FIXIN' UGLY HOUSES FOR MONEY...

How Small-Time Real Estate Investors Can Earn $1,000,000 And Lots More

by

Jay P. DeCima

KJAY PUBLISHING

Opinions and recommendations given herein are based on the author's actual experiences and on research believed to be reliable. It should be noted that investments mentioned in this book have been chosen only to demonstrate a given point. The reader must necessarily act at his own risk. This book is sold with the understanding that the publisher and author are not engaged in providing legal accounting or other professional services. If legal advice or other expert assistance is required, the reader should seek competent professionals in those fields. If you do not wish to be bound by this statement, you may return the book in good condition to the publisher for a refund.

Library of Congress Catalog Card Number 97-75993

ISBN 0-9621023-1-8

Published and Distributed by:
KJAY Publishing Co.
P.O. Box 491779
Redding, CA 96049-1779
1-800-722-2550

Acknowledgment to Kathy

Special thanks to Kathryn P. Balkovek, my best friend and business partner, who's help was absolutely essential in producing this book. Kathy labored for many hours typing, revising, shuffling and rearranging pages—eventually turning them into something we could both read and understand.

Kathy sandwiched the work on this book between her regular everyday chores of paying bills, making bank deposits, answering my 800 telephone service, writing paychecks and keeping track of almost 200 tenants.

If it weren't for Kathy's help, I seriously doubt that these pages would have ever been organized enough to produce a book. As it turned out, it took a lot of effort on both our parts to interpret my original draft, which had been scribbled out longhand on 600 sheets of loose leaf binder paper.

Using my own advice—"Never take on a business partner unless you can't do the whole deal by yourself"—I realized I was not able to do this project by myself and, thus, needed a partner. That's precisely the reason I've dedicated this page to Kathy—she is that partner. Without her help, there is simply no way this book could be in your hands right. There's even a good possibility it might *never* have been completed!

Preface

I don't know how you think, but let me tell you something about myself. Real estate investing truly fascinates me! People often ask me, "Why don't you quit buying more properties? You've already got all the houses you'll ever need for one lifetime. Why don't you sell them and take the money to enjoy yourself? After all, you can't take the houses with you, and you're not getting any younger, you know!"

It's true, I do own more houses than I'll ever need. And certainly, I could sell out and travel around the world if I wanted to. The problem is that I ENJOY DOING WHAT I'M DOING. If I were forced to retire tomorrow and someone asked me, "What do you plan on doing now that you're retired?" My answer would be, "I plan to buy rundown houses and fix them up."

Real estate investing is a very exciting business the way I do it, and it would be difficult for me to stop doing it. Each new deal is an exciting challenge because every single transaction is different. Not only are the deals different, but the techniques I use keep changing. I'm continually learning more about investing, and, thus, becoming a better "deal maker" each time I go after a new property. Once you get hooked on the excitement that comes with making new deals, it's easy to become addicted, at least it certainly turned out that way for me.

During the past twenty years, I've read at least 30 books about real estate investing; but I must tell you, few, if any, present the ideas and strategies you'll find in this book. The reason for this, I believe, is because I practice real estate investing and landlording every single day. That's all I do, except write about it! The point is, what you read in this book is not necessarily the unanimous opinion of every "so-called" real estate expert, and it is not written about strategies that might work when the moon is full! Rather, it's about blood, sweat and tears. All mine! What I have written is a "How To" guide for do-it-yourself investors, just like myself. I will show you how to start with very few dollars in your pocket and acquire cash-producing real estate in less time than it takes the average person to get promoted at their regular job.

I have devoted many years and thousands of dollars to acquire the knowledge I will share with you in this book. If you are presently working for a salary and want to become your own boss in the future, I will tell you how to do it. If you wish to earn more income for yourself and family, I will teach you about the highest paid "part-time" job you'll ever find anywhere. If you're looking for an exciting new career, like I was, with absolutely no limits to how much money you can make, you don't need to look any further. This is your lucky day, because I'm about to show you exactly how you can do it for yourself.

If you're the kind of person who can start something new and stick with it, then you're the kind who will benefit a great deal from this book. Furthermore, I can promise you this much—If you begin using some of the money-making strategies I teach in this book, vigorously applying the ones that fit your personality and style, you can become a very successful real estate investor of whatever size you choose.

You can quit after four rental houses or continue full-steam ahead until you own a hundred properties! Whichever way you choose, you'll discover something inside you has changed. Never again will you feel dependent upon an employer, and you won't have to put up with a boring dead-end job. This book will give you another option to provide for yourself and family. Perhaps even more important, you'll never again be at the mercy of recessions, employer downsizings, bankruptcies or any other economic ills that beset most people. In short, you will have complete control over your financial destiny.

A housing entrepreneur is what I am! Entrepreneurs are the true capitalists. Entrepreneurs make America the financial powerhouse of the world. They are creative people who will not accept no for the answer. If banks won't finance your real estate deals, I will tell you about someone who will. If you need cash quickly and you can't sell your real estate fast enough, is there something else you can do? I will show you how to get the cash another way! During the Great Depression of the '30s, nearly half the total work force in the country was unemployed. Most entrepreneurs were not among them. The reason is because entrepreneurs are survivors. They learn how to prosper in good times and cut their losses when things go bad. Once you learn how to survive as a housing entrepreneur, you'll never again have fears when the economy suffers a down-turn. You'll know exactly what you must do to survive.

There are three important ingredients you must have in order to be a successful housing entrepreneur. The first is KNOWLEDGE. That's where I can help you, because my book will be your guide. My personal experiences can be your teachers and my profitable transactions will provide convincing evidence that my techniques work.

The second thing you must have is SELF-CONFIDENCE! This will come with knowledge and from actually completing your own real estate deals. When you know how to do something successfully, your confidence level rises by leaps and bounds. The final ingredient you must have is, without question, the most important of all three. It is PERSEVERANCE! Failure to stick in there until you reach your goal is nearly always a fatal mistake for entrepreneurs.

I've watched many investors become discouraged and stop on the one-yard line. Quite often someone right behind them will pick up the ball and make a fortune by simply going one more yard. I am so convinced about this third, and most important, ingredient that I've included a homily, below, that was a favorite of the late Ray Kroc, entrepreneur and founder of the world's largest hamburger store—the McDonald's Corporation. (This man has always been a guiding light and inspiration to me.) These words have great application here—

Press On: Nothing in the world can take the place of persistence. Talent will not; nothing is more common than

*unsuccessful men with talent. Genius will not; unre-
warded genius is almost a proverb. Education will not;
the world is full of educated derelicts. Persistence and
determination alone are almighty.*

Source: *Grinding It Out*, autobiography Ray Kroc, Berkley Medallion Books, New York (1977).

Fixing rundown houses is an excellent starting point for investors and career changers alike, who are short of money and experience. Buying rundown properties from motivated sellers is much easier and a lot cheaper than buying nice looking real estate. The competition is significantly reduced, plus it's much less sophisticated in terms of deal-making.

Small multiple-unit properties like I recommend often fall between the cracks when it comes to investor interest. The big guys, especially partnerships and syndicators, generally ignore them because they're simply too small and troublesome. Most small-time investors rule them out because a large percentage of them are "single house" thinkers. They often conclude that older rundown properties are worn out, obsolete and more trouble than they'll ever be worth. Many lack the vision to spot hidden gold covered by dirt and neglect.

For these reasons, I often refer to my small multiple-unit properties, both houses and apartment buildings, as "leper properties." The reason is that not many investors are willing to touch them. Instead they fight the crowds, competing to purchase the nicer looking, sweet smelling properties, often referred to as "Pride of Ownership."

What many investors fail to understand is the value of cash flow *now!* While I will certainly agree that tomorrow's equity is an important consideration for down the road, my top priority is always *GETTING* DOWN THE ROAD. Cash flow producers must always come first, which means earning profits as you go along. "LEPER PROPERTIES" are the best cash producers for getting you down the road. If you will follow my advice in the beginning and acquire the kind of properties that earn you consistent profits, then one day you'll have enough money to buy any kind of real estate you want!

x

Table of Contents

Introduction by Jay 1

Chapter 1: How to Make $1,000,000 Working Smarter................5
   Making Serious Money Requires Extra Helpers.................................6
   White Picket Fences Provide Big Payback.......................................6
      Fix-Up Profits vs. Wages at the Sawmill....................................8
   Leverage and Compounding Are Your Silent Helpers.........................9
      Why Invest in Fixer Houses...10
   Double Your Income, or Earn a Million...10
      Where Does All the Money Come From......................................11
   Making Big Bucks From the Yucks...12
   Why Fixers are the Perfect Place to Start..15
   Properties Must Have Potential for Adding Value.............................16
      Seller Financing All or Part..18
   Buying Back Your Own Debt Worth Big Bucks................................19
      When You Up the Income, You Up the Value 20
   Long-Term Profits and Cash Flow...22

Chapter 2: The Haywood Houses—A Textbook "Fixer-Upper"...23
   Classified Ads Can Sometimes Lead to the Gold Mine....................23
      Find What You're Looking For and Act Quickly.........................25
   Fixing People-Problems is Worth Big Bucks 26
      Flexible Sellers Provide High Profit Opportunities.....................27
   Good Financing Sets the Stage for Big Profits.................................29
      Looking for Loans in all the Wrong Places................................30
   Fixer Skills Turn Ugly Duckling Into Beautiful Swan 31
   To Make Big Money You Need a Profit Plan...................................34
      Waiting for "Mr. Good Buyer" 35

Chapter 3: The Profit Advantage Using Fix-Up Skills 37
   Only Two Methods to Make Money..37

Selecting Right Strategy is Key to Success 38

Adding Value—A Practical Method for Average Folks....................38

Fixing Houses is Equal Opportunity for All..............................40

Knowledge is What Builds Fortunes 41

The Women's Advantage Shows Up 42

Changing the Looks Always Adds Value Faster...........................42

Sizzle Fix-Up Provides High Investment Returns.........................43

Biggest Opportunities Often Found Where You Least Expect..........44

Renovators, Remodelers are a Different Breed.........................45

Fixing Up Bad Management Pays Equally Well..........................45

How to Determine What to Pay................................46

Still Waiting for the Tooth Fairy—Forget It!..........................50

Forget About Investing in Unfamiliar Places.........................51

Adding Value Strategy is Perfect Opportunity 52

Key Ingredients to Look for When Making a Bargain Purchase.......53

**Chapter 4: Specialization is Quickest Path To Earning Big
Profit** **55**

Getting Started Ranks First......................................56

It's Important to Position Yourself to Make Money....................57

Forget the "Blue Ribbon Deals"—Get Started 58

Looking for Mr. Right—Not Perfect.....................................58

Finding Sellers Who Truly Want to Sell................................60

Ugly Property Buyers Get the Best Terms...............................60

Personal Effort Creates Instant Equity and Long-Term Wealth......62

Don't Buy Until You Know How Much to Pay 63

Beware of Forked-Tongue Sellers....................................63

Owners Seldom Overstate Income on Tax Returns.........................64

"For Sale" Often Means "I Need Help"...................................64

Don't Wanters—They Don't Know How to Manage....................65

Violation of Habitability Laws—A Motivator.............................65

Looking for Ugly Deals 67

Property Management—Key to Success................................67

Looking for Just the Right Property.................................69

Diversification Later on is Best Strategy 69

The Best Odds for Your Success 71

Ten Must-Do's That Will Speed Up Your Success....................72

Specialized Training Materials Available 74

Chapter 5: Good Realty Agent Plus Thorough Analysis Speeds Your Success..75

Good Agents Don't Cost You Money—They Help You Make It 76

How to Find an Agent That's Right for You 76

Five Important Benefits an Agent Provides....................................78

Thoroughly Analyze the Deal Before Making an Offer.....................81

Jay's Income Property Analysis Form..81

The Basis for Negotiating a Purchase..87

The Most Controversial Expenses..87

Real Estate Agent Can Help You Build Wealth 88

Chapter 6: Substitute Personal Skills When You're Short of Cash..91

Getting Started With the Right Properties 91

Buying Properties the Old-Fashion Way..92

Develop a Reasonable Plan—Then Follow It............................94

Sellers Seldom Tell You Like It Is 95

If the Numbers Work—Write Your Offer Immediately.....................96

Double Your Pleasure and Your Profits....................................98

The Key To Wealth Is Solving Bigger Problems 99

Finding the Right Properties 99

How to Develop Instant Equity..101

Beware of Properties that Don't Provide Cash Flow.....................102

Fix-Up Skills Earn Wages—Knowledge Builds Fortunes........102

Fix-Up Properties Require Multiple Skills 103

Big Bucks Come From Understanding Economics..........................104

Take Advantage of Old Fashion Charm 106

The Biggest Mistake is Over-Fixing.......................................106

Work That Shows Fast—Earns Fast 108

Chapter 7: The Price is Determined by Income and Location 109

Complex Formulas are Not Necessary..109

Don't Count Pennies Doing Fix-Up.......................................111

Jay's Profit Strategy: Up The Rent—Increase The Value.............113

Looking at Profits and Cash Flow..115

Stay Away from Weird Stuff—Stick to Basics 115

Learn Where Your Profits Come From...116

Without Cash Flow Today - There Might not be Tomorrow 117

Buying at Right Price is 90% of the Battle...................................117

You Can't Pay the Bills With Equity 118

Operating Your Own Property is Always the Best Teacher............119

Know Where Profits Come from Before You Buy............................119

Two Numbers Make all the Difference.....................................120

Location, Location, Location...121

    Income Properties are Supposed to Provide Income.................122

    Predicting the Future is Only a Guess, at Best 122

Investing Long-Term for Future Growth...................................127

**Chapter 8: Finding the Right Properties and a Motivated
Seller...129**

The Four Basic Methods of Finding and Buying Fixers................129

    Watching the Daily Classified Ads 130

    Using a Real Estate Agent 131

    Working the Multiple Listing Book................................132

    Initiate Written Cold Calls...133

Cold Calling is a Proven Technique 134

Finding Profitable Deals is the Goal 135

    Finding Sellers Who Truly Need to Sell 140

Learn to be a House Detective..141

"For Sale" Often Means "I Need Help".....................................141

    "Don't Wanters"—Most Don't Know How to Manage.............142

Motivated Sellers Seldom Tell Their Secrets.............................143

    Vision Plus Long-Term Thinking Worth Big Bucks.................144

The Courage to Look Where Others Don't 145

Buying Properties from the Bank..146

    Timing is Everything From Wine-Making to Real Estate........147

Determining How Much to Pay is a Must...............................148

The REO Man Rings Twice Sometimes.................................151

    Where Does all the Fix-Up Money Come From 152

Chapter 9: Fixing Rundown Houses for Money........................155

Don't Fix Things That Don't Pay You Back.............................155

    Big Bucks Come From Understanding Economics.................156

Investments Measured on How Well They Pay.........................157

    What You See Counts for Everything...............................158

You Must Position Yourself to Make Money.............................160

    Rundown Properties Offer Most Possibilities......................161

Cash Flow and Equity Opens Many Doors..............................163

Diversification Okay—But Only After Cash Flow 165

Specialized Help is Available...166

Chapter 10: Jay's Moneymaker "Foo-Foo" Fix-Up Strategy........167
The "Foo-Foo" Cover-Up Strategy Exposed 168
Lawns, Shade Trees and Picket Fences are Hot......................169
The Fix-Up Revolution—Made to Fit and Ready to Use 170
The Majority of Fix-Up is Cream-Puff...171
You Won't Get High on Paint Fumes Today 173
The Handy-Person Fixer University...174
Window Coverings are Top Sizzle Items...............................174
Fix-Up Skills Often Worth More Than Cash..................................176
Houses With all the Right Things Wrong..............................176
Biggest Payday Comes From Knowing Where to Kick 182
Changing the Looks Adds the Quickest Value.............................183
Sizzle Fix-Up Offers the Biggest Profits.......................................184
Profitable Fixing Boils Down To "What Does it Cost?" 186
Passing the Savings on to the Landlord—That's You!............188
Painting is a Drag—But Also a Top Money-Maker 191
It Pays to Invest Where the Money's At................................192
Always Get a Second Opinion..192

Chapter 11: Secrecy, Plus Adding Value Creates Profits.........193
The House Detective Approach 193
The Secret Path to the Gold Mine...194
Getting Started on the Right Foot 195
Forget Sellers Who Can't Make You a Good Deal.........................196
The More You Know the Better You Can Negotiate.................198
Offer Relief From Pain in Lieu of Cash...198
Free and Clear Properties—No Mortgages.............................201
My Favorite Profit-Maker is Adding Value.....................................202
Setting Up "Worry-Free" Investments.....................................203
Three Value Factors When Considering Investment.....................204
Good Location Equals Higher Rents and Top Multiplier..........205
A Two-Point Gain Nearly Doubles the Value.................................207
Over-Paying—The Deadliest Investor Sin.....................................209

Chapter 12: Where Do All the Profits Come From......................211
Playing the Appreciation Game..211
The Magic of Compounding..212
Not Just Any Real Estate Will Work.......................................214
Tax Shelter Benefits 215
Leverage Lets You Soar With the Eagles................................217

Brain Compounding Can Increase Your Wealth............................219

Adding New Profit Bulbs On My Money Trees..............................220

Chapter 13: Negotiating Deals That Earn Big Profits.............225

Stripping the Puffery from "For Sale" Ads....................................227

What is Needed for Maximum Control..228

What the Classified Ads are Telling You......................................229

    You Must Learn the Facts Before Making Offer.....................230

Detective Story Offers Excellent Training....................................232

    Winning Over the Seller Leads to Winning Negotiations.........233

Don't Win the Negotiations and Lose the Property.......................235

It's Always Best to Let Seller Participate.....................................236

    No One Manages Property and Does the Repairs for Free.......238

Successful Negotiations Puts Money in Your Pocket.....................240

Additional Help is Available...241

Chapter 14: Fixing Million Dollar Problems...........................243

Creating Equity With Very Little Cash...243

Hillcrest Cottages—A Million Dollar Problem.............................244

    Knowing the Real Reason for Selling is Big Advantage...........245

Key Ingredients for a Super Deal..246

    Fixing Up Hillcrest Cottages..248

Removing the Risk From a "No Down" Sale.................................250

Extra Profits with Wrap-Around Financing..................................252

    Wrap-Around Mortgages Gives Seller Better Control.............255

Benefits of Seller Carry-Back Financing......................................257

    A Profit Opportunity You Shouldn't Overlook.......................260

Chapter 15: Investing With Others—Small Partnerships.........261

Why Would Anyone Want a Partnership.......................................261

    Partnerships Must be Based on Mutual Needs.......................262

Jay's Rules for Finding a Money Partner......................................264

    Partnerships Can Solve Your Money Problems......................265

Benefits Must be Totally Equal for all...266

    Looking for Partners in all the Wrong Places.........................267

    The Partnership Promise...274

Important Terms—Rules of the Partnership.................................274

    No Money—No Problem—Once You Prove Yourself.................276

You Must Develop Good Track Record..276

Chapter 16: Sell Half the Property to Increase Your Income....279

50% Sales Turn Negative Income Into Positive..............................279

The Task Is To Quickly Fix Up the Property and Add Value.........282

How To Market a Fixed-Up, Fixer Property..................................283

Joining Together for Profit Opportunities 286

Never Invest Without a Written Agreement..................................288

Chapter 17: Jay's 90/10 Money Partner Plan for Wealth...........293

Small Ownership Cost Buys Big Returns...............................294

High Returns and Buying Power are Keys to Plan.........................295

Never Forget the Golden Rule of Investing...........................297

Leverage—the Investor's Hamburger-Helper.........................297

Proof of Skills Needed First..298

In Search of Investor Cash...299

Selecting the Right Partner is Critical....................................300

Handyman Skills are Worth Big Bucks..301

Buying Right Sets the Stage for Making Profits 303

Beware of No-Down-Payment Transactions................................304

Chapter 18: 100% Financing With Seller Subordination...........307

Loan Terms are More Important Than Interest Cost 308

Jay's 30-30 Seller Subordination Plan.......................................310

Where Does All the Money End Up? 311

Real-Life Case History of South-Side..315

Please Tell Me—Where's The Beef! 316

No Limit to Creativity in Real Estate...317

Making Yourself a Better Borrower...319

How to Build Your Financial Integrity 320

Jay's Five Basic Financial Documents For Borrowing...................321

Fixer Jay's Loan Kit for Borrowers...324

Chapter 19: Free Fix-Up Money From Uncle HUD.....................325

"Wait and See" Will Take You Nowhere.....................................325

More Than One Way to Profit...326

Uncle Sam Provides Money for Fixing Affordable Houses..............326

Dealing with the Local Housing Authority 330

No Money Down Deals are Very Possible 331

Housing Authority Needs Landlords to Participate......................332

More Help for Broke Owners With Knowledge 335

City Loans Work In Tandem With Grant Funds............................335

The Extra "Red Tape" is Grossly Overstated..........................337
Understanding the Motivation at City Hall.........................340
You Help Yourself Most When You're Helping Others 342

Chapter 20: Buying Back Mortgage Debt for Bonus Profits......343
Look for Property With Private Mortgages.........................343
Jay's Red Mustang Strategy...347
    If At First You Don't Succeed—Try Harder 347
What You Should Know About Buying Debt.........................350
    Making Money by Accident Hooked Me 352
    Don't Throw the Gold Away With the Sand.........................353
Reasons Why Mortgage Holders Sell for Discounts.....................354
Value, Like Beauty—is in the Eye of the Note Holder.................356
Jay's "Christmas Letter" Generates Profits Year-Round 356
Investors Need a Healthy Financial Diet.............................358
Add a Professional Touch...358

Chapter 21: Managing Tenants is Key to Investment Profits....359
    We Do It for the Money...359
The Dream—Working for Yourself.....................................360
    Tenants Can Make or Break a Property.............................361
Success Means Wearing Many Hats...................................362
Fix-Up Skills Worth Big Money......................................363
Proprietorship—A Must...364
Giving Fair Value for a Fair Rent...................................366
    Don't Underestimate Dangerous Situations 368
Landlording Can Be Learned Very Easily.............................368
Quick Action Keeps Landlords One Jump Ahead......................370
    Don't Bother Asking the Dumbest Question.........................372
How to Manage Your Tenants By Mail 373

**Chapter 22: Landlording Skills Can Make You Very
Wealthy 375**
Horror Stories are Caused Mostly by Ignorance.......................376
    Operating Properties Like a Business...............................377
    Fewer Rules are Best—But be Sure to Enforce Them 379
    Emotions Should Not Control Landlord Decisions 379
Many Landlords Help Tenants Go Haywire.............................381
The First Rule of Business is to Define Your Customer 382
Keys to Good Management are Action and Enforcement................384
    You Must Always Get the Money First.............................385

Good Tenant Records are Essential...386
Rental Contracts Don't Need to Be Complicated....................389
Landlording Should be About Profits...................................391
Tenant Urgency—Not My Urgency.....................................391
Obey the Laws of Habitability...395
Long-Winded Contracts Don't Mean They're Better......................397
Discrimination DOESN'T Protect Deadbeats 398
Collections First—Love and Kisses Later.............................399
The Value of Tenant Cycling...400
Fairness Always Counts the Most for Everything 401
Jay's 60/40 Strategy Pays Big Dividends 402

Chapter 23: Cash Flow Keeps You Green and Growing 403
Decide What You Are—Investor or Speculator...........................403
Sorting Out Dreams From Reality..405
Borrowing From the Bank is Good For Bank.............................406
Real Estate Investors Must Think Like Business Folks........408
Jay's Formula for Making Money in Real Estate 409
Plan Must Be Simple With Achievable Goals 411
Write Your Profit Plan—Then Get a Second Opinion...................414
Compounding—The Secret To Building a Fortune 414
Matching Your Offer With the Property 415
Weak Sellers Make Weak Offers Work.................................416
Easy Path to Cash Flow...419
How to Make Lemonade Offers Work 420

Chapter 24: The Big Picture and Long-Term Wealth................421
Don't Get Bogged Down With Routine Stuff.............................421
It Takes a Workable and Realistic Plan..............................422
The Wall Street Journal Dream 423
My Best Profits Come From Dumb Landlords............................425
The Dream Alone is not Enough...426
High Rent-to-Value Ratio is Tip-Off to Profits 428
Selling for What You Paid, and Still Making a Profit.................429
I Didn't Grow Up to be a Landlord....................................430
Change Brings On New Opportunity 432
Investing in Real Estate is Like Kissing Frogs....................432
Avoid Doom and Gloom Like the Plague 433
Positive Cash Flow Makes it all Worthwhile..........................435

xx

Appendix: Exhibits and Resources..437

Index..451

Introduction by Jay

Starting my house-fixing career in Northern California, back in the 1970s, seemed like a perfect opportunity at the time! I had no idea back then that one of the worst real estate recessions since the end of World War II was lurking around the corner. I had no way of knowing that interest rates would suddenly shoot up to 22% and completely close the doors on traditional real estate financing.

What a terrific way to start out, I thought! I couldn't have picked a worse time if I had planned it. When interest rates began to climb, nearly all the real estate activity in town came to a screeching halt! Hardly anyone was interested in buying or selling. Looking back now, I realize it was probably a good thing for me that I couldn't predict the future, otherwise I would have likely kept on punching a time clock at the telephone company—and figuring how long before I could draw Social Security.

I still remember my early struggles. Friends and business associates kept telling me the same thing, "Jay, there is absolutely no way you can buy rundown houses today to fix up and expect to make any profits for yourself. To begin with," they told me, "Most banks are not willing to finance real estate today, especially the kind you're buying. Besides that, you won't be able to sell your properties because there's no appreciation anymore! Obviously that means no future profits! You've got to face reality Jay—The days of making money in real estate are over, the bubble has finally burst!"

I remember reading a book by William Nickerson, *HOW REAL ESTATE FORTUNES ARE MADE* (Simon & Schuster, 1963). In his book, Nickerson says:

Although opportunities are much greater during boom times, I have come to the conclusion that opportunities are always present in good times or bad! Anyone who really wants to can make a fortune in real estate. To succeed one requires only the initiative to start and the determination to keep applying the three R's of renovating, refinancing and reinvesting.

Nickerson's words gave me the courage to ignore my critics.

Quite often I've found that things you don't understand too well can end up helping you more than the things you do understand! For example—I didn't understand why it was the wrong time to buy FIXER HOUSES, So I kept on writing offers and buying those kinds of properties anyway! I didn't understand that borrowing fix-up money at 20% interest was way too expensive, so I borrowed the money and fixed the houses anyway. Nearly everyone told me I couldn't sell the houses because it was such a terrible seller's market! However, in just thirteen months time, I sold my Haywood houses (details in Chapter 2) and made a $150,000 profit.

Over the years, I have learned that it's far better to be a little bit dumb about things and to act, than it is to be super intelligent and never accomplish anything! It may sound like I'm a little dumb when I tell you this, but I promise you, it's true! GOOD OPPORTUNITIES NEVER DISAPPEAR, PEOPLE SIMPLY FAIL TO RECOGNIZE THEM! Action is the magic ingredient that separates successful people from those who can't figure out what to do.

My seminar students are always asking me this question, "Do you honestly believe there will always be an opportunity to make big money fixing rundown houses?" Let me answer this way—according to the latest government survey conducted by the U.S. Department of Housing and Urban Development (HUD), the need for decent, affordable rental housing exceeds production by at least 250,000 units annually. Adding to this problem is that more rental houses are deteriorating below habitability standards than are being rehabilitated. TRANSLATED, this means that fixing rundown houses is truly a golden opportunity for do-it-yourself real estate investors like myself. Indeed, the future is brighter than ever and there's no end in sight.

People often say to me, "Jay, you sure are a lucky devil! You jumped head first into real estate investing at exactly the right time! Your timing was perfect, but tell me truthfully—Do you still think the same things you've done for yourself can be done by others in today's economy?" My answer is a loud and clear, "YES IT CAN!" Furthermore—as you will discover by the time you've finished reading this book—the economy and timing have hardly anything to do with fixing houses for profits. Profits will come from ADDING VALUE and your own PERSONAL SKILLS. That's the real beauty of fixing houses. The only limits are your willingness to learn how and, of course, getting started!

Beginning with the first chapter, I'll share with you an exciting strategy about making big money! I'll show you how to set yourself up for life FINANCIALLY. You must be willing to learn a few new techniques and develop some special skills. There's no question, I can teach you how but, obviously, you must jump in and get the job done.

You'll be pleasantly surprised, as I was, to learn that money is not what you need most to be a successful real estate investor! Unless, of course, you're counting the cash expense to buy this book! Forget that right now, because I promise you'll earn it back many times over. To begin with, pay very close attention as you read the first chapter, because it only takes one property like Hillcrest to get your book cost back a thousand times over! *Hillcrest is a 21-unit fixer property that you will read lots more about in Chapter 14.* Real estate profits can multiply like rabbits by using leverage, but they seldom get much better than my Hillcrest property.

Chapter 1 is first in order because it shows you that *small-time investors* can **earn big-time profits** doing fairly simple fix-up jobs. However, all chapters are important, because each one will teach you new and exciting ways to make money. By the time you're done reading the book, you should have enough knowledge to start turning "ugly duckling" properties, like my

Hillcrest, into beautiful "swans." When you do, your beautiful swans will start producing those lovely golden eggs I call CASH FLOW!

If you're the kind of reader who highlights important information with a fluorescent marker, I fully expect this book to look like Walt Disney's doodling pad when you're done! If it doesn't, you should back up and start again, because you're skipping over way too much good stuff.

There's one final point I wish to make before I send you on a Money-Making Education through these next 24 Chapters. Do not expect me to tell you if a 10% loan is good or bad—or which bank will loan you money—or even where you should invest in "Fixer-Upper" houses—Its my hope you'll be able to tell me the answers by the time you've finished reading.

What I will show you are techniques and strategies that work anywhere, anytime—with or without bank loans. What you'll learn from me has been working for at least 100 years— And I'll guarantee you—It's going to work at least 100 more. I'm a firm believer in the age-old wisdom that argues, *"Feed a man a fish dinner, and you've helped him for a day! But, when you teach him how to fish, you've helped him for a lifetime."* If you agree—let's get on with our fishing lessons.

CHAPTER 1

How to Make $1,000,000 Working Smarter

*M*ost people are too busy earning a living to make any serious money! I'm talking about the kind of money that can make you wealthy enough that financial problems will no longer be your biggest concern. Unfortunately, most folks simply don't know what to do or how to begin! The reason for this lack of knowledge is that MAKING MONEY 101 is not taught in traditional places of learning. The fact is, most educators are still preaching the age-old proposition that hard work, long hours and a steady job at the mill are your best guarantee for a happy life and financial success. The problem is that today few facts support this theory!

To begin with, working harder and longer hours has strict limitations! For example, suppose you had a job that pays $200 per day for a regular 40 hour work week. No matter how hard you work or how many hours you work, you can't possibly earn more than 2 or 3 times your normal paycheck!

Even if your employer would allow you to work another full 40 hour shift at double-time pay, it's likely your earnings would only be about 2 1/2 times your regular pay after tax deductions! I would agree it's much better pay, but still pitifully short of what I would call serious money. To earn that, you need 10 or

20 times more income. Obviously, there's not enough hours in the week to earn this kind of money the old-fashion way.

MAKING SERIOUS MONEY REQUIRES EXTRA HELPERS

What it all boils down to is this—if you want financial independence, which is what serious money brings, and if you wish to have it during your lifetime, you'll need some extra helpers of the non-human kind! Their names are LEVERAGE and COMPOUNDING.

As is the case with any helpers, your job will be to provide them with a project to work on, plus skilled leadership! With the assistance of leveraged real estate and compound earnings, you can far exceed the limitations of a regular paycheck. It sounds a bit strange if you've never heard this idea, but just bear with me. I guarantee you'll have a far better understanding as I take you on a magic money-making tour through these pages.

One of the questions I'm most frequently asked at my seminars is, "HOW MUCH MONEY CAN I MAKE FIXING UP RUN-DOWN HOUSES THE WAY YOU SUGGEST?" Obviously, there's no single answer, because everyone who invests in FIXER-UPPER PROPERTIES will do it differently. For example—some investors may decide to do only limited cosmetic fix-up, such as painting and clean-up. Others, with more hands-on experience, like builders and contractors, may choose to upgrade foundations, add rooms and revamp the walls.

WHITE PICKET FENCES PROVIDE BIG PAYBACK

Quite frankly, foundations and walls are just a bit more than I care to tackle. I've discovered that my biggest paydays come from repairing things that need fixing and cleaning up! Hauling away junk and painting nearly everything that shows is always a top money-maker—and rejuvenating dead or dying yards by planting new shrubs and lawns is quite inexpensive compared to the profits you'll earn. As a finishing touch, to bring out the charm, I always like to add my signature improvement—a three-foot-high white picket fence enclosing the front yard. A white picket fence gives any house the "homey look," and from a pure economics standpoint, fences will return $10 for every dollar you spend to build them.

That's exactly what I did to my Hillcrest Cottage property, which you will be reading a lot more about as we go along. Hillcrest and my five Hamilton Avenue houses were sold together in a single package installment sale! I earned as much money for just this one sale by itself, with only two year's worth of fix-up work, than most people will earn during their entire working careers.

I never dreamed this would be possible, but let me assure you, it is! In fact, I'm still collecting payments to prove it. Let me show you why fixing rundown houses will beat the pants off working your life away down at the local sawmill. The chart below will help you see the big money difference between working for wages and working for yourself, like I do.

Typical Wages for Sawmill Worker - 40 Years

Worker's Age in Years	Term in Years	Average Wages Per Year	Total Wages
21 to 25	5	$21,000	$105,000
26 to 30	5	23,000	115,000
31 to 35	5	26,000	130,000
36 to 40	5	28,500	142,500
41 to 45	5	31,000	155,000
46 to 50	5	33,500	167,500
51 to 55	5	37,000	185,000
55 to 60	5	40,000	200,000
Totals	**40**		**$1,200,000**

Fix-Up Profits vs. Wages at the Sawmill

As the chart shows, working 40 years at the sawmill will earn you $1,200,000 in my town. You will spend approximately 80,000 hours on the job (2,000 hours per year for 40 years equals 80,000 hours). By dividing the total wages by the hours, you can see that sawmill workers average $15 per hour for working a lifetime at the sawmill. Naturally, income taxes will reduce their take-home pay.

By way of comparison, my Hillcrest sale earnings were $1,200,022, paid to me over a period of 26 years and one month. Obviously, I didn't work anywhere near 80,000 hours to earn my money, since I only owned the property for two years before I sold it! I have calculated that my fix-up work took about 2 years from start to finish. However, not all of my regular workdays were spent at Hillcrest. I was also fixing up several other properties during the same period of time.

Assuming that I had worked two full years at 2,000 hours per year, you can see rather quickly that my hourly rate of pay would be a little over $300 per hour. THAT'S 20 TIMES MORE EARNINGS than the mill worker. Plus, I spent only two years of my life to earn the same amount of money it will take a mill worker 40 years to earn.

For the sake of comparison, I've shown you what the average sawmill worker in my hometown can expect to earn working 40 hours a week for the next 40 years. That's assuming the mill stays open! I don't believe there's any question which career you'd choose if you knew about fixing houses the way I do it. If there were some way the mill worker could increase his hourly pay to $300, like I earned fixing my Hillcrest property, he'd take home 24 MILLION DOLLARS in wages by the time his 40 year career was over.

My point is this—we all get exactly the same number or hours in a work day, the same number of days in a week and so forth. THE BIG DIFFERENCE BETWEEN FOLKS WHO EARN MODEST WAGES AND THOSE WHO MAKE MILLIONS IS HOW THEY SPEND THEIR TIME!

LEVERAGE AND COMPOUNDING ARE YOUR SILENT HELPERS

Obviously, there's no way you can earn $300 an hour working at the sawmill. In order to make big money you must spend your working hours doing the kind of things where your earnings can be tax sheltered and leveraged. Also, you must work at jobs where you earn the biggest profits for the least amount of time spent. LEVERAGE and COMPOUNDING will be your silent, but powerful, helpers when you fix up rundown properties like my Hillcrest Cottages. Probably the best news I can pass along is that there's absolutely no dollar limit on how much you can earn doing this. The sky's the limit! We'll discuss more about leverage and compounding and how it helps you build wealth as we go along, but for now let me tell you about your chances for success.

Almost anyone can be successful fixing houses if they truly set their mind to doing it! Education is available from folks like me who have already done it. Practice will make you an expert in less time than you might imagine. On the other hand, more practice at the mill will only give you a sore back, I'm afraid! If you agree with me so far, keep reading! I'll show you why it's not necessary to have a lot of money in order to make a lot. That should perk you up!

Why Invest in Fixer Houses

To begin with, fixer real estate offers one of the best opportunities for making a great deal of money in the shortest possible time!—also, with the least amount of personal risk involved! Furthermore, the opportunity is available to nearly everyone, because fixers can be acquired for a minimum amount of up-front cash and with the best possible terms.

Fixer properties offer do-it-yourself investors a unique opportunity to substitute their personal handyman skills in lieu of a normal cash down payment—often this means a 20-40% cash savings right up front. The popular myth, which says, "IT TAKES MONEY TO MAKE MONEY," simply doesn't hold true in the fix-up business. This advantage will become increasingly clear to you when you read several actual case histories of high-profit deals I've done over the past few years.

DOUBLE YOUR INCOME, OR EARN A MILLION

It is well within the reach of ordinary working-class folks, assuming they have the desire to learn, to become very successful and financially independent fixing up rundown houses. Earning a million dollars, if that should be your goal, is not an unreasonable target. Many achieve the goal in 10 to 15 years. Naturally, it goes without saying, you'll earn every nickel you make, but there is no limit to what your earnings can be! If your goal is to double your present income, that's easy enough to do. If your sights are set on becoming a millionaire, I suggest you just keep on reading and find out exactly how it's done.

WHERE DOES ALL THE MONEY COME FROM

Many book writers seem to be very vague on this point! Some will tell you, "If you'll just follow the formulas in my book, the profits will take care of themselves." To me, that's simply not clear enough. I insist on knowing where the money is coming from, so I'll know exactly where to concentrate my efforts. Let me take the mystery out of the money! Profits and paydays come from three primary sources in this business. Naturally there are variations and combinations, such as selling partial payments on seller carryback notes and payments received from partnership buy-ins. We'll discuss these later.

Three Major Sources of Money Coming In

1. *Monthly Rental Income* – Net cash flow.

2. *Property Sales*
   a. Cash money from escrow at closing.
   b. Seller carryback financing. Monthly installment payments (receivables).

3. *Borrowing*
   a. Seller subordination at the time of purchase.
   b. Equity loans during period of ownership.

We'll discuss each of these sources in much more detail later. But first, let me tell you how important I feel it is to have all three sources available as options at all times. You need to understand there are times when it seems like you can't give real estate away! You try your best to sell, but even when you lower the price, there are still no offers. Oh sure, someone might offer to buy you out for a song and dance, but that's no good! Besides, didn't you get in this business to make money? Why in the world would you decide to give all your profits away?

On the flip side, there are times when real estate is hot! That's when buyers will pay almost any price without question. It's frenzy buying like a supermarket sale with tomatoes advertised at 8¢ apiece or 3 for a quarter and everyone buys three. No one seems to pay any attention to the math! When people feel good about buying real estate, the price will often take a back seat in their haste to buy something quickly. Obviously, that's the proper time for selling.

Monthly Rental Income Keeps You Green and Growing

Most important to real estate wealth-building is the steady flow of green cash. That means money you can depend on coming in monthly to pay the bills! Most bills must be paid every month. Cash flow is often overlooked by novice investors who think only of profits from selling. Borrowing the words from a famous hamburger millionaire, Ray Kroc, founder of McDonald's Corporation: "IT'S MOST IMPORTANT TO STAY GREEN AND GROWING." Staying green means having enough income to pay all the bills. Net cash flow from rents is your best guarantee to stay green.

Lack of cash flow is the biggest problem I have with trying to mix investing with speculation! Speculators are all too often willing to tolerate short-term difficulties, like little or no cash flow, in the hopes that they'll soon strike it rich from a big sale. More often than not, the big sale never happens and the speculator goes bankrupt or out of business.

It's for this reason that I strongly recommend owning and operating KEEPER RENTAL UNITS. By doing so you'll always have cash flow generators to pay the bills. I also strongly recommend that cash flow rental units be your very first investment. It's most important to make investments that produce cash flow as quickly as possible. **Cash flow** must always be your **number one goal** if you intend to stay in business and earn big profits doing this stuff.

MAKING BIG BUCKS FROM THE YUCKS

Buying and selling fixer houses is not seasonal type work—nor is it a fad! It works well anytime, and it can be an extremely profitable business when you do it right! Many investors get very good at fix-up, but fall dreadfully short when it comes to developing a good management plan and marketing strategy. Some still make money, but not nearly as much as they could with decent planning.

The reason that fixing rundown houses offers such a high potential for making big profits is because investors can purchase them for only a fraction of their fixed-up price! By quickly adding value—primarily from fixing and cleaning—ugly rundown properties can be transformed into attractive houses that renters and buyers are willing to pay big dollars for. Transforming UGLY DUCKLINGS into beautiful SWANS is not complicated or scientific, and it pays handsomely once you get the hang of it!

A PROFITABLE SELLING PLAN REQUIRES PROPER TIMING

Giving away hard earned profits is not good business, so you must make plans to avoid it. There's a time to sell and a time to hold on. Cash flow is what allows you to hold on until the right time to sell. That's why it's so critical! Proper timing is very important when it comes to making big money in real estate. It's something like waiting to catch the biggest wave for a surf-boarder.

In real estate, we call this "selling during an up-cycle!" Ideally, you should plan for selling properties when great multitudes of buyers are out shopping for them. When buyers outnumber available properties, it drives up the selling prices. That's called a "seller's market." This is the time to sell for the highest price and with the most favorable terms for you, the seller.

When you have sufficient monthly cash flow coming in from rentals, you are in the best position to wait for exactly the right time (up-cycles) to make your sales. That's worth big bucks even though it means you must learn landlording while you wait. Landlording and property management will take time to learn, but the benefits will far exceed the troubles of learning—I will guarantee you that!

Lump-Sum Cash vs. Monthly Payments

Whoever said "Cash is King," was absolutely right! I would never disagree with that! However, there are many ways to skin a cat—and, likewise, there are many ways to become wealthy without waiting around for large cash payments. It's well known that cash sales made without proper tax consideration, or a plan for re-investing, can often cause a loss of capital. The loss of capital for any investor, especially in the early stages, can result in a serious growth problem. You could even end up going backwards—worse yet, going broke!

I have sold properties for thousands of dollars above the going market prices, because I've given excellent terms to buyers. When you consider those extra dollars earning 2 or 3 times more interest for me than most banks earn from their loans, it's easy to see why carryback paper is very good for your financial health. Obviously notes don't appreciate, so you'll need to keep real estate in order to keep growing. A good balanced diet of rental houses, and carryback notes with occasional equity borrowing sprinkled in, provides a well-nourished investment program with a guaranteed monthly cash flow.

It Doesn't Cost a Ton of Money to Start

Contrary to the old saying: "IT TAKES MONEY TO MAKE MONEY," I must respectfully disagree! In fact, I intend to show you that fixing rundown houses and small apartment buildings can earn you lots of money with very little up-front cash invested—sometimes none at all—when you learn to buy properties the way I teach you.

All of my early purchases had to meet two important investment criteria:

1. They had to be properties that I could acquire with minimum cash down payments—no more than 10% of the purchase price and sometimes less, whenever I could convince the seller.

    This first rule is not nearly so difficult as you might imagine, once you get targeted on the right type of properties.

2. The properties must generate positive cash flow within six months to a year after I acquire them. This doesn't mean positive cash flow on paper!—it means that green "foldin' money" I can stuff in my pockets every month, after I pay all the property expenses.

My goal was to acquire properties that would start producing earnings quickly, so I could quit my regular 9-5 job to invest in real estate full-time. If you learn to invest my way, but still wish to continue working at your regular job, rather than changing careers, that is perfectly all right! I'm sure the extra money you earn will prove you made a wise decision.

WHY FIXERS ARE THE PERFECT PLACE TO START

The simple explanation is that fixers are easy to purchase and they offer the best potential to earn quick profits without having to wait for appreciation to help you. Also, fixers can be acquired with very liberal terms, in most cases, and with very minimal personal risk to the buyer, if you structure the deal properly. As a result, fixer buyers can enjoy the biggest profits with the least amount of risk and have almost total control over their investments.

In case you're wondering what I mean by total control, I'll explain it more as we go along. However, consider an all-to-common purchase agreement where the buyer signs a promissory note that's all due and payable in five years! If you are the buyer and you don't have the money, or can't borrow it five years from now, chances are you could lose the property you've spent your blood, sweat, tears and money on! That's an example of very poor control over your investment. Fixer buyers can do a lot better than that. The primary reason is that sellers are forced to make more concessions in order to sell rundown real estate.

Adding Value Builds Profits Much Faster

There are many different investment strategies for making money in real estate, but almost all of them depend on future appreciation for the lion's share of profit-making! Appreciation

is worth big bucks when you're fortunate enough to own properties during inflationary times! However, when you own real
estate during a stagnant economy, you need a technique that
makes money without appreciation if you intend to stay in
business very long. Let me tell you about my strategy where
profits are not totally dependent on appreciation or even a
growing economy. I call it the "ADDING VALUE STRATEGY."

PROPERTIES MUST HAVE POTENTIAL FOR ADDING VALUE

In order to make a property more valuable, the property must
have the potential for improvement. There are many properties for sale that don't have the potential for adding value.
When you acquire these kind of properties, all you can do is
keep them operating efficiently, collect the highest income you
can and hope the value increases someday. In other words,
once you buy the property, you're more or less held hostage by
the economy—if it's good, you'll probably do all right; if it's not
so hot, chances are you'll be a victim of poor timing.

Investing in this manner is not nearly as profitable nor as
safe as using my adding value strategy! There are several reasons why my plan is better, but the biggest reason is that you
can achieve almost total control over what happens with the
property financially. This is made possible by purchasing the
right property to begin with. Remember what I said—YOU
MUST ACQUIRE PROPERTIES WITH THE POTENTIAL TO ADD ON
VALUE. Rundown properties with fix-up potential and properties which are poorly managed are the best candidates for
adding the most value quickly!

LESS COMPETITION ALWAYS EQUALS BETTER BARGAINS

When you set your sights to acquire rundown properties and
poorly managed real estate, you are automatically putting
yourself in the "profit mode" right from the start! The reason
for this is—there's far less competition. Most buyers are
turned off by properties that are ugly or rundown and have
management problems. This means there's a reduced number

of buyers for these kind of investments. Naturally less competition allows you to control the PURCHASE PRICE and TERMS—especially when no one else is making offers at the time you are! There have been many occasions where my offer was the only offer to purchase a rundown property. Obviously, sellers are receptive to most any reasonable offer under these circumstances if they're really serious about ridding themselves of their problem!

The following profit-making terms and conditions are generally always available to buyers of problem properties:

1. *Low purchase price (20-40% below fixed-up market value).*

2. *Minimum cash down payment required is normal.*

3. *Liberal seller financing for all or most of the mortgage debt.*

4. *Opportunity to increase income quickly (under-performing properties).*

5. *Immediate chance to reduce operating expenses and improve the bottom line.*

6. *Improve cash flow quickly by eliminating dead-beat tenants.*

Let me explain why these six TERMS and CONDITIONS are worth big bucks to investors who have the skill and know-how to fix the problems!

Low Purchase Price

Obviously when you can purchase a rundown property for 20-40% under the potential market value, you are building in a sizable profit to start with. It also means your debt service (financing) will be much less than comparable non-fixer properties and most likely can be held to 50% or less of the gross income. If you can acquire properties with 10-15% down payments (high leverage) and keep the monthly payments less than 50% of the gross income, you'll be in the positive cash flow mode right from the very outset.

Minimum Cash Down Payment

Sellers of fix-up real estate and properties with management problems are in no position to hold out for normal down payments, if they expect to sell their problems. I have seldom paid more than 15% down for any property! Also, many of my down payments have been for less cash because they were "lemonade down payments." LEMONADE DOWN PAYMENTS are part sugar—which is the cash, and part lemon—which is something else of value, like my old ski boat or used camping trailer! Even junk furniture stored in the garage will work sometimes.

For example, my offer on a $100,000 fixer property might be 15% down—consisting of $5000 cash and $10,000 worth of ski boat, motor and equipment. Chances are, my boat set-up would not sell for a nickel more than $5,000 through the classifieds, but to a motivated property seller, a value of $10,000 seems reasonable. Besides, how many "burned-out" property owners are boat appraisers? For the best results, execute this plan during hot sunny months—near a lake, if possible.

SELLER FINANCING ALL OR PART

Seller financing is the Cadillac of all financing when you learn to negotiate good terms like the following list:

1. Long-term payoff (15-30 years)

2. Low interest rates, 6-9% range, today's market.

3. No "Due-On-Sale" clause in note or mortgage.

4. No pre-payment penalty in note or mortgage.

5. No late fee in note, unless the seller insists on having one.

6. No other restrictive terms or conditions, such as buyer agreeing to repave common roadway when holes or ruts appear.

> *THIS CONDITION WAS ACTUALLY ONE OF THE TERMS IN A PROMISSORY NOTE I ASSUMED. IT'S NOT REALLY ENFORCEABLE, HOWEVER.*

Seller financing is better than FHA loans, GI loans or any other type of institutional financing when you structure it properly! Naturally, fix-up property sales are perfect for this, because most banks simply won't write loans for this type of real estate. Sellers must finance the sale themselves or they can't sell in many cases. Motivated sellers who own properties that won't qualify for bank financing have no choice other than carry back a mortgage or sell for cash (which is not too likely).

BUYING BACK YOUR OWN DEBT WORTH BIG BUCKS

One of the most profitable opportunities missed when seller financing is not a part of your investment strategy, is the chance to buy back your own mortgage debt later on from the beneficiary (that's the seller you make payments to). During the course of 20 years, or so, many things will change! For example, a seller who is only to happy to receive monthly payments for the next 20 years at the time of the sale, may suddenly find himself in a cash bind several years down the road. Money shortages are quite commonplace for all of us. Things like death, divorce, college funds, lifestyle changes and loss of employment or income can quickly create a serious need for immediate cash!

When you design your seller carryback mortgages with the good terms (for yourself), like I showed you above, they have much less market appeal to professional note buyers. Note buyers like to have a late payment clause, pre-pay penalty, high interest rates and much shorter terms; they won't pay very much for mortgages without them. This means if they do make an offer to buy the note, the price they offer will generally be so low the seller will be insulted—and, probably won't accept!

What this means is that the mortgage you purposely designed with very good terms for yourself is worth much less if it's sold before the pay-off date. Now you can buy it back much cheaper, because the seller probably can't sell it to anyone who would pay as much as you will. This strategy is worth big bucks when you do it right. I purchased my own $77,000 note for $41,500, just three years after I signed it. You'll never get this opportunity when dealing with banks or hard money lenders.

Under-Performing Properties

Investment properties that are not producing the amount of income they should are truly "gold mine opportunities" for investors who can spot the problems and fix them. The reason this technique is so lucrative is that the purchase price is based on current income production, which is low (under-performing). This price will generally be much less than its fixed-up value—the key here is to be able to clearly understand what is wrong and have the knowledge to fix it! It takes several properties (practice) to get good at this, but when you do, it's like taking candy from the baby.

When you have the ability to increase the income stream by whatever means you use, you will automatically increase the property value. This is the essence of my ***ADDING VALUE STRATEGY***. It's routine business for me to up-grade small multiple unit properties by doing physical clean-up and fix-up work to the buildings. At the same time, I'm gradually moving in new tenants who are willing to pay higher rents for a clean, fixed-up apartment.

WHEN YOU UP THE INCOME, YOU UP THE VALUE

Rundown apartments or junky houses that rent for $300 per month in a $400 marketplace are perfect examples of an opportunity for adding value. To start with, I would probably be willing to pay about six times the gross rents for houses renting at 25% below market value! Let's say we have eight units renting for $300 per month, for a total of $2,400 per month, or $28,800 annual. My purchase offer would be six times $28,800, or $172,000.

The value of a property that commands top market rents of $400 per month doesn't stay valued at six times gross (as when it's under-rented). Instead the value will increase to something like eight times the gross rents when the property is fixed up and looks good. As you will see, that can represent a big value difference!—eight units renting for $400 a month, equals a total of $3,200 per month, or $38,400 annually. Eight times $38,400 equals a new value of $307,200. When you learn to acquire under-performing properties like this example, you can quickly make yourself $135,000 richer. Suppose it takes you a year or so to complete the work! It still beats working for $15 an hour at the sawmill—don't you agree?

Excessive Operating Expenses

Start-out investors will often rush out and purchase leveraged properties at retail prices. Next, they hire professional property managers to run them. It doesn't take long before they learn a painful lesson about excessive expenses! To begin with, inexperienced investors and professional managers are almost the perfect recipe for bankruptcy! If there's a single most important experience that every new investor needs under his belt, it's the experience of operating his first investment property HANDS-ON. This is the best way to learn first-hand how much things cost and where the biggest savings can be found. Most smaller income properties with a 50% or greater monthly debt service will not support hired services.

As a general rule, operating expenses for older fix-up properties, which include management, taxes, insurance, repairs and maintenance, will cost anywhere between 45-60% of the total monthly income. It's not uncommon to find new investors making mortgage payments (debt service) in excess of 60% of the total monthly income. It shouldn't be very difficult to see there's a serious problem here! The first rule is: DON'T DO THIS! The second rule is: When you find properties with these kinds of financial problems, they may very well be excellent opportunities for ring-wise operators who know how to reduce the expenses. Also, you can often re-negotiate high mortgage payments with private note-holders, once you are able to determine that they really don't want the property back. Many will take lower payments instead!

Occupied With Deadbeat Tenants

Unruly tenants will often frighten potential buyers away from high profit deals. Investors who will spend the time necessary to learn local landlord-tenant laws can put themselves in a money-making mode. EDUCATION IS WORTH BIG BUCKS HERE—it arms you with the special know-how to handle tenant problems that often scares away most of your competition.

Many years ago I acquired a seven unit property filled with hostile-looking bikers. The purchase price was about half of what I felt the value could be. The owner was even afraid to show me the property—he was scared to death of his own tenants. I simply filed eviction papers and had the Marshal serve all the tenants. Several weeks later all the bikers were seen rolling down the interstate only minutes ahead of the Marshal, who had gone back out to evict them. All I had to do was clean up the jumbo-size mess they left in order to earn a handsome profit of almost $60,000. The evictions were hardly more than a short-term inconvenience for me, which unlocked the doors to LONG-TERM PROFITS and CASH FLOW. It's a very worthwhile trade-off—believe me!

LONG-TERM PROFITS AND CASH FLOW

The key to making big money—both monthly cash flow and long-term profits—starts with paying the RIGHT PRICE and NEGOTIATING THE RIGHT TERMS when you buy! You must negotiate until you achieve these goals, otherwise you are likely to have serious financial troubles after you become the new owner. Naturally, as in any other business or investment, there is no single rule or activity that will guarantee your success! However, BUYING RIGHT will go a long way toward making it happen.

CHAPTER 2

The Haywood Houses—A Textbook "Fixer-Upper"

*M*y Haywood houses were a textbook example of the kind of property that can make poor investors a whole lot richer in a reasonably short period of time! As investors often say, "This property had all the right things wrong with it."

CLASSIFIED ADS CAN SOMETIMES LEAD TO THE GOLD MINE

I found Haywood in the classified ads one Saturday morning. The ad read as though it were written especially for me. The described property sounded almost perfect! I responded quicker than flies to a picnic! By the way, let me pause to emphasis an important point here—***speed pays off***! When you hear about a deal that sounds really good, check it out quickly, especially if other people will know about it—like in the classifieds! Here's how the Haywood ad was worded:

> **INCOME PROPERTY FOR SALE—2 DUPLEXES PLUS 7 OLDER COTTAGES ON 2 ACRES IN CITY LIMITS. GROWTH AREA—FUTURE COMMERCIAL ZONING. PROPERTY NEEDS WORK—LOW DOWN PAYMENT. OWNER WILL FINANCE FOR 10% INTEREST. PRICE $189,000. CAPITAL REAL ESTATE CO. 413-4567**

First thing I did was to call the real estate office. The agent was off that day, so I called him at home! Somewhat reluctantly—after I promised that, if I liked what I saw, I would immediately call him back—he finally gave me the property address. He wanted to represent me if I decided to write an offer—he was the listing agent and had high hopes of representing me, too! That way he would get the total (100%) sales commission.

After looking at the property and deciding it was definitely what I was looking for, I called him back to set up an appointment. I was ready to make an offer—but first, allow me to "flash-back" and tell you what I saw at Haywood.

The location was excellent! It was in the east area of town where all new growth was headed. Although nothing exciting was going on at the time, I could sense future commercial zoning. Naturally, I'm always happy to have commercial potential when I buy properties; however, I won't pay extra for "Pie in the Sky"—and, neither should you. That's called speculating. What I'm doing is investing.

My first interest in a deal is always cash flow! Speculators, on the other hand, are willing to bet on a "Big Hit" sometime in the future—but for many, the big hit never comes, so there is no future. When you buy for cash flow first, like I do, you'll guarantee a future for yourself. Don't ever forget this point!

I was pleasantly surprised to discover that the "duplexes" in the ad were really four individual, two-bedroom houses. Detached houses have more appeal to tenants because they offer more privacy and individual living. Also, the older cottages were not actually cottages! Rather, they were older houses of various shapes and sizes. Some even had garages! On the issue of "PROPERTY NEEDS WORK"—no one would question that. The tall weeds and brush growing between the houses and around them was so high it nearly hid them. Several junk cars were scattered about and, fortunately, they were hid by the weeds.

FIND WHAT YOU'RE LOOKING FOR AND ACT QUICKLY

The one word that best described the Haywood houses would be "neglect." And, as you shall soon learn NEGLECT is worth big bucks for us do-it-yourself fix-up investors. Fixing neglected properties is mostly what I call "Grunt Work?" It taxes your back, but not your brain. Physically, it's better than a membership at the local gym, because, instead of spending money, you'll be earning it. Grunt work, on average, represents about 85% of most fix-up projects I've done. As you might guess—it's easy to procrastinate or fiddle around too long with an offer on ugly properties like this. Don't—because you'll lose them!

Besides the high weeds at Haywood, there were a host of other things that made the property ugly! Let me list them, so you'll know what to look for next time you visit a potential money-making opportunity.

1. *Unsightly yards, dead grass, unkept trees, high weeds.*

2. *No painting done for many years. Bare wood, peeling paint or repulsive colors on buildings.*

3. *Broken down fences, porches, sheds, carports.*

4. *Ugly roofs that distract from looks of houses or buildings.*

5. *Broken and bent, non-working garage doors—or lack of doors.*

6. *Exposed "pier-type" foundations with accumulated junk shoved underneath the houses.*

7. *Falling-down fencing and ugly entrance porches that looked like tenant "add-ons."*

8. *Non-running automobiles. Worse yet, vehicles sitting on blocks with no wheels.*

9. *Piles of junk strewn about property, including stolen shopping carts.*

10. *External fixtures falling off houses—like gutters, fascia boards, gates, window trim, shutters, screen doors, porch lights, ugly amateur-built add-ons and broken windows covered with cardboard.*

11. *Unsightly pens built with chicken or hog wire and scavenger materials.*

12. *Unsupervised dogs and stray cats running around every-where.*

13. *"Spider-web" wires, overhead electrical and telephone, running in all directions. Very unsightly!*

FIXING PEOPLE-PROBLEMS IS WORTH BIG BUCKS

Besides the physical things wrong with the Haywood property, there were also people problems! FIXING PEOPLE PROBLEMS PAYS BIG BUCKS—same as fixing house problems. In many cases owners become so "fed up" or intimidated by their tenants that they are willing to sell out for much less than the potential value. For fix-up investors who can handle people problems, big profits can be earned rather quickly! Here are the most common types of people problems I've encountered.

1. *Scary looking people with tattoos who hang around property drinking beer and working on junk cars.*

2. *The motorcycle crowd—where one or two legal tenants move in, then all their biker friends become permanent guests.*

3 . *Loud-mouth renters who constantly yell, fight and scream, causing good tenants to move out.*

4. *Deadbeat tenants who pay only when you chase them down or catch them with cash. Most are always behind with rents.*

5. *Renters who attract a constant stream of visitors, especially nights and weekends (dopers do this).*

6. *Uncontrollable tenants who routinely violate the owners' rules. Examples: Allowing unauthorized live-ins to occupy premises. Hauling junk cars onto property. Doing substandard alterations to living units (houses or apartments) without owner's permission.*

I'm sure there are several more I could think of, however, the six I've listed here should be enough to acquaint you with the basic people problems! They are the main sources of fuel for what I call; "The Fed-Up Factor." Many owners become sellers because they get *FED UP*. They simply get sick and tired of non-conforming, deadbeat tenants robbing them of the earnings they anticipated. Their dreams are shattered.

The Haywood owner had a mild case of "People Problems" when I arrived—and, it helped me a great deal in negotiating an excellent price and terms! Sellers will make big concessions when they lose interest in their property. In this particular case, the seller had moved to another town and asked me if I would mind over-seeing the property while we waited for escrow to close. He had no interest in even visiting his property again. Whatever his real estate salesman and I decided to do was perfectly all right with him.

Smart investors in this situation can fix short-term people problems in exchange for receiving valuable long-term benefits. For example, I could remove every tenant and replace them with new ones within a 6 month period of time—maybe less! That's a short-term problem easily corrected by a knowledgeable landlord.

In exchange for knowing how to solve tenant problems, I will expect to purchase the property anywhere from 20 to 40% under the normal market price, with a small down payment (5 to 10% range). I am also expecting long-term owner financing (10 to 20 years). These are excellent benefits for a buyer— plus, they are long-lasting. They're exactly the right ingredients for making a bundle of money with a fixer property.

Flexible Sellers Provide High Profit Opportunities

The seller wanted $30,000 cash down, which I didn't have. I offered $195,000, with $15,000 down and two additional principal payments of $5,000 each. The first payment due one year after close of escrow, the second due two years after close. You'll notice my offer is $6,000 more than the asking price in the newspaper ad. I was hoping that in exchange for paying a higher price, the seller would accept a smaller down payment and give me good terms on the seller financing.

The owner counter-offered asking $20,000 cash down payment. He wanted three $5,000 lump payments instead of two. He also agreed to carry back a note for $175,000, for 15 years, at 9.5% interest. One big concession I asked for and got was that my payments for the first three years would only be $1,200 per month. After three years, I would increase the payments to $1,500 per month, which was the amount the seller wanted initially. I explained that $1,200 was all I could afford based on present rents—later, after I upgraded the property and increased the rents, $1,500 per month payments would be acceptable to me. The seller agreed and we signed the deal.

Seller Financing Fits Needs of Both Parties

One reason I've had such good success with my fixer projects is that I've purchased older properties where sellers COULD and WOULD provide the financing. Buyers and sellers can design creative terms that work to solve each other's problems. Banks and most institutional lenders simply cannot do business this way. Their rigid lending policies and strict by-laws will not permit the kind of creativity we often need to make these deals work.

Many good buys would have been lost for me if creative financing had not been an option! Always look for properties where owners can provide all, or most, of the financing. Quite often new investors become totally baffled when they can't find a bank loan. I cannot over-emphasize the importance of seller financing, especially for do-it-yourself investors who need to keep all their options open—like buying the note back at a discount price in the future (see *Chapter 20*).

FLEXIBLE FINANCING MADE HAYWOOD WORK

My Haywood houses were a perfect example of seller flexibility. He originally wanted a $30,000 cash down payment, which I didn't have. I was lucky to scrape together $20,000 cash. To even come up with that amount, I had to borrow on my overloaded VISA card. You'll notice the seller allowed me to make a smaller down payment if I agreed to make three future principal payments. I agreed to pay $5,000 payments at the end of the first, second and third years, following the close of escrow.

The seller originally wanted 10% interest on his carryback note—I offered 9.0% and we compromised at 9.5%. That seemed fair enough at the time. However, even at 9.5%, the interest payments alone on a $175,000 promissory note were more than I could afford starting out. Here again the seller was very flexible. He allowed me to pay a reduced payment of only $1,200 per month for the first 36 months. I agreed to increase the payments to $1,500 or more per month starting at 37 months after close of escrow.

Obviously, during the first three years the monthly payments were less than the 9.5% interest stated on the note. There was no accumulated payments or add-on interest to this deal—we simply agreed to reduce the payments to $1,200 per month during the first 36 months. The seller would not start receiving his full 9.5% interest rate on the note until after the first 36 months. This concession alone saved me over $6,600 in interest. Now you can understand the reason why I agreed to the extra $5,000 principal payment.

If you will recall, my original offer specified only two $5,000 principal payments. As it worked out, the interest rate I paid during the first three years was slightly over 8.0% and not the 9.5% interest stated on the note. The lower mortgage payment ($1,200 per month) was extremely beneficial to me because my rental income was quite low to begin with. I knew that after I cleaned up and fixed the property, it wouldn't take long to increase the rents to market rates. I felt the current rents were about half of what they would be after two years of my ownership.

GOOD FINANCING SETS THE STAGE FOR BIG PROFITS

The good seller financing I was able to obtain was the key that would allow me to unlock big profits later on. Price-wise, I did not try to beat the seller down on his asking price. In fact, I paid him more than he was asking, but the terms he gave me were well worth the extra.

The terms of the financing you get when you purchase a property will directly affect your profits when you sell. They will also determine whether or not you'll get cash flow from the rental income. A 15 year mortgage term is about my average seller carryback. In my view, anything less than ten years should be avoided, unless there are special circumstances involved. Examples might be purchasing property for a very substantial discount—say 40%, or perhaps zero interest financing.

Clobbered Financing Won't Work

A common mistake that far too many investors make is becoming overly concerned about how much they pay, rather than how much time they have to pay it. Here's what happens: It's very difficult to sell properties with "clobbered" financing (that means lousy terms). Buyers will shy away from lousy terms. Typically, buyers want to pay so much cash down and assume some "nice and easy" long-term financing. If you cannot offer that, it generally makes your property less desirable to the buyers. It also means your profits will be less than they could be! When the financing is short-term, it's likely the monthly payments will be so high that it's impossible to ever achieve cash flow—also, most buyers are very leery of big balloon payments that are due three to five years after they purchase the property.

Looking for Loans in all the Wrong Places

Many deals I've witnessed will go like this—the investor buys an older property for $50,000. He's able to buy it for $1,000 cash down, because it's a real "dirt-bag" property. However, the seller wants the $49,000 balance in five short years. Older properties are always tough candidates for acquiring conventional bank financing, even when the loans are variable-rate with extra high points and low loan-to-value (LTV) ratios. The seller in this case has agreed to take $500 monthly payments, including principal and interest (mostly interest). Obviously, there will be a large balloon payment of over $40,000 at the end of five years.

The buyer figures he can "clean up" the property and fix whatever needs fixing. After that, he intends to sell or refinance the deal to make his profit. The problem with this strategy is that few buyers will purchase an older property with a $40,000 loan due in just five years. That means refinancing is necessary and that can be a very expensive proposition.

If you sell to a homeowner, he or she will need to qualify personally for the loan, as well as the property itself. This can be a real hassle and it often takes a long time! Worse yet, there can be many added expenses before everything is done. Obviously this means less profit for you.

If you sell to an investor, you can almost forget a refinance with non-owner occupancy lending rules. Naturally, it's tougher with older properties. Even if you find a lender, chances are the only type of loan you'll get is one where you are personally liable, in addition to the loan being secured by the property! It will also be a short loan, which means 50 to 65% of their conservative appraisal.

Many lenders will not allow sellers to carry back a note for any part of the financing. This means the buyer must have a larger down payment. Investors don't usually pay large down payments. Before you're through, this seemingly simple straight-forward transaction can turn into a real horror story. Worse yet, with $500 monthly loan payments, there's a good chance you'll never be able to rent this property high enough to have cash flow, should you keep it. *MY ADVICE*: Don't do deals like this—it's much too risky for such a limited profit potential.

FIXER SKILLS TURN UGLY DUCKLING INTO BEAUTIFUL SWAN

Let's get back to my Haywood houses. I want you to see how just one property can be enough to give your real estate career a tremendous financial boost. In terms of plain old profit-making from start to finish, I earned a respectable $8,000 a month for my "hands-on" fix-up skills and management.

I want you to understand that the skills I'm talking about are skills easily learned by anyone. Mostly, my skills involve clean-up, hauling junk and various other jobs that make a property look bright and shiny when I'm done. I call it *grunt work*. Equally important is my technique of rearranging the tenants. "Tenant Cycling" is the term I use for this particular management technique.

When I buy a property in rundown junky condition, I most always inherit tenants who are similar in condition to the property. You can't fix one without the other, because they go together like a pair of shoes. They are different parts of the same problem. For example, if you rent a clean property to dirty junky tenants, they soon have the property dirty again. Conversely, if you rent dirty property, you won't attract clean tenants. *ONLY DIRTY TENANTS WILL SHOW UP TO RENT DIRTY PROPERTY*. That's one of the unwritten rules of landlording. Don't try to change it. Those who do often end up drinking too much and run the risk of becoming dirty tenants themselves.

Before I leave the subject of clean-up and tenant cycling, let me offer some profitable advice. You must learn landlording skills the same way you learn how to fix up your properties. You will severely limit your potential for making profits in this business if you don't learn to operate your properties efficiently while you own them.

Getting a Tan and Building Your Bank Account

I spent the better part of one full summer doing outdoor work on my Haywood property. Sometimes, I would hire a handyman to help me. Several houses were in desperate need of major clean-up from top to bottom. They all needed painting. I started my inside fix-up on the first vacant house after a nonpaying renter decided it was better to move.

The most dramatic change was the cleaned-up appearance of the property as you drove in—a job that took about two months to complete. Mostly, this process was comprised of lots of clean-up and cutting down the overgrown trees and bushes. First, we chopped down the high grass and weeds around the houses, then we started watering ALL DAY. We sprinkled on new lawn seed to speed up the process—and after four months, all the yards were green as the city park. Like I tell my seminar students, it's extremely important to make the kind of improvements that show—*FIRST*. That means you should always haul away the junk, clean up the yards, evict the junky tenants, paint the building exteriors and fences, and anything else that shows, *FIRST (when you first drive up)*.

People passing by the property will notice your work immediately, especially if the property has been a real "scumbag" for a long period of time. Often passer-bys have stopped to tell me how nice the property looks since I started working on it. Everyone appreciates seeing ugly rundown houses fixed up and made attractive again. Occasionally a deadbeat tenant I've just evicted for non-payment of rent will stop by and say to me, "This place really looks great. I don't guess I'd mind paying the rent now." Unfortunately most deadbeat tenants seldom mean what they say.

Most Buyers Judge a House by Its Cover

If you forget some of the money-making tips I'm sharing with you here, try very hard to remember what I tell you next. It goes right to the heart of my fix-up strategy—it will also add thousands of dollars to your bank account, if you do as I suggest.

BUYERS WILL PAY TOP DOLLARS FOR LOOKS. More than just a few times I've watched real estate agents sell clean, attractive rental properties—while comparable ugly houses don't even draw a look! Often the ugly properties were far better values with more income and cheaper price tags. The thing I want you to understand is that a ton of money can be made in this business simply on the basis of how property looks after you fix it up!

You will discover that the buyers you eventually sell your rental houses to will be more than willing to pay higher prices for attractively painted, squeaky-clean properties. If you can offer them decent terms and provide most of the financing you'll learn just how easy it is to really "clean up!"—and I'm not talking about the yards this time.

I would estimate that at least 95% of all buyers make their buying decisions based primarily on the appearance of the property. Knowing what buyers think and do has proven to be very profitable knowledge for me. It provides me specific directions for exactly what I must do to earn my biggest profits as a fix-up specialist. It will work exactly the same way for you.

To Make Big Money You Need a Profit Plan

Before I completed my Haywood purchase, I had already penciled out my profit plan several times. My "ball-park" cost estimate for fix-up was $20,000. That amount included all the painting, fencing for several front yards and installing wood panel siding on three houses to improve their looks. I also planned to replace four garage doors and install new carpets in six houses. I always install new window coverings, either drapes or mini-blinds, when I "rehab." There were several leaking water lines to fix and several new sewer clean-outs that had to be installed to open up sluggish lines.

Most of the work that was needed was plain old grunt work! Naturally as with most fix-up houses, the maintenance and repairs were long overdue. According to the tenants, no one, including the owner, had ever been seen fixing anything. The tenants who stayed with me (only three) were delighted with the change of ownership.

The End of a Very Profitable Season

Two wonderful things happened to me during that summer. First, I got one of the best suntans I've ever had—and secondly, I earned about $75 an hour while I got the tan. By

Christmas time, I had eight new tenants and my rents had increased from $1,650 a month to $2,940. In case you need help with the math, that's $1,290 a month more than when I started. I considered that to be a very sweet return for my fix-up work. By the way, my actual fix-up costs eventually totaled about $18,000, and that even included the interest charges on my VISA cards. I was very pleased with the way things turned out—and even more satisfied that I accomplished the task for less money than I had estimated.

WAITING FOR "MR. GOOD BUYER"

Haywood survived the winter nicely and suddenly it was summer again. Income properties in my town were scarce, and it seemed like the timing was about right to make a decent sale. I ran the following classified ad to test the water:

> ### 11 HOUSES 2 AC. COMMERCIAL
>
> **GOOD TENANTS—ANNUAL INCOME $38,400—LOW DN. GREAT TERMS—10% OWNER FINANCING $350,000 CALL JAY TODAY—WON'T LAST LONG.**

The ad was right, it didn't last long! The second offer turned out to be the buyer. The selling price was $345,000!

I sold Haywood with the following terms: $50,000 CASH DOWN PAYMENT with an "All-Inclusive" note and trust deed (wrap-around) for the balance of $295,000. Payments to me were $2500 or more per month, all due in 13 years. You may recall the payoff date on my $175,000 note when I purchased Haywood was 15 years from close of escrow. I timed my carryback note to match the balloon payment I would face in 13 years.

I hope you can see the financial benefits of buying the right kind of property. Properties that can be improved quickly, inexpensively and with owner financing. Haywood was almost the perfect handyman property. When it sold, I got my $20,000 cash down payment back, along with the $18,000 fix-up costs and a little extra cash to boot! Also, I was now the beneficiary (the one who gets the payments) on a wrap-around promissory note that paid me $1,000 a month net income for the next 13 years.

Remember, Haywood next time you drive by a junky property with ugly houses and high weeds. *Think about the hidden profits*. Would you take on a summer's worth of work for $150,000 profit? I've had many jobs in my lifetime that paid me substantially less.

CHAPTER 3

The Profit Advantage Using Fix-Up Skills

Many people have a serious mental block when it comes to buying ugly rundown houses. Somehow they just assume that the present looks of a house has something to do with its long-term value. Nothing could be further from the truth! I'll explain as we go along.

I have watched many seemingly smart investor-types, including real estate sales professionals, walk away, or advise clients to do so, from so-called junker deals. Quite often that can be a bad financial decision, especially for the smaller "mom and pop" type investors who usually have very limited funds to begin with. Limited funds can earn much higher profits and earn them a lot faster when they are wisely invested in ugly ducklings. Let's discuss basic money-making techniques.

ONLY TWO METHODS TO MAKE MONEY

There are basically two ways to make big profits in real estate! The first is to locate and purchase a quality property at substantially less than its current market value—not necessarily the appraisal value, but rather its true value in terms of how much money it can earn for you! Earnings would be monthly cash flow now or as soon as possible, in addition to longer-term profits from a sale or exchange in the future. My personal view is that the highest priority should always be **cash flow now**. The reason is rather simple—investors must buy groceries on what they earn now, otherwise, they won't be around to reap any profits in the future!

Big Sharks Can Bite Harder Than Small Ones

The difficulties with this first method becomes painfully clear the minute you hit the marketplace in search of a bargain! What you'll first notice, right off the bat, is that there are a lot of investors out there with you. The competition is fierce! Not only that, but you'll also discover there's a bunch of dummies out there, with a lot more money to buy real estate than you have. I don't need to remind you, cash always wins! Just in case that doesn't bother you too much—I think you should also understand there are many ring-savvy "real estate sharks," who continually watch the market like buzzards circling over a dying cow. They are smart investors who can move lightning quick when they spot a bargain. Some are real estate agents who buy up listings before the general public gets "out of bed!"

SELECTING RIGHT STRATEGY IS KEY TO SUCCESS

I think you can understand what you're up against, if you are just starting or don't have much money! The odds that you'll find decent quality properties at a profitable discount and be able to snap them up quicker than the real estate sharks seems a bit presumptuous, to say the least. But, don't despair, there's still hope for poor investors who don't know much about the game yet—it's the second method for making big money and it works just as well. It's my method of choice and certainly the one I recommend for do-it-yourself investors. It's called "The Adding Value Method."

ADDING VALUE—A PRACTICAL METHOD FOR AVERAGE FOLKS

The good news is—the adding value method doesn't cost near as much money, on average, and the competition is much easier to deal with! That's because a large percentage of potential fixer-buyers wouldn't know a good deal if they were sitting on it! Amateur buyers have a tendency to rely on emotions rather than on buying skills. One common tendency is "playing house" with investment properties. Such things as a small dining room, purple toilets, black wall paneling and orange carpets have little to do with the investment quality or profitability. Yet, I've seen many a good bargain passed over because of these things. You'll find the competition is much less sophisticated in the smaller property fix-up market!

FIXING UP LOOKS AND MANAGEMENT EARNS PROFITS

The fix-up adding value method of investing comes in two flavors. The first is where the property needs physical fix-up work—things like painting, landscaping, fixing leaky roofs and building white picket fences. Also, you may need to upgrade inside, things like plumbing fixtures and new flooring. Obviously, these will add immediate value to any property. The second flavor is changing the operating procedures generally brought about by poor management. A good example is when out-of-town owners allow the tenants to manage themselves. Tenants who manage themselves will eventually manage the owner right out of business. Once an owner realizes he's losing serious money every month, it doesn't take long before a real estate bargain is available on the market.

Quite often, what unwary owners learn the hard way is that poor management not only robs them of everyday profits and cash flow, but it also costs them most of their long-term profits to boot!

Folks who have little experience in the fix-up business will often incorrectly judge a property by how it looks rather than by the benefits it will provide. You should stop and ask yourself every now and then, "*WHAT AM I DOING IN REAL ESTATE?* Do I want properties that look good, or do I want properties that pay good?" There must be no confusion about what your goal or purpose is if you wish to be successful.

One of my first recommendations, especially for new investors, is to sit down and plot out exactly how much money you'll start getting back each month once you become the owner of a property. Do this exercise before you buy, not afterward! People who purchase single family houses often skip over this drill. If they were to "plot out" negative cash flows month after month on a sheet of paper, where they could stare at the depressing negative numbers, I'm sure many would avoid the heartaches of doing "dummy deals."

Fixing Houses is Equal Opportunity for All

After doing fix-up for so many years, I must make a confession—fixing is much more satisfying and rewarding to me than buying the newer sweet-smelling and more expensive houses, although I do own some now. Let me simply make this observation about newer houses. They provide an excellent vehicle to reduce excessive cash flow. Also, if you have a problem with too much cash stacking up in your bank account, newer houses, with high monthly mortgage payments, can quickly eliminate the problem.

Someone recently reminded me that newer properties have other qualities just as important as cash flow. "They instill a great sense of 'ownership pride'," they told me. I'm a bit reluctant to buy that argument, but I will concede that money's not everything, so let's move on to some other stuff!

There are those who incorrectly believe that fixing houses is a job that only experienced carpenters or contractors can do. Nothing could be further from the truth. Almost anyone can do this job. In the final analysis, it matters very little who performs the physical "fix-up" work, as long as the right things get done.

Owners doing their own fix-up work will only enjoy a money saving advantage if they fix the right things at the right time. Both are very important. As you shall learn, KNOWLEDGE is what makes the big money. Swinging a hammer or swishing paint brushes will merely earn you average wages or save a few bucks in the short run.

The following two questions are frequently asked about fixing up houses.

1. ***Does fixing up houses require any special licenses or professional skills?***

2. ***Can women succeed in the fix-up business just as easily as the men can?***

These questions will be answered in the following paragraphs.

FIX-UP SKILLS EARN AVERAGE WAGES

As a general rule, no licenses are required by owners who fix up houses for themselves. However, if you do it for someone else—for example, as an employee or independent contractor, that's different! It's very likely you will need a license to be perfectly legal. Remember this book is not exactly the best source of information for doing everything perfectly legal. After all, you've never seen a rich cop, have you!

Having fix-up skills can certainly be an advantage because it's one less thing you'll need to learn about. But, having said that, I'm going to tell you something you should definitely underline and never ever forget. It's very important information because once you clearly understand and accept it, you can direct your efforts where the money is. BIG PAYDAYS DO NOT COME FROM FIX-UP SKILLS; THEY COME FROM REAL ESTATE SKILLS AND SPECIALIZED "HOW TO" KNOWLEDGE. When I first started out in this business, I had no idea how important this really was—although, now it makes perfect sense to me.

KNOWLEDGE IS WHAT BUILDS FORTUNES

To illustrate my point, just think about all the thousands of licensed building contractors who can do almost anything to a building—they have plenty of skills! Now think about what most of them don't have plenty of—if you answered money, you're right! Believe me when I tell you fixing skills alone are not enough if you intend to make big money in this business. Don't decide that fixing houses is not for you because you can't stop a toilet from running. That isn't where the money is! Let it run while I tell you what to do.

The Women's Advantage Shows Up

Fixing houses is definitely equal opportunity! The job is not gender sensitive, with perhaps one small exception—I think women understand living space better than most men I'm acquainted with. For example—cupboards and closets, cabinet space and electrical outlets in the bathroom—my male fixer team often ignores or overlooks the importance of these items. They are often called to our attention when a lady renter calls to complain about only one cabinet or not enough electrical outlets for all her bathroom goodies. Women seem to have a natural instinct when house remodeling is involved and I suspect this comes from their homemaking abilities. I have caught my male fix-up crew building a bedroom closet just large enough for three wire hangers, women fixers seem to know better than that.

Most everyone can eventually stop a leak in the toilet. Likewise, everyone can do a halfway decent paint job if they really try. Knowing how to turn those chores into cash money is a horse of a whole different color!

CHANGING THE LOOKS ALWAYS ADDS VALUE FASTER

It's not by accident that I always begin my fix-up project in the front yard. I've seen professional appraisers value identical houses as much as $10,000 difference because of plain old filth and junk on one property. In other words, a clean house is worth $10,000 more, simply because the owner hauls away the trash and keeps the house looking nice. Think about that for a minute. That's a lot of money for ordinary clean-up skills. Suppose it takes a whole week (40 hours) to haul away garbage and clean up a property. That's $250 an hour—nearly as much as a brain surgeon gets on his days off!

Scrutinizing Your Fix-Up Plan

All fix-up work should pass some financial scrutiny! Does it really need doing? I believe most improvements should be justified on the basis of paying for themselves. I expect the payments to come from higher rents or bigger profits as my reward for doing the work. Fixing or changing things around purely on the basis of personal likes and dislikes will seldom provide a justifiable "mark-up" (profits). Those kind should be avoided. This happens to investors who quickly charge forward without a plan. It also happens to folks who fall in love with investment properties. I advise you to be very careful and avoid these common pitfalls. Remember, fixing up dumpy-dirty houses is not glamorous work; but if you do it right, it will pay you better than anything else I know of.

Fixing Average Rental Houses In Average Locations

I haven't mentioned location before—that's another entire discussion. However, let me just say this—don't buy property near the Beirut Airport or in locations where you'll need a Bradley armored vehicle to drive through the neighborhood. It's not that you can't make money in a combat zone, because you can! The reason is this—houses like I'm recommending are not scarce in decent areas once you develop your "star search" network. Bad areas are simply not worth all the hassle. Save your energy for painting.

SIZZLE FIX-UP PROVIDES HIGH INVESTMENT RETURNS

Let me take a moment to say that all fixer properties must be brought up to basic minimum building code standards before you can expect them to generate income. That's a must rule for all properties.

When I talk about fixing for dollars, I'm primarily referring to what I call "sizzle items"—things like white-picket fences, fresh paint, window coverings, ceiling fans, wallpaper, new plastic countertops (Formica), attractive floor coverings, planters, shower curtains, decorative porches or entrance doors, trees and shrubs, green lawns, new faucets, modern toilets and new plastic shower enclosures.

The reason I call these sizzle items is because they are attractive and eye appealing, as well as useful. Sizzle items seldom have anything to do with code problems. For example, old dingy carpets will pass a code inspections same as bright new carpets; trees and shrubs have nothing to do with codes or safety; and neither does curtains or ceiling fans. What these items have is lots of customer appeal. This appeal translates into big dollars at the box office (a.k.a., my rental office). The very same appeal makes selling your properties much more profitable because they look better.

BIGGEST OPPORTUNITIES OFTEN FOUND WHERE YOU LEAST EXPECT

I'm constantly asked, "How do you find these kind of properties? 'Diamonds in the Rough,' as you call them!" Many folks tell me, "There's no properties in my hometown like you write about." If you will spend time looking, as I suggest, you'll find the properties are there! You just haven't found them yet. Obviously, there's a host of reasons why you haven't—but I've found most folks haven't been looking for them. Most investors I know do the traditional kind of looking!—if they like apartments and multiple units, they tend to look for traditional apartment buildings; single house buyers generally drive through the residential subdivisions looking for investments. Most investors would never think about looking at slaughterhouses or old motor lodges. Let me suggest that you expand your vision a bit. You might find that "non-traditional properties" are a lot more fun, in addition to being more profitable. At least, give it a try!

Creating Homey Atmosphere is the Solution

Since older houses are not the same as newer ones, don't try to make them so, instead try to capitalize on the marketable features not found in the modern day construction. Older houses quite often radiate charm—a homey feeling with high ceilings, woodwork, large porches, yard space, old windows (dressed up), evaporative cooling with separate heating, storage sheds, separate garages and, more often than not, mature shrubs and trees. All these items can add to the charm of older buildings. Add a freshly painted white picket fence after everything else is cleaned and spruced up and you'll have lots of customers!—Either renters or buyers, depending on your investment plan.

RENOVATORS, REMODELERS ARE A DIFFERENT BREED

I've discovered there is no inexpensive method to turn older houses into new houses. Many amateur fixers try to accomplish this task only to find their bank account disappears faster than the house changes. Herein lies the most important difference between what I do and what remodelers and renovators do. Believe me, *it's a very expensive difference, too!*

Often remodelers will replace entire plumbing systems with all new piping; sometimes they have the entire house rewired. They tear out old flooring and replace floor joists and girders. They replace wood windows with new metal frame styles. Some will even jack a house up to level it. That means they must also fix all the cracks and often redo the stucco exterior. Don't do fix-up this way unless money is not the object—you'll lose your shirt if you do.

FIXING UP BAD MANAGEMENT PAYS EQUALLY WELL

A good example of this would be buying a building that's always half empty because not enough effort is made to rent the vacancies. The seller of such a building will probably not be able to obtain the highest price possible because his income is artificially low. Income properties or rental units generally sell for a price that's largely determined by the income they generate. Most often a factor called *gross rent multiplier* is used to establish a selling price.

For example, if a six-unit building—fully occupied most of the time—is earning $400 per month per unit (six times $400, or $2,400, monthly), and—if the gross rent multiplier for the local area is eight—all other things being equal, the building will likely sell for about $230,000. It's figured like this: six units times $400, or $2,400, per month, times 12 months equals $28,800 annually, multiplied by the gross rent multiplier of eight equals a $230,400 selling price ($230,000 rounded).

Obviously, $230,000 is a "ball-park" number. Many factors—such as location, economic climate (is anyone buying?), financing, condition of building and motivation of the seller—will come into play during sale negotiations. However, remember this point—skilled buyers will pay pretty much in line with the building's actual earning record, not its potential earnings in the future! Higher earnings in the future should be your reward for better management!

HOW TO DETERMINE WHAT TO PAY

Using the example above, let's say an average of two units were not rented all last year. In that case, two times $400, or $800, per month times 12 months equals $9,600 in lost income. The reason doesn't matter! What is important is that the actual income earned is only $19,200, rather than $28,800 when all units are full. Now, when we calculate the selling price, it will be eight times $19,200, or $153,600—that's a big difference!

Real estate agents will often do their best to conceal or down-play the actual income because it makes a tremendous difference in the asking price of properties. In the example above, the difference is $80,000! Most real estate salesmen will talk to buyers about *SCHEDULED GROSS INCOME*, meaning that every room or apartment unit is filled up. Sometimes they'll even factor in future rent increases, too. You'll find that agents don't lack for creativity when it comes to setting a selling price. Of course, that's what commissions are based on!

To add some reality to the asking price, I always insist on seeing a copy of the seller's current tax schedule, *FORM 1040-E*. Quite often I find a big difference in the income reported to IRS and what the salesman tells me. I imagine it's similar to looking at the real estate agents automobile mileage reported on his personal tax return, then comparing it with the mileage he tells his auto insurance agent. Quite often numbers reported for the same things are not really the same at all!

SELECTING THE RIGHT PROPERTY FOR ADDING VALUE

Location – Several considerations are important! Ask yourself these five basic questions about location. If you answer yes to all five, it's okay to proceed.

1. Would I personally feel comfortable (safe) working in this part of town?

2. Do I think this particular location is safe at night, relative to other sections of town or areas?

3. Does this area look like it's reasonably stable (i.e., not going down hill, in terms of renter desirability)?

4. Are city or private services—like buses, police protection, schools, fire department, hospitals and decent shopping— available within a reasonable distance from the property?

5. Does this location look like a normal rental area where I can easily attract the kind of customers (tenants) I'm looking for? For example, low income families, HUD assisted, working class or rich tycoons with Cadillac Sevilles and speed boats.

CHEAP PRICE DOESN'T NECESSARILY MEAN GOOD DEAL

In determining whether the location is suitable for purchase, many inexperienced investors develop "tunnel vision!" They only "zero in" on what they perceive to be a *cheap price.* Remember, if the price is cheap, compared to the surrounding area, there's always a reason. The secret here is to not get "sucked in" on a cheap price. Make sure the answers to the five basic location questions, above, are "Yes" before you jump in!

To illustrate what I'm telling you—say you negotiate to purchase a duplex building in a particular area for a price of $70,000. We'll say that comparable duplexes in the area are selling for somewhere between $85,000 to $90,000. On the surface, without knowing anything else about the deal, it would seem as though you've found a property 20% below the average market. That part sounds great!—20% is an excellent discount, under normal circumstances. However, do remember this—everything that glitters isn't necessarily gold. Fool's gold glitters too!

Think About Who Will Live There and Pay Rents

If anyone of the five location questions are answered "No," you could be buying a problem, regardless of the 20% discount. For instance, a 20% discount is not nearly enough, in my opinion, if you don't feel comfortable working on your own property. If you're not comfortable being there, and you're the owner, I assure you no one else will either. Worst of all, neither will any potential customers (the renters). If the property is in a rough area, only rough tenants will live there.

In the case of a "No" answer on Question 5, you should ask yourself, "How could I ever make money with the property if I'm not able to attract the kind of tenants I want to live there?" Generally speaking, working class tenants desire to live around other working folks. That's mainly what they have in common! Conversely, they would not likely be happy for very long living next to non-working renters who stay up listening to the screaming music until 3:00 o'clock in the morning. In case you are "gearing up" to rent duplexes to the BMW crowd, you better figure on garage doors that open without anybody touching them.

I need to emphasize one point before I leave the five location questions. Nowhere have I mentioned anything about making predictions, judgments or forecasts about future appreciation. Many inexperienced buyers get themselves all "honked up" on this issue. My view is this: The future is speculation at best! If you know how to use a Ouija board, that's fine—it's as good as any other method. What happens quite often is this: A buyer will become convinced that the property is going to be much more valuable *SOMEDAY*. This thinking or guessing leads to paying too much. Don't pay for the future now. Stay away from guessing, you'll be better off! All you must do to make a profit is answer "Yes" to Question 3, that's good enough!

Fixing up houses, physically *DOING THE WORK,* is much easier to do than most people imagine. It's also much more profitable than most folks realize at first. It's not the least bit uncommon to earn back double the amount of money invested for a down payment and fix-up costs in a very short period of time. Quick, 100% returns are commonplace among experienced fixers. Obviously, you'll need to get the hang of it first! Fixing houses is the best "fast track" wealth strategy for common everyday people. It's also the most inexpensive and safest way to invest in real estate today.

You Can't Get There (Wealthy) Starting From Here

In my town, basic run of the mill, three-bedroom starter houses cost $105,000. Rented, they'll bring in $750 per month, tops. With a 30 year, 8 1/2% fixed FHA mortgage, the payments are $692.03 per month for principal and interest. That's after paying $15,000 cash down, plus an additional $5,000 for all the other costs. When you add in the taxes and fire insurance policy, we're talking $820 per month without a single nickel for repairs and management. It's easy to see that total monthly expenses could average somewhere around the $900 range if I bought this house!

I don't know about your financial means, but to me this kind of deal doesn't make much sense! Worse yet, it may not make any profits for years to come. The only thing I can see happening here is continually paying out hard cash every month, waiting for the value to go up or hoping my rent can be increased enough to cover monthly operating expenses.

To borrow the over-worked slogan from a past presidential campaign—"Where's the Beef?"—I tell you this is not investing, it's speculating! I have no problem waiting patiently to earn *some profits*, but I'm not the least bit interested in waiting for all of them.

THE BIG MONEY IS THERE, BUT DON'T FOOL YOURSELF

You can't build big bank accounts on what you think you can do, only on what you really can do! You can save yourself a lot of time and wasted effort by sizing yourself up correctly to start with. For example, when I first began buying houses, I was not really a very skilled investor. I was actually a telephone man! I had over 23 years experience doing telephone work. People still thought of me as a telephone man even when I owned 18 houses. It took three or four years after I quit my telephone job, before people began to think of me as a real estate investor

What does all this have to do with investing, you're probably thinking. The answer is "plenty." You can benefit yourself a great deal by doing an honest self-evaluation. Figure out what you're really capable of doing right now, not what you imagine you can do! In other words, level with yourself. It's not wrong to know almost nothing about real estate investing when you're starting out. What is wrong is to fool yourself into thinking you do! This is a serious mistake that can hold you back, permanently. Also, it makes you an ideal target for any "snake oil salesman" with a "Get Rich" program to peddle.

It's most certainly true, you can make yourself a ton of money investing in real estate. You can do it gradually at home part-time on your kitchen table. You can build an unbelievable amount of wealth and security for yourself and your family. You can also be your own boss and insulate yourself from the ups and downs of a shaky economy that besets most ordinary people. I'm convinced you can do all of this, but you must be completely honest with yourself when you start.

STILL WAITING FOR THE TOOTH FAIRY—FORGET IT!

I used to see a popular TV infomercial showing a "late-night" real estate guru skimming across the waves in a speed boat surrounded by bikini-clad beauty queens. Million dollar mansions dot the shoreline in the background. It's still unclear to me whether real estate or bikinis were the point of focus. However, I can tell you this much—If you have any delusions about the houses you see on the TV screen being the kind you should be trying to purchase, call me quickly because I've got a bridge I'd like to sell you!

It's totally unrealistic to think that "new guys" can simply jump in and beat out the "experienced," old-pro investors, buying choice real estate at bargain prices! Their superior skills, knowledge of the marketplace and probably tons more money, make for a very uneven "playing field." It would be about like matching a high school boxer with a ranking professional. The odds are simply too long!

FORGET ABOUT INVESTING IN UNFAMILIAR PLACES

It's not the least bit realistic for inexperienced investors to think that somehow they'll have better success acquiring properties away from home. Many "wanna-be" investors seem to have a far-fetched notion that they can just stroll into town and pluck out bargain properties right under the noses of local sharks. That idea, my friends, is total nonsense! Investing in another area might play well on TV cable shows or sell a few books, but in most cases it's a huge mistake. It will cost you, rather than pay you! The idea sounds attractive because it conjures up visions of travel, adventure and romance. I call it the "greener pasture itch." Small-time real estate investors looking for a duplex or two, are much better off investing in the boring town they live in. At least, they're familiar with the hometown turf! Judging by what's happened to many out-of-town investors I know, I'm convinced the chances of making any serious money are about a thousand to one and that's with good luck! Smart investors do it at home!

Pause for a Word From the Announcer

I have purposely side-tracked the "how to" part of my adding value strategies for a moment in order to stimulate your thinking. It's worthwhile here, I think, to recall the catchy "one-liner" from the Clint Eastwood flick, "Dirty Harry." Harry said, "A man's just got to know his limitations." Nowhere does this line have more truth than for the do-it-yourself real estate investor. Investors must understand their own capabilities, both strengths and weaknesses. Only then can they develop a plan that fits within their personal limits.

ADDING VALUE STRATEGY IS PERFECT OPPORTUNITY

I have used exactly the same advice I'm passing along to you! The day I started investing, I knew there were two important things I was short of—KNOWLEDGE and MONEY. Do-it-yourself investors just can't lack a whole lot more than these two things.

On the positive side, I had a couple of good things going for me. First, I had a very strong desire to learn everything I could about real estate investing as quickly as I could do it. Secondly, I was 100% ready and willing to substitute my own personal efforts, whatever was needed, to make up for my lack of money. Those two things can level the playing field if you don't renege on them! Also, they work absolutely perfect with my ADDING VALUE STRATEGY. The following is an actual classified ad from my local newspaper that seems to fit my plan:

WILL TRADE, SELL OR WHATEVER

Small duplex unoccupied on 3/4 acre. Partly remodeled. Ran out of money. Each unit 2 bedrooms and detached garage. City services. Good potential for fix-up. Assume $21,000 - $167 per month. Will take vehicle - Lot paid for - Motor home, boat, etc. For equity, call 227-3414 Leave message for owner.

When I read this ad, I can pretty much tell that the two things I don't have much of, KNOWLEDGE and MONEY, will not likely be too important in terms of doing a deal. Conversely, my two strong suits would seem to fit this property quite nicely—namely a STRONG DESIRE TO LEARN and 100% WILLINGNESS TO CONTRIBUTE PERSONAL EFFORTS. My thinking went something like this.

Agreed, I'm not very knowledgeable about real estate transactions because I'm just getting started here! It looks as if I'll be dealing directly with an owner who probably doesn't know much more than I do, otherwise why would he be stuck with such a mess? Also, I don't have much money, but it doesn't seem like I need very much to acquire this property. The owner says he is willing to take almost anything for his equity! Obviously, my personal efforts (saw, hammer and labor) can be substituted for lack of fix-up cash.

Pending further investigation, this property sounds very exciting to me because it seems almost "tailor-made" for what I am able to contribute. If an ad sounds interesting, you should immediately make contact and start talking with the owner. This is how "high profit" deals start out. I realize we're merely discussing the ad at this early stage, but try to vision what you might quickly uncover here! Some things are written, others you must interpret by reading between the lines. Also important to note is the ad appears to be written by the owner, rather than a real estate broker office. Most generally, owners have a tendency to state the truth because they are not professional ad writers.

KEY INGREDIENTS TO LOOK FOR WHEN MAKING A BARGAIN PURCHASE

1. EASY TO PURCHASE. VERY LITTLE CASH REQUIRED, TRADES ACCEPTABLE.

2. IT'S OKAY TO ASSUME EXISTING FINANCING, IF THERE IS ANY.

3. PRIVATE NOTES (MORTGAGES) ARE MUCH BETTER THAN IN-STITUTIONAL LENDERS (BANKS AND MORTGAGE COMPA-NIES).

4. WILLINGNESS OF SELLER TO CARRY BACK FINANCING ON GOOD TERMS (MINIMUM OF 10 YEARS).

5. MORTGAGE PAYMENTS (TOTAL OF LOANS) ARE 65% OR LESS OF TOTAL INCOME.

6. LOW RENTS THAT CAN BE INCREASED WITH PROPERTY CLEAN-UP.

7. DIRTY, UGLY PROPERTY, WHICH MEANS LESS COMPETITION.

8. OUT-OF-TOWN OWNERS WITH HIRED PROPERTY MANAGER OR TENANT MANAGERS.

9. "GIVE UP" OWNER WHO HAS DECIDED TO DO SOMETHING ELSE.

10. "OUT OF CONTROL" PROPERTY, SUCH AS THAT WHICH HAS BEEN TAKEN OVER BY "BIKERS."

11. YOU HAVE THE ABILITY (SKILLS) TO FIX UP PHYSICAL PROBLEMS.

12. YOUR JUDGMENT. THE PROPERTY WILL FUNCTION OKAY AS IS, ALLOWING ADEQUATE TIME TO UPGRADE AND FORCE THE VALUE UP.

These 12 items are certainly not everything to consider, nor will any of them necessarily guarantee the property is a bargain. They are, however, excellent ingredients that go into the making of a profitable deal.

CHAPTER 4

Specialization is Quickest Path To Earning Big Profit

SPECIALIZATION is the quickest way to learn this job. It's also the best way I know to rise above your competition. When I first started buying rundown houses, most of my competition was from agents and owners who looked at every property listed for sale. I only looked at fix-up properties and could therefore concentrate all my efforts on just one specialty—MY SPECIALTY.

After a few deals, I became very knowledgeable and soon learned how to beat my competitors on many of the best money-making properties. Later on, agents started calling me as quick as they listed good fix-up properties. Needless to say, you're in the driver's seat when you are the first one to know about a good deal. It took awhile to build my credibility; however, by concentrating in just one area of the market, I soon became recognized as the local authority of *FIXER HOUSES*.

Think Like An Investor

Before you start looking at the specific types of properties to purchase, it's time to pause and think. It's most important to start out by adjusting your thinking to fit the situation. What situation do I mean? Ask yourself this question: "Who is most likely to sell their property at the best price and with the best terms?" Let me tell you who—SELLERS WITH SOME KIND OF A PROBLEM. Generally, they won't tell you about their problem.

You will need to find the problem on your own. That's the situation and that's the kind of seller you should be looking for.

I can assure you, sellers like that are out there. You must purchase properties from sellers who are really serious about wanting to sell—NOT FROM SELLERS WHO *MIGHT* WANT TO SELL. That little five letter word, *MIGHT,* makes a very big difference. The seller you must find should have a strong reason for selling. It can be one of many different reasons. However, I've found the most common of these to be financial problems, management problems, divorce and job transfers or moving. It's not uncommon to find all four of these problems existing at the same time.

GETTING STARTED RANKS FIRST

Before I tell you how to buy your first property, or first several, let me remind you of a very important fact of life. Everything we do begins with "the first time." It's not necessary to be fully trained and have years of experience before getting started. On-the-job training has long been recognized as one of the most effective methods for achieving journeyman status quickly. Getting started is always top priority. First, find out what you must do, then go out and do it. Keep learning how to do it better as you are ACTUALLY DOING IT. Live combat training with real bullets will speed up your education by tenfold.

NO MONEY DOWN IS NOT THE GOAL

First, let me explain why it is not always best to buy for no money down. Quite often a little cash, inserted at just the proper moment, will hasten the desired results and will more than make up for the problem of finding the money somewhere. You normally get far better terms. Obviously, you must have some cash, or you must find it somewhere.

Many "No Down" investors attempt to use what I call the "Christopher Columbus technique." When they begin their investment voyage, they have no idea of where they're going or what they should buy. When they finally do acquire a property, they seldom know much about it until after they close the deal. Finally, they seem to think the entire transaction should be financed, using someone else's money. Quite often they go broke before they ever realize where they've been or what they've done. Hopefully, by learning a few basics you can avoid such a fate. A little early planning can make your maiden investment voyage smoother sailing and more profitable.

IT'S IMPORTANT TO POSITION YOURSELF TO MAKE MONEY

For "hands on" investors, like myself, I have long held the notion that the best type of real estate to buy, when you are just starting out, is something between rundown and ugly to a complete "junker." The degree of rundown or ugliness will be dependent mostly upon how brave you choose to be. Personally, I think the braver you are, the better off you'll be with your first several acquisitions.

EVALUATE YOUR PERSONAL FINANCIAL STATUS

As you might suspect, how you invest will depend on your present financial situation. Do you have a regular job right now? How long will it last? Are you planning to quit or retire? Do you have some cash available? Do you own a home with borrowing equity? Are you flat broke with nothing but determination and a pretty face? Each individual's situation will be different to start. However, do not despair if you don't have much money right now, many have done fine without any.

GUTS, DETERMINATION and DESIRE to succeed are the top priorities. They are necessary with or without money. Money simply adds some comfort. Remember, that just as many flat broke investors make it big as do those who can easily borrow "start-up" money from their mothers-in-law.

The selection methods for purchasing your first property will be the same whether you are currently employed and hoping to change careers down the road or presently unemployed and desperately in need of work that will produce some spendable income right now.

FORGET THE "BLUE RIBBON DEALS"—GET STARTED

Here's some good advice for you: Don't over-complicate getting started. Getting started ranks higher than finding the best deal in town, when you first start out. Don't look forever trying to find a "blue ribbon" property. When you first start, chances are you will stumble past the right deal on the way down to see it, without even realizing you were there. The point I'm making here is most of us don't really know enough when we first start to distinguish a good bargain from a "Grade B" dirty deal. (By the way, that's poultry jargon from my old chicken farmin' days.) The best approach in the beginning is to SPECIALIZE in whatever type of property you decide works best for you. I recommend the older fix-up types, because profits can be earned much faster than with the nicer looking units. Making profits is much faster when you *add value*.

Continuing Education—A Must

In my own particular case, I specialize in fixer (rundown) houses and apartments. That's how I started and I'm still doing the same kind of deals today. I've gotten used to the money. I also invest in other types of properties now, but only if the projects are good enough to meet my financial objectives. Also, remember I've gained lots of experience from buying properties over the years. I have already paid my dues. I'm an educated buyer now, thanks to what I've learned from others, attending many educational seminars, reading several hundred books (many over several times) and spending literally thousands of hours practicing my first love—*FIXING RUNDOWN HOUSES*.

LOOKING FOR MR. RIGHT—NOT PERFECT

The first property you buy should be the largest one (in terms of the number of rental units) that you can acquire based on your available resources. Resources can include cash, trades, or sweat equity (working on the property in exchange for the normal cash down payment). Trades can be boats, trailers, cars—even your personal residence, bare land or whatever. If you're a good negotiator, perhaps a no-cash-down deal is possible. However, be very careful with large mortgage payments and short-term financing with large lump sums or balloon payments due in a year or so. Don't get yourself trapped by either one, especially when you're just starting out and you don't have extra cash reserves.

Zeroing in on the Target

The target purchase should be a rundown property with management problems and a very discouraged owner. The location and condition are not nearly so important as seller motivation and the potential for property improvements or upgrades. *Improvement means the ability to increase the property value and in turn increase the rental income.* Obviously, there are properties in some locations that really need tearing down and starting over. However, the vast majority of older rental units suffer only from poor management and neglect by owners who simply milk the property and do little or nothing about upkeep or improvements.

SELECTING A PROPERTY THAT'S RIGHT

Your first property purchase and the next several after that will be extremely important to your investing career, although it's very likely at the time you do it, you probably won't realize just how important it is. There are right properties and wrong properties available in the marketplace. Choosing the kind that earn *LONG-TERM PROFITS* and *SHORT-TERM CASH FLOW* is what I'm talking about.

Making a good solid "first property selection" can easily spell the difference between establishing a strong permanent launching pad for your investment career or creating a deep financial hole that will quickly bury you. Your dreams of wealth can quickly fizzle out and die if you purchase properties that cost you more than they'll ever be worth. Buying the right property is the first step, and it's extremely important because it's usually make or break for new investors. What I tell you next will provide valuable information on how to buy your first properties right and what to look for before you sign the deal.

FINDING SELLERS WHO TRULY WANT TO SELL

The most difficult part of buying property right has nothing to do with contracts, paperwork, escrow instructions and presenting offers. It's finding sellers who really and truly desire to sell their properties. We are looking for sellers who have compelling reasons to sell. Sometimes the fear of losing their property in foreclosure creates pressure. That's an obvious reason. Many reasons are hidden or not so visible; however, always remember this—*A REAL NEED SITUATION MUST EXIST FOR THE SELLER, OR YOU WILL HAVE GREAT DIFFICULTY MEETING YOUR PURCHASE OBJECTIVES* (that is, buying the property with the kind of terms you need to make the transaction a profitable one for you). Again, I will repeat myself here because it's so important.

You must locate sellers who, for whatever reason, have a real need or a very strong desire to sell. Nothing short of that will work!—don't forget this part. Also, don't pursue sellers who don't have some urgency. You'll be wasting your precious time. In the long run, your time is more valuable than money.

UGLY PROPERTY BUYERS GET THE BEST TERMS

To start with, sellers of these kinds of properties are not in a position to be very picky about who they sell to. They can't play "hard ball" with the price and terms like owners of higher quality, nicer looking properties. The main reason is because 95% of all potential buyers are "turned off" by the rundown and ugly condition. Consequently, lack of buyer competition will greatly limit the owner's ability to sell.

What this means is you can almost always buy these properties for much less cash up-front (lower down payments). Also, it's likely the seller will be forced to accept much weaker terms. Lower equity payments and carryback notes are very common. Some of these deals can be 100% owner financing. Ugly, fixer-type properties are generally older properties.

Many of them no longer have conventional mortgages (i.e., bank loans or savings and loan mortgages) to pay off. When they do, they are most likely low balance loans with lower interest rates and nearly always assumable to new buyers. For "beginning" investors especially, let me say this loud and clear—**OWNER FINANCING IS THE KIND YOU WANT!** Owners are almost always more flexible to deal with than banks.

Property purchases where the owners will carry back low interest financing are the kind of transactions that allow you to buy real estate with minimum cash down payments and, yet, still be able to generate cash flow. Bank financing with higher interest rates and variable rate mortgages are not the kind you want. Bank financing will seldom be much of a problem when you buy older rundown type properties like I recommend. The reason is—original bank loans, if there were any to begin with, have long since been paid off—and most banks won't write new loans on older properties today.

Know Yourself and Develop Your Hidden Talents

One very important reason to start out with this type of property is because it's a good experiment for you. It provides you with an opportunity to learn what you can and cannot do. If you purchase a rundown house or small apartment building that's already an existing eyesore when you buy it, ask yourself this question: "How can I possibly make it worse?" Even with very limited handyman skills, your efforts are still likely to make some worthwhile improvements. If you don't do things exactly right the first time, so what! Who cares! No one, but you, will probably even notice. Simply do it over again until you get it right.

Doing ordinary clean-up work, which almost everyone can do fairly well, is likely to result in a major upgrading. Most certainly it will improve the looks. When you tackle the more sophisticated improvements or repairs, take your time. Read a book or two and look at the how-to pictures. I promise, you'll be pleasantly surprised to find out how many things you can actually do if you make the effort.

A Plan That Forces You to Learn

As a rule, investors will not develop these do-it-yourself skills when they buy an ordinary "non-fixer" type property. Most people are generally reluctant to tackle jobs they've never done before. They're afraid they might ruin something or make a bad situation worse than it already is. That's a major advantage to buying and working on your very own "guinea pig." With fixer properties, most folks are not nearly so afraid to experiment. I will promise you this much—you'll learn to do some things you never thought you could. More importantly, the experience you get will be invaluable as you continue to build your real estate wealth.

PERSONAL EFFORT CREATES INSTANT EQUITY AND LONG-TERM WEALTH

Another important reason for buying rundown ugly houses is that you can add value very quickly to these types of properties. What this really means is that you will be able to buy cheap and sell for a profit in the shortest amount of time. You can also increase rents and develop a positive monthly cash flow much more rapidly with these properties. Unlike buying pride of ownership properties at top market prices where you must wait for appreciation to profit, fixer properties are different. You can force values up by making improvements. Upgrading the property automatically increases its value. You don't need to wait nearly so long to earn your profits this way.

Clean-Up is Not Difficult and Cash Flow Comes Faster

With fix-up real estate, often a simple clean-up and painting, together with changing the tenants, will earn you more money almost immediately. With increased income from rents, you now have a more valuable property. Several of these deals can "pump up" your equities quickly and allow you to build wealth much more rapidly than with non-fixer properties. I call this *"fast track investing."* The chapter on fast track investing (Chapter 11) tells more about adding value.

Finally you get another important benefit, which I rank just as high as the others—it's called confidence in yourself. Once you actually experience positive money coming in, you begin to realize, *"HEY, THIS IS ME!* I'm really making good things happen."* Confidence in yourself is worth more than the money. And, strangely enough, money builds confidence faster than anything I know of.

DON'T BUY UNTIL YOU KNOW HOW MUCH TO PAY

If you use my INCOME PROPERTY ANALYSIS FORM (see Chapter 5), it will force you to dig out the information you need. It's an excellent checklist! And, of course, everyone should use some kind of checklist so expenses don't get overlooked. Perhaps the most important use for my form is as a negotiating tool. With all INCOME and EXPENSE numbers on the table, it quickly becomes apparent if the property will make any money. This form tends to take away the emotional tendencies of negotiating. For example, if the seller argues that the expense numbers I'm showing are too high, I say, "Fine. You show me what they really are, then prove it." When the form is properly filled out, it will show you and the seller exactly how much money is needed for operating expenses and how much is left over to pay the mortgage debt. It helps you avoid paying too much for a property when you can see all the income and all expenses in one place, at the same time.

Beware of Forked-Tongue Sellers

Do not buy any property based on average expenses for the area. Sit down with a blank piece of paper and figure them out for yourself. If you're a beginner, go heavy on expenses and light on income. You'll find out quick enough why I advise you to do this. If you're already operating your own real estate, then you know the reason why!

Here's what helps me the most. I consider all financial data (income and expenses) to be A BIG LIE. I never believe the financial information given to me by the seller or his agent. Don't misunderstand me here, I'm not calling sellers and agents liars. However, I want you to think, like they are thinking. This will help you "dig out" the correct numbers and get the proof to justify all expenses. For example—utility bills—the seller can easily get a computer printout showing several years worth of billings by meter numbers. Obviously, this is excellent proof. It's exactly what you need to verify what the utility costs have been over a period of time. Remember, only the current owner can get this information. When you figure the annual taxes, don't forget your new tax bill will be based on how much you pay for the property. It won't be what the seller is paying now.

OWNERS SELDOM OVERSTATE INCOME ON TAX RETURNS

For income verification, I think it's best to see the seller's federal tax returns. *The 1040 Schedule E* is what you need, in most cases. I have never seen lower income figures published anywhere for any property than on the owner's tax returns. Since income property sale prices are largely determined by the income generated, it's obvious how you'll benefit by having the correct income figures. I have found from personal experience the first income numbers I'm given are generally higher than those shown on the seller's tax return.

Find Out What the Seller Really Needs

Initially, it's quite likely you won't know the seller's real needs because he won't tell you about them. Eventually, as you gain more knowledge from practice and experience, your detective instincts will improve and you will learn how to discover the information you need more easily. Knowing the real reason a seller wants to sell can be a tremendous advantage when negotiating the kind of terms that you must have to be successful. *Successful means profitable*—that's why it's important!

"FOR SALE" OFTEN MEANS "I NEED HELP"

I know from experience there are many owners of small rental properties—the kind you need to buy—who have their properties listed for sale and what the "FOR SALE" sign really means is "I NEED HELP." These sellers have lots of different problems. However, many of them need some serious help. Their solution to getting help is to sell the property. These owners can provide big financial opportunities for investors without even realizing it, unless, of course, they read my strategies, same as you are doing right now.

DON'T WANTERS—THEY DON'T KNOW HOW TO MANAGE

Many real estate lecturers refer to these types of sellers as "DON'T WANTERS." The number one reason most sellers of small income properties are DON'T WANTERS is because they cannot manage their properties. Mostly, they can't take the hassle of dealing with their own tenants. Also, most of them have never taken the time to educate themselves in preparation for this most important task. They are victims of the "Columbus" technique. They don't know what they're doing or where they are going. They are simply drifting with no direction. They have no plan and, therefore, are unable to fix what's wrong. You don't want to invest like that, believe me—it's not much fun!

VIOLATION OF HABITABILITY LAWS—A MOTIVATOR

Many owners become very motivated to sell when needed repairs stack up or when the tenants "gang up" and refuse to continue paying rents until something is done. Quite often city or county housing officials add extra tension when habitability laws are violated. Housing authorities can provide tremendous motivation to owners who neglect their properties. Many of these owners become highly motivated sellers when they find out what building officials require them to do. Out-of-town owners often panic and decide to sell at the first chance they get. You can find these properties by attending abatement committee meetings in your local buying area.

Owner's names and mailing addresses can easily be obtained by checking the county parcel maps, locating the property, then matching the parcel number with the tax rolls to obtain names, assessed value and mailing addresses for individual owners, partnerships or corporations.

Watch Out for the Hole in Your Doughnut

It's very important in pursuing our goals, that we always keep our eye on the target. Our target here is to become a successful investor and operate cash flow properties. Nothing else matters once you are started. You must be totally determined to succeed. To emphasize this point, I'll share a poem I learned in my third grade classroom years ago. It has great application here, I think.

"AS YOU TRAVEL THROUGH LIFE'S JOURNEYS, WHATEVER BE YOUR GOAL—KEEP YOUR EYES UPON THE DOUGHNUT AND NOT UPON THE HOLE."

I don't mean to imply that your first rental property purchase has to be a hole. I'm just saying that the best deals will come from motivated sellers who don't know how to manage their property. Quite often the property becomes an ugly eyesore and the owner becomes a serious don't wanter. This type of owner is very likely to sell with excellent terms for the do-it-yourself investor—that's the doughnut you want!

A Fireman is in the Strongest Position at a Fire

It is very important for you to understand that bad management and bad tenants are only temporary problems (and they are curable). You want properties with curable problems. On the other hand, short-term, high interest mortgages and negative cash flows are serious long-term threats that cause permanent problems. More often than not, they are non-curable. Permanent problems can be fatal when you're just starting out.

Most owners of rundown properties must sell cheaper with flexible terms (very good for us fix-up buyers), because they don't receive many offers. They become very lonely sellers. After awhile, with no offers, they finally realize they'll only get help if they sell at FIRE SALE PRICES. Of course, we are the firemen when this happens. It's up to us to provide the ladder so that discouraged owners can climb out of their fire trap without burning themselves too badly.

LOOKING FOR UGLY DEALS

Most buyers (at least 95%) place very heavy emphasis on LOOKS. Therefore, we immediately have one major advantage buying rundown properties. Ninety-five percent of the buyers don't show up. That's less competition! It's knowing what to buy and how to go about buying it correctly that makes the difference between profit-makers and negative deals. Underline knowing how to buy correctly, then circle the sentence and read it again. Remember, it's the BOOKS (the dollar amount) not the LOOKS, that really count in this business. Unless, of course, you want to operate your real estate as a hobby. I am assuming you're like me; however, I don't mind working in the dirt if it pays well.

I think you get the idea now, so let's try it out. First, find an ugly, "beat up," rundown, poorly managed property with a "Don't Wanter Owner." Also, don't forget that—based on your available resources, dollars to buy with and the time you have available to spend working on the property—you should be looking for the largest number of rental units you can physically handle, either by yourself or perhaps you have a partner to help you. You must determine this before you purchase, not afterwards. That's the goal! Now you must do it!

PROPERTY MANAGEMENT—KEY TO SUCCESS

Management is a difficult job but, like most other jobs, there is an easy way and a hard way to do it. It's most important to learn the easy way right from the very start. The combination of owning and managing rental property is not just my personal formula for achieving success. It's a well-tested and proven method for acquiring great wealth. The great wealth part is not a mystery. It's the profits we earn from adding value to our real estate and providing a place to live for many tenants who need our product.

A San Francisco investor friend of mine writes about our task. The following is a quote from his book about managing properties:

> "If you are comfortable with the role of owner and land-lord and are not intimidated by the responsibility for setting out the rules by which people may live in your property, you will find a vehicle for self-employment and self-expression that is difficult to match in our society."

I am in total agreement with my landlord friend. Hopefully, the information I give you will help prepare you for your role as manager and owner.

Real estate investing can be a highly profitable and rewarding experience for you, if you know what you're doing and learn how to do it right.

PRICE PER UNIT IS KEY INVESTMENT NUMBER

The largest number of units has to do with financial leverage and percentages. For example, it's often just as easy to purchase ten rundown houses with the same size down payment, as you would have to make to purchase a four-unit, nicer-looking apartment building. The seller with the nicer looking units can obviously be more selective (i.e., he can hold out longer for top price and terms). That's because a lot more buyers are interested in buying nicer looking properties. Assuming all the important numbers (price and terms) fit, it's always better to pay the least amount down per unit as possible in order to obtain maximum returns on your up-front money (down payment). If you buy 10 rundown houses, a $20,000 cash down payment would equal $2,000 per unit for the 10 apartments. The same down payment would equal $5,000 each for just four nicer looking units. Again, always try for the least amount down and the largest number of units.

Buying Properties is a Numbers Game

Look how percentage applies to vacancies. The more units, the better. One vacancy in a ten unit group will equal only 10%

vacancy, while one vacancy in the four-unit property equals 25% of the total units available for rent. Obviously, this percentage is not conclusive by itself because other important details have not been considered. I'm merely trying to show you that bigger numbers means spreading the risk. More units mean it's cheaper to operate per unit, as well as cheaper to purchase. You'll be getting a lot more bang for your investment dollar. If this doesn't seem clear now, I promise you, it will later on. If I can help you buy the first property correctly, so that you can actually experience success, I will have achieved my purpose. After that, you'll do just fine by yourself, same as I've done.

LOOKING FOR JUST THE RIGHT PROPERTY

I talk to people almost daily who have decided that investing in income-producing property is a great idea, but they can't seem to figure out what type of property to invest in. Many folks do nothing while trying to figure out some easy magic formula that will offer instant wealth and guaranteed success. Take my advice here—stop looking for perfect properties with guaranteed profits—*there are none*. Most people who wait for "the perfect deal" are still waiting.

DIVERSIFICATION LATER ON IS BEST STRATEGY

Once you've become established financially and, even more importantly, have become more knowledgeable about real estate investing, you may then enjoy the luxury of diversification. You might even venture out and buy several "pride of ownership" properties. After all, money's not everything! Besides, you might want to own a good looking property to drive by with friends or your mother-in-law so you can show off a bit. It's important to remember that most of the world judges by looks. Most of my advice is about making money. Once you have all the money you need, who cares what you do. You might even consider buying property that just breaks even.

The Greatest Seven-Inch Fisherman

TV Minister, Robert Schuller, tells the story of a young boy watching an "old man" who is fishing off the river bridge and doing quite well at catching his limit. The young boy couldn't help noticing something very strange about the old fisherman. It seems that ever time he would catch a real nice size trout, 10 or 12 inches long, he would very methodically take it off the hook and toss it back in the river. Each time he would catch a small fish, he would take it from the hook and carefully place it inside his fishing basket, making sure to close the lid each time.

After several hours of watching the old man fish, the boy's curiosity got the best of him. He walked out on the bridge and told the old fisherman what he had observed. "Why?," he asked, "Do you keep all the small fish and throw the big ones back in the river?" The old man looked at the boy, slightly amused and smiling, then answered, "Because I only have a seven-inch frying pan. The bigger fish won't fit in my pan." The old fisherman had already determined ahead of time what worked best for him. Now he was merely working his proven plan.

Most beginning investors would be much better off using a single size pan to begin with—in other words—specialize! Nothing fits seven-inch pans better than seven-inch trout. Once you become famous for your seven-inch trout dinners and understand exactly how to make money serving them, only then would I recommend getting another frying pan. It's far more rewarding and profitable to be good at doing one thing very well than to be just mediocre trying to be good at everything. Besides, it's a lot easier to learn just one thing at a time. Your life will be much less complicated, believe me!

If, by now, you get the feeling I want you to specialize in the early stages of your investment career, then give yourself an "A" for perception. You may also move up to the front of my investor training class. Who knows, with any luck at all you might even become the teacher's pet someday.

ALWAYS INVEST WHERE DEMAND IS HIGHEST

It's very important to stay within your means financially and to acquire the type of properties you can handle yourself. If you do that, you can almost write-off the risk factor. Inexpensive rental houses and small apartment buildings will always have long waiting lists of qualified tenants if you keep the properties looking attractive and in good repair.

According to U.S. Housing & Urban Development (HUD), 50,000 lower income rental units are disappearing in this country, annually. They are torn down for urban expansion, condomized, and some just fall down. The reason doesn't matter much. The point is that they are becoming scarce as hens' teeth. First thing you know, the Federal Government might start subsidizing landlords who own what's left. Certainly that would make as much sense as paying farmers to plow their tomatoes under so they don't flood the market and bring the prices down.

Because they are scarce and in such high demand, the risk of owning and operating inexpensive rental houses is almost non-existent. That is exactly what new investors need, NON-EXISTENT RISK. There are plenty of other things to worry about.

THE BEST ODDS FOR YOUR SUCCESS

William Nickerson, author and rehab-millionaire, began fixing rental houses over 50 years ago. Bill quite accurately concludes in his best-selling book, *HOW I TURNED $1,000 INTO THREE MILLION IN MY SPARE TIME*, "The chances for success are 1,600 times better owning and operating rental properties than for starting another type of business." If you need more convincing, I suggest you write to Bank of America Business Services Dept. or Dun and Bradstreet Credit Rating Service, Inc. The information they provide about starting other businesses and the odds of success are quite gloomy by comparison.

A Good Plan Always Requires Action

Once you have developed your financial success plan, you've decided where you want to go and selected a vehicle (fixer property) to get you there, based on your personal resources and abilities, you've completed the first step. Now you are ready to implement your plan. Don't wait until you know everything about investing. You will never get started! Investment education comes mostly from actual investing. Obviously, you must learn all you can from books, classrooms and seminars. However, unless you actually do some investing yourself, you'll be missing the most important part of your training. For Fixer Jay Seminars and Fixer Camps, see addendum "Training Products."

TEN MUST-DO'S THAT WILL SPEED UP YOUR SUCCESS

1. First, **develop a total investment plan** from start to finish. I recommend specializing to begin with. Say you plan to buy four rental fix-up properties each year for five years. Each one must produce $100 per month cash flow. That's a reasonable plan.

2. **Learn your local market**. Know what properties should cost, and what they can reasonably sell for. Learn how much rent you can get. Do this step before you buy, not afterward.

3. **Develop a business sense**—think like a retailer. That will help you to pay wholesale prices when you buy. Buying at retail prices and selling for retail prices simply won't work.

4. Learn to spot or identify hidden bargains quickly, then act fast to acquire them when you do! Remember, competition is keen. You must develop a sixth sense for "sniffing out" hidden money-makers. **Become a house detective**!

5. **Learn landlording, firsthand**, from doing it. Manage your own customers (tenants). Many inexperienced investors farm this function out to professional property managers. I consider this a serious mistake for new

investors—maybe it's okay later on, but owners should know the job inside and out first.

6. ***Invest, don't speculate***! Investing is a plan to make money. You must be able to identify exactly how you will do it. That's why step one is necessary. Spectators are guessing without a plan.

7. ***Establish local trade accounts***. Most building supply stores will give you 10% discounts. It's not automatic—you must ask for the discount. Besides saving money, you build a solid credit history. Another benefit is you don't need to carry a pocket full of money around. Paying monthly statements is better for bookkeeping.

8. ***Learn to live on tax-free or tax-sheltered income***. Rents you collect are normally tax sheltered. Rehab loans, like Title Ones, are tax free, same as borrowing on equity or refinancing. When you collect $100 rent, you get to keep $100. When you earn $100 in wages, you keep only $60. Taxes eat up the rest.

9. Learn how to ***do deals where you have as close to 100% control as possible***. Basically, this means owner financing and doing the management yourself. Avoid short pay-back notes and variable-rate mortgages offered by institutional lenders.

10. Once you have developed a plan that works well and consistently makes money for you, stick with it until it stops working. Most investors suffer from a common weakness— they are suckers for a better mousetrap. ***Avoid the "Too Good To Be True" temptation—it generally is!***

Throughout this chapter, I've told you about many different things that will help point you in the right direction, when you purchase properties. Try to acquire as many units as you can with the money you have and, obviously, as many as you can handle. I've told you about the value of planning and developing your hidden talents! These things are the necessary building blocks for becoming a successful fix-up specialist. That's what I recommend for you—at least in the initial years of your investing. Later on, you'll have many more options, but never forget that monthly net income is what you want first! With it, you'll continue to grow and become successful; without it you won't!

SPECIALIZED TRAINING MATERIALS AVAILABLE

If I've convinced you, so far, let me recommend my specialized home study training materials! I'll guarantee this program can help put your education in the "Fast Forward Mode." The reason I say this is because everything it teaches you are things I currently practice myself. You won't find any untried or unproven techniques in the workbooks and audio tape products. Instead, you'll find that this training material will move you quickly toward the profits! Naturally, you'll have to do the work yourself, but I'll show you strategies to speed up your success and keep you from getting bogged down with non-profit projects.

My deluxe home study fix-up course, "FIXING RUNDOWN HOUSES AND SMALL APARTMENTS" (Product #2100), will keep you going in the right direction. Additional details can be found in the Appendix, along with descriptions of other training materials to help you be more successful. Order forms can be found in the back of this book, or call *Jay's toll free order line*, **1-800-722-2550**, for faster service.

CHAPTER 5

Good Realty Agent, Plus Thorough Analysis, Speeds Your Success

*L*ike a good husband or wife, a skilled real estate agent can be a valuable asset to every investor! And, much like building a strong and successful marriage, you will need to spend some time and effort finding exactly the right agent who fits your style. Chances are you won't find Mr. or Mrs. Right on the very first date, but don't be hesitant to dump a few along the way. That's how it works for everyone during the hunt.

The worst part is always breaking the ice, introducing yourself to complete strangers who you think might be able to help you. There's no way to short-cut this process, so it's best to simply charge ahead and do it! When you eventually find an agent who can appreciate your goals and is willing to spend time helping you achieve them, you'll suddenly realize you've added a powerful new tool to your "wealth-builder's" kit.

REAL ESTATE AGENTS ARE THE EYES AND EARS OF THE REAL ESTATE BUSINESS. Approximately 96% of all sales and trades involve licensed sales persons and their brokers. It would be very foolish, indeed, to harbor any serious notions about excluding them from your investment plans. The best thing you can do for yourself is to begin diligently searching for a good one.

GOOD AGENTS DON'T COST YOU MONEY—THEY HELP YOU MAKE IT

Before I get started on the *HOWs* and *WHYs* of selecting the right agent for you, let me first discuss the issue of commissions. Those are what agents earn when they help you buy and sell real estate. You must first understand there is no such thing as a standard *ONE SIZE FITS ALL* commission! They come in different sizes and are often negotiated.

For example, most agents in my town charge about 6% commission for selling houses. However, that's not the market I'm a part of. House buyers generally represent one-time transactions for most agents. I'm a buyer and sometimes seller whenever the opportunity arises. Unlike the guy who buys a house and that's it, I'm good for many transactions. Naturally, I represent a greater source of income for my agent.

Understandably, agents can work for smaller percentage commissions when multiple transactions are anticipated! To give you an idea of what I'm saying here, I'll share my own experience—the average commission I've paid out for all my transactions during the past several years is slightly less than 3% of the purchase price. That's roughly half of the so-called going rate. I find it best to negotiate the commission fee in advance of every deal I make. Just remember, it must be fair for the amount of work involved.

HOW TO FIND AN AGENT THAT'S RIGHT FOR YOU

Just like a budding romance, the first thing that has to work is chemistry! It will do you no good in the long run to force yourself to work with an agent who disgusts you, no matter how good you think they may be. Obnoxious agents are best left to service obnoxious home buyers, since there's no shortage of either one out there! I mention this because it's very important to develop a relationship that can last a long time. A lasting relationship is much easier when you like each other. I've had just two agents in the last 20 years, or so.

If you are just starting out in your search for an agent and are relatively new to investing, you need a more experienced person to help you. You also need to seek out an agent who's specialty is income property or investment sales. In the larger brokerages, you'll find more specialists. In the smaller offices, sales people take on everything that comes through the door. Many times you'll find the more experienced agents working by themselves or in smaller offices. Larger firms always have a ton of new agents cycling through to get experience.

WHAT ABOUT AGENTS WHO SHOP FOR THEMSELVES

I often hear seminar leaders tell investors to seek out real estate agents who buy and sell properties for their personal investment accounts. This advice is akin to leaving your fiancée at an all-night bachelor party, hoping your friends will always do what's right for her!

I can see no advantage to having my own agent competing with me to acquire bargain properties. I once accused my current agent, Fred, of doing exactly that. We were both looking at the numbers on a small subsidized (HUD) apartment building for sale, when I unexpectedly had to leave town for a week. When I returned Fred had purchased the building. He said that since I hadn't advised him that I was interested, he assumed that I wasn't. What he couldn't explain very well was why he didn't wait until I returned before he made the purchase. Fred started treating me extra nice and bought me many lunches for a year or so trying to make me forget. Finally, I did and it's never happened again.

Meeting an Agent Face-to-Face

The best way to begin your search for Mr. or Mrs. Right is to visit the local real estate offices in the area where you plan to invest. Ask the receptionist if she knows which agent sells the most apartment buildings or income properties. Get yourself introduced and tell the agent exactly what you're trying to accomplish. The interview will be your opportunity to learn what the agent thinks he can do for you! But, remember, he must know exactly what you want done in order to respond in a meaningful way.

Try to be as specific as you can. In my situation, I might tell them, "I'm interested in 'fixer-type' rundown properties. I prefer detached houses on a single lot. For example, 10 or 12 older houses on an acre of land would be something I'm very interested in. I'm also looking for properties with owner financing. I'm particularly interested in properties that have several private notes or mortgages that I can assume. Deferred maintenance (rundown), problem tenants and properties that need major clean-up are my specialty. I wish to avoid new bank financing whenever possible and I'm trying to purchase properties with 10% cash down. I can also add notes or other properties in trade."

This description of what I'm looking for and my special preferences should be enough to give most any agent a fairly decent idea of what I'm in the market for. Naturally, the price will have to fit the deal, but don't ever forget that the asking price seldom has much to do with the final purchase price, particularly in the fixer business.

FIVE IMPORTANT BENEFITS AN AGENT PROVIDES

Once you find an agent who seems like he or she talks your language and, of course, demonstrates some honest action like jumping right in and finding a few properties that seem to fit your "looking for" instructions—you'll be off to a running start.

One thing to remember here—both you and your agent are new to each other! Don't make the agent do all the work. You should help in every way you can, especially in the "getting acquainted" mode. For example—if the agent is showing you properties in good condition, and you have told him you want junkers, reiterate your instructions so you get what you want. By helping your agent, who is trying to help you, you'll end up the big winner.

The benefits you'll receive by taking the time to develop this relationship will be worth big bucks to you in the long term! Here are five of the most important benefits:

1. Your agent has immediate "pipe-line" knowledge of when a bargain property is listed for sale, either as a member of the multiple listing service or by networking through his associates and contacts. You'll get the information quickly so you can write an offer fast if the property is what you're looking for. Being first, or near first, is very important in this business.

2. A good agent will automatically do the "weeding out" for you once he or she becomes accustomed to what you really want. My agent Fred, always brings me everything I need for making an educated evaluation on each deal. The information provided is normally a property profile, copies of existing promissory notes and either a filled-out INCOME PROPERTY ANALYSIS FORM (see the last page of this chapter), or at least the necessary data to fill one out. This is valuable "time saving" work for an investor, yet it's needed before any intelligent buying decision can be made. Obviously, it puts Fred closer to a commission if his help results in an offer to purchase.

3. My agent provides a middleman "buffer" between the buyer (me) and seller. This is a valuable service to me or any real estate investor who owns multiple properties already. "Mom and Pop" real estate owners (the kind I buy most properties from) often feel intimidated negotiating with me one-on-one. They seem to feel that because I own so many properties and I'm successful, they'll automatically end up on the short end of the stick! It's a perception that's hard to overcome regardless of whether it's true or not. Fred can generally reduce this problem for me in his capacity as a neutral third party. Sellers often feel that a licensed person will be more sensitive to their needs, as opposed to direct, face-to-face negotiating with a "ring-savvy" buyer.

4. A good agent will never let the commission block a sale! The good ones are creative, and will take a fraction of what they have coming in order to close the sale. They'll let you pay the balance later on, perhaps in monthly payments.

My first agent, Merv, allowed me to pay 50% of his commission at closing, and then agreed to take a promissory note for the balance, with monthly payments anywhere from $50 to $250, depending on my projected cash flow. At one point, I was paying monthly commissions of $1,250 to Merv. In addition to helping me, Merv was very happy to have the steady monthly income, plus 8% interest.

5. A good agent can put you in contact with money lenders, both private "hard money" guys and institutional lenders with programs that fit what you're doing. Lenders shop real estate offices looking for qualified buyers among their clients. This is a valuable benefit to investors who are always in need of funds for upgrading and acquisitions. Naturally, you and your project must qualify in order to take advantage of this opportunity. Nonetheless, money is always the ammunition that keeps investors in the hunt.

No-No's to Avoid if You Expect Loyalty

As you might guess, the benefits must flow both ways between the agent and investor. Here's what you *shouldn't* do if you want to develop a profitable relationship!

1. Don't jump from agent to agent! Use your agent for all your transactions, unless you have special exceptions previously agreed upon. Loyalty will move mountains.

2. Don't try to squeeze the commissions. If you are like me— that is, you negotiate commissions—do it before the agent goes to work, not after the deal is written up and in escrow.

3. Don't send agents on "wild goose" chases! The veterans will dump you if you try, but don't even do it to the inexperienced dummies. They will soon catch on and will have nothing more to do with you. I've heard "so-called" real estate experts tell novices to instruct their agents to draft up and present "low-ball" shotgun offers to purchase properties from the multiple listing book. Ask any respectable agent what he thinks of that advice! Take it from me, don't waste your agent's time with such nonsense.

4. Don't make a ton of offers without ever closing anything. No agent can survive without "paydays" any more than you can. Take better aim so you'll hit the target and close the deals!

5. Don't deal directly with just any agent who calls you! Always refer them to your own agent. Also, some sellers will insist on dealing directly with potential buyers and they don't want to involve an agent! That's fine, but tell your own agent about the exception. Sometimes, I pay Fred a small fee, say $500 to $1,000, to help me do the legwork in the background. If you take good care of your agent, you will benefit in the long run, believe me!

THOROUGHLY ANALYZE THE DEAL BEFORE MAKING AN OFFER

Throughout this book, I will refer to average costs—plus, several "Rules of Thumb" I use to determine whether a property is a good candidate to consider purchasing. *RENT MULTIPLIERS, COST PER UNIT* and *ANNUAL RENT RETURNS* are the most common ones I use! However, when it comes to the important financial examination of a property, I always use my INCOME PROPERTY ANALYSIS FORM to help me uncover all the answers.

The property analysis form helps me discover many easy-to-overlook property expenses! It also provides an excellent tool for negotiating with sellers. I consider the negotiating value of the form just as important as its use in developing the income and expense data. I'll explain this in more detail as we go along. You will find my form quite simple to use, filling in the blanks is all you'll need to do.

JAY'S INCOME PROPERTY ANALYSIS FORM

Even though my form is quite simple to use and filling in the blanks shouldn't require a MBA degree, it is still easy to overlook important information, believe me! Therefore, I'll guide you, line-by-line, down the form and explain my reasoning, as well as how I came up with the numbers, as we go along. The example I'm using is an actual case history for an eight-unit property I negotiated to purchase several years ago. Let's call the property "JAY'S RUNDOWN HOUSES," Elm Street. (A blank copy of my Property Analysis Form can be found in the Appendix.)

The information I'll need is the actual INCOME and EXPENSES that will transfer to me if I should purchase the property. Notice that the income and expense data is shown on a monthly basis. You may refer to the income property analysis form on Page 89 as we discuss each line. Remember, the information I'm looking for is the actual INCOME and EXPENSES that will transfer to me should I purchase the property.

Line No. 1—Total Gross Income

This figure is the total income from all sources, including rents, laundry, storage or whatever else is earned by the property when it's fully rented! That is, 100% of all the income.

Line No. 2—Vacancy Allowance

Allow a minimum of 5% of total gross income for this line, even if the units are 100% rented at the time of purchase. The reason is—no rental property stays 100% rented all the time. If the seller suggests it does, request a copy of his tax forms, *1040 Schedule E*, for the past two years. Simply divide the income each year by 12 to establish the average monthly income he reports to Uncle Sam. This double-check proves very enlightening to most buyers.

Line No. 3—Uncollectable or Credit Losses

Allow a minimum of 5% of total gross income for this line. Generally it's more for rundown type properties, like I purchase, because *people-problems* are likely to be present with *property problems*. This line reflects losses from tenants who "skip out" without paying and evicted tenants who are forced to leave after a judgment. Chances of collecting any money from them are slim and none. Forget that, just start over and do better next time.

Line No. 4—Net Rental Income

This line represents the actual net amount of money you have left to operate the property—sometimes referred to as gross operating income. It's the actual money you get to hold in your hand each month.

Now let's talk about all those nasty expenses.

Line No. 5—Taxes Real Property

This line needs to show what the property taxes will be (per month) on the day you acquire the property! First, you will need to estimate approximately what you think you'll end up paying for the property, then apply your local county tax rate to the estimate! For example, in my case, I anticipate paying approximately $195,000 for the Elm Street houses. My county taxes will be slightly over 1% of the selling price. Therefore, $195,000 times 1% equals $1,950 in annual taxes. This amount divided by 12 months equals $162.50 (rounded to $165) per month.

In California, the seller may have been enjoying low property taxes if he has owned the property for a length of time. However, because of Proposition 13, the purchaser's taxes will immediately jump to 1% of the selling price as soon as the property transfers title. Don't get trapped into using the seller's lower tax bill when computing your expenses on this line.

Line No. 6—Fire and Liability Insurance

You must apply the same logic on this line as you did for Line 5 taxes! Ask yourself—what will insurance cost me based on the condition of the property and the amount of fire and liability coverage I will need for proper protection? Some sellers I've bought properties from have been grossly under-insured because it's cheaper!

Fire policies are based on footage construction costs to re-build the property in case you suffer a loss. If actual construction costs are 60¢ per square foot in your area and you decide to take-over or assume the seller's existing policy calculated at 30¢ per foot, you'll quickly become the "stuckee" if you ever have a loss! Line 6 should be the monthly amount for insurance that provides you with adequate protection if you become the new property owner. Don't be caught under-insured after the fire trucks leave.

Line No. 7—Management

I personally charge 10% for my management services because I'm worth it! Professional managers charge anywhere from 5% to 10% for the job! At any rate, properties don't manage themselves. People must do it and, obviously, it does cost money. It's a very legitimate property expense. Here's the problem you will always encounter when you negotiate to buy properties from owners who do their own management—they don't understand this expense and will always ask, "Why are you showing a property management expense of $250? Aren't you planning on managing the property yourself?" I always tell them, "Yes, but I don't plan on doing it for nothing!" They usually respond by telling me, "We do our own management and we don't take a fee."

I ask, "Are you trying to tell me you will manage property for nothing! If you are, I'd be happy to let you continue managing for me at the same price!" Managing property and tenants is not an easy job and whoever does it should receive compensation. I allow 10% on this line. However, that also includes legal and accounting, too! I do my own evictions when necessary—*that's a bargain!*

Line No. 8—Maintenance

I always allow at least 5% maintenance expenses. It might be a bit less for newer properties, but it could go higher for older houses and apartments. Maintenance is the routine up-keep that's necessary to keep the property in good condition, so that it can continue to earn income. Maintenance does not necessary improve the property. It merely keeps it running in proper order and at the same level.

Line No. 9—Repairs

When the doorknob falls off, you must put it back on—that's a repair. When the roof leaks and causes damage to the ceiling sheet rock, you must repair the roof—then repair the ceiling! It's important to remember that all properties will need repairs because things fall off and break. Obviously, older properties will require more repairs. I always allow a very *minimum* of 5% for older houses, but 10% is not unreasonable.

Line No. 10—Utilities Paid by Owner

It will pay you to research and verify which utilities are paid by the owner. Sellers have a tendency to state their expenses on the low side! Sometimes they'll produce receipts for heating bills in the summer months. Quite often they will group all their utility expenses and give you an estimate like $100 per month for everything. What you need to know is whether or not each house or apartment unit has its own separate meters for gas and electric service. Quite often apartments will have a separate owner's meter for the laundry room and exterior lighting, such as porches, hallways and parking areas.

Utility companies will furnish printouts of electric, gas and water usage. The owner of record is the only person who can obtain this information. My advice is to ask the seller to provide you printouts for the past 12 months of service so you can average out the costs for a full year. As a general rule, apartment owners must pay water, sewer and garbage service. The key here is to check everything out for yourself. You don't want any surprises after you purchase the property. Again, the most important thing you need to verify is if each house or apartment has its own separate meters—*meaning the tenants pay for their own utility services.*

Line No. 11—Total Expenses

This line is the addition of lines 5 through 10. If the total of this line is more than 40% of line 1, re-evaluate your purchase offer to make sure you are not over-paying for the property.

Line No. 12—Operating Income

This line is the amount of money that's left to pay the mortgage payments (debt service), after all operating expenses are paid.

Line No. 13 and 13A—Mortgage Balances and Payments

Show each of the existing mortgages, monthly payments and dates when each mortgage will be paid off (due dates). Totals on 13 and 13A—remember, these are totals for just the existing mortgages on the property—*before you buy it.*

Line No. 14—Monthly Cash Flow Available

Subtract line 13A from line 12. This is the amount of money you'll have left each month after paying all expenses and all existing mortgage payments. It's the cash flow available to pay the remaining balance of the seller's equity—in this example $29,600.

In this particular transaction, I propose to pay the seller a $20,000 down payment and assume both existing mortgages (a total of $145,400). The balance I will still owe the seller for his unpaid equity is $29,600. As you can see, on line 14, $66 is the amount of cash flow I'll have left each month to pay the seller's equity.

Proposed purchase price	$195,000
Proposed cash down payment	20,000
Mortgages (two) buyer assumes	145,400
Balance of seller equity	29,600
Totals	$195,000

THE BASIS FOR NEGOTIATING A PURCHASE

Armed with a properly filled out INCOME PROPERTY ANALYSIS FORM, your negotiations will be much more objective! That's because they are based on accurate and reasonable projections.

All income should be verified by you and the seller. This can be done using current rent receipts or last year's tax return (1040 Schedule E). Either way, the figure can be established and agreed on by both parties.

Vacancies and credit losses are difficult to argue against! That's because all rental properties have vacancies from time to time. They also have tenants who don't pay rent for one reason or another. The combined total of 10% for both of these occurrences is not the least bit exaggerated! Some properties consistently operate at much higher losses.

Property taxes and insurance can easily be checked out! For determining the tax expense, call the tax assessor's office. Ask what the taxes will be if you purchase a property for $195,000. Don't tell them the exact location of the property, just the dollar amount and the proposed purchase date will be enough.

To find out the insurance cost—call an agent and ask him or her to give you a "quickie" cost estimate for a policy covering both fire and liability on the proposed purchase. The agent will need to know the property address, but he needn't go inside the houses to give you a "ball-park" figure. Provide him with the approximate square footage of each unit and an estimate of the age.

The Most Controversial Expenses

Management, maintenance, and repairs are the most frequently argued expenses by sellers. They will tell you it doesn't cost anything to manage or do repairs and maintenance because *"we do it ourselves."* I always ask them, "Would you be willing to continue doing the same job for me at no cost if I buy your property?" You already know the answer to that question. Still, the issue needs to be settled so that both parties can agree on what the expenses for these items should be. Assuming the seller is not happy with percentages I use, I simply tell him that I don't wish to purchase his property—*and then do all the work for nothing.*

Again, I ask for income tax records to substantiate the expenses. Two full years of 1040 Schedule E tax returns will do the job. I have never received tax returns that show the same numbers given to me by the seller or his agent. In almost every case, the income figures shown on the tax returns are less and for some odd reason, the expenses are always much higher. I think it's something like the unexplained difference between the miles people drive, as stated on their automobile insurance forms and the mileage they report for the very same period on their 1040 tax forms—it would seem that many cars will never wear out based on the auto insurance mileage; but for the purpose of tax deductions, they travel more miles than a Greyhound bus.

REAL ESTATE AGENT CAN HELP YOU BUILD WEALTH

Real estate agents are the eyes and ears of the real estate business. If you harbor any serious thoughts about excluding them from your investment plans—**MY ADVISE IS DON'T**. A sales commission is "peanuts" compared to the value of a good agent working to help you achieve your financial goals.

I have heard seminar instructors tell their audience—Its best to avoid agents whenever you can to save commissions! Only use them for free appraisals and advice, or to write up dozens of stupid offers from the multiple listing book for a mass mail-out! Shotgun offers, they call them. In general, the suggestion was to get free service, but avoid commissions! I must tell you friend—this is poor advise—if you think it's right—pretend you're the agent for a moment! Would this treatment seem satisfactory to you?

INCOME PROPERTY ANALYSIS FORM

Property Name _"Jay's Rundown Houses" - Elm St._   Date _Several Years Ago_

LINE NO.	INCOME DATA (MONTHLY)	PER MONTH
1	Total Gross Income (Present)	$ 2500
2	Vacancy Allowance Min. 5% LN-1 Attach copy of 1040 Schedule E or provide past 12 months income statement for verification	$ 125
3	Uncollectable or Credit Losses (rents due but not collected)	$ 125
4	Net Rental Income	$ 2250
	EXPENSE DATA (MONTHLY)	
5	Taxes, Real Property	$ 165
6	Insurance	$ 150
7	Management, Allow Min. 5%	$ 250
8	Maintenance	$ 175
9	Repairs	$ 125
10	Utilities Paid by Owner (Monthly)	$ 76

Elec	$ -0-	
Water	$ 25	
Sewer	$ -0-	
Gas	$ -0-	
Garbage	$ -51	
Cable TV	$ -0-	
Totals =	$(76)	

11	Total Expenses	$ 941
12	OPERATING INCOME (LN 4 - LN 11)	$ 1309

Existing Mortgage Debt

1st Bal Due	$102,800	Payments	(Montly)		Due Mo/Yr	
2nd Bal Due	$ 42,600	Payments $	806		Amortized	(2011)
3rd Bal Due		Payments $	437		Balance Due	(2005)
4th Bal Due		Payments $				
5th Bal Due		Payments $				
13	Totals	$145,400	(13A)	$		
14	MONTHLY CASH FLOW AVAILABLE		$	1243		
	(LN-12—13A)	(Pos or Neg)			$66	

NOTE: Line 14 shows available funds to service new mortgage debt from operation of property.

REMARKS: All lines must be completed for proper analysis. Enter the actual amount on each line or Ø.

Here's the deal gang—try to find an agent who shows some interest in working with you. Take your time, you don't have to choose the first one you meet. But when you find one that clicks—you'll both likely know it. Clearly explain in detail—**what you will buy** if the agent finds it for you—remember, no agent can work for you very long without a payday. If you need a few more helpful pointers on this subject—write to Fixer Jay and ask for my free audio type "**Select the Right Agent**." It's an interview with my personal agent Fred. Send $2.00 cash for postage and handling.

CHAPTER 6

*Substitute Personal
Skills When You're
Short of Cash*

GETTING STARTED WITH THE RIGHT PROPERTIES

I intend to show you how to make more money than you ever thought possible—fixing up rundown real estate! Naturally, it's you who must be willing to roll up your sleeves and do the work! If that bothers you, let me suggest you start looking for another book.

In this chapter, I will teach you some techniques that will profit you very handsomely, but you will first need to be willing to pay your dues. If I knew an easier way, believe me, I'd tell you what it is, but since I don't, I'll teach you the same way I learned.

The first order of business is to seek out and locate the kind of real estate that has good profit potential. If you're anything like me, when I started out, money was never any problem!—it was absolutely clear from the start *I had none*. When you have no money, you have a serious restriction when it comes time to buy. However, in the long term, you could be better off than your richer friends because you're forced to become a better shopper. Also, when every single dollar is important, you'll learn to evaluate expenses very carefully. Reducing expenses is the same thing as making profits.

Don't Wait—Nothing Ever Happens

Right after willingness to go to work—knowing what you must do is the highest priority! Many folks with high hopes and the best intentions seem to always be waiting for something to happen. Whatever they think is supposed to happen seldom does. Waiting to save enough money to start investing is perhaps the world's most popular excuse for doing nothing. It's also one of the lamest that I know of.

NEVER LET LACK OF MONEY STOP YOU

No money for a down payment should never stop investors who are willing to learn a few new tricks and then apply them. Quite often new investors will tell me, "I'm willing to spend everything I've got to become successful, but my real problem is I just don't have any money right now." I say, "Okay, that's fine. Let's just spend what you do have right now." Generally, what "start-out" investors have most of is TIME and WILLINGNESS. Assuming you have both, I've got some good news. TIME and WILLINGNESS, along with the techniques I'll show you in this book, will allow you to get started right now, without further delays or any excuses!

BUYING PROPERTIES THE OLD-FASHION WAY

Years ago, I read a book called "100 WAYS TO PURCHASE REAL ESTATE WITH NO MONEY DOWN." As it turned out, the author had merely assembled 100 blank real estate purchase deposit receipt forms, filled in the blanks with every conceivable purchase arrangement he could think of, good or bad. He claimed the book represented 100 proven ways to buy property with no money down.

Since many of the same purchase terms were repeated over and over throughout all the proposals—I'd guess only eight or ten of the bunch could be called true originals. The other 90, or so, were merely repeats or modified versions of the same thing.

I mention this book for a reason. There's good news and bad news. We'll start with the bad. Most of the offers I read would not stand much chance of getting accepted in the real world unless the buyer's I.Q. was around "30-something." The good news is—You will never need 100 different ways to buy real estate anyway! So, don't waste your time reading about it.

YOU ONLY NEED A FEW METHODS THAT WORK

I'm absolutely certain I'll never live long enough to need anywhere near 100 different ways to buy real estate! I've never had to use more than four or five different ways in the past 30 years. When you consider all the real estate I've acquired— more than 200 houses and apartments—it would seem apparent that four or five good methods might be all you'll ever need, as well. The point I'm making here is—you don't need 100 ways to buy property. It's not a question of quantity, it's the success of the methods you use that counts!

When I write about this subject, I'm always reminded of the young man who boasted about knowing 39 different ways to kiss, but—the problem was—he could never find a girlfriend! Don't forget, the object is to purchase properties. You only need several good methods that are likely to be successful and stand a good chance of closing deals! Successful offers must also meet the needs of sellers. It's very important to structure offers that stand a good chance of being accepted, otherwise you'll waste a lot of valuable time, plus you'll lose a lot of good deals in the process.

I once listened to a so-called "Real Estate Guru" tell his audience to, "Write 100 different offers every month, then let your real estate agent present them to every potential seller." He called this plan "The Shotgun Method." He claimed it really works great because all you need to do is get two offers accepted every month and you'll soon own all the real estate you want! In case you think this method has some real merit, I'd say—try it out on your favorite real estate agent. I'd be very interested in his opinion, assuming, of course, it's printable.

Some ideas you hear about at seminars will sound almost brilliant when presented by a smooth-talking pitchman. However, I've found brilliant ideas are a lot like kissing—the real test doesn't come until you can find a girlfriend to try it!

DEVELOP A REASONABLE PLAN—THEN FOLLOW IT

The first plan will be to acquire income or profit-producing real estate with little or no cash using what you can contribute, *YOUR TIME AND YOUR EFFORT!* Let's assume you have the time and you are also willing to contribute your personal labor. In short, you are ready to begin right now. First, let's do what I call my *"COMMON SENSE QUIZ."* I want you to ask yourself two questions. The answers are intended to get you focused, as well as maximize the use of your time and effort. This idea might seem a little strange to you at first, but let me assure you, once you find a few good deals, strange ideas will begin to seem just like normal, I can promise you that!

Ask Yourself These Questions—Memorize My Suggestions

Question 1: What kind of a property might you expect to acquire without any cash down payment? Here are a few suggestions to help you decide:

   a. *Rundown ugly* (no maintenance being done).
   b. *Tenant problems* (motorcycles, junk autos, deadbeat tenants).
   c. *Empty property* (tall weeds, broken glass, garbage everywhere).
   d. *Financial problems* (foreclosure, bankruptcy, bank repossession).
   e. *Partially completed building* (all activity stopped).

The reason you are looking for ugly, distressed-type properties is because they are most likely to be big problems for their owners. The strategy here is to use your time and your efforts to fix these problems in lieu of the normal cash down payment.

Fixing seller problems has a cash value. Often the value of fixing big problems will add up to a higher dollar amount than the normal down payment might have been. The point is, your willingness, plus your ability to fix problems for others, can create a very lucrative opportunity for you.

Many investors, including myself, have used this wealth-building technique to quickly develop large real estate portfolios. It is an excellent strategy for investors who have very little cash for a down payment, which includes most of us when we're just starting out.

Question 2: What kind of sellers are likely to own the type of property described in Question 1? Here are some likely candidates:

a. Out-of-town owners.

b. Owners with financial problems.

c. Family problems (divorce, death, life-style change).

d. For sale by owner ads (all newspapers).

e. Owners who have lost jobs.

f. Elderly or disabled.

g. Inherited property owners.

h. Owners who advertise lease-option or even houses for rent.

i. Job transfers, owner moved, now has two house payments.

SELLERS SELDOM TELL YOU LIKE IT IS

An important point to remember is that people don't always say what they mean. For example, "LEASE WITH OPTION TO OWN" might really mean the owner doesn't want to be bothered for at least a year or more. Most out-of-town owners don't like long distance renting, especially with problem tenants, but will seldom say it that way. Owners who inherit properties often think rental houses are too much of a hassle. Obviously, they weren't the ones who bought the property in the first place. Owners with financial problems or those with two mortgage payments will seldom tell you they're hurting when, in fact, they sometimes are. You must learn to dig out this kind of information in order to make offers that will generate a reasonable level of interest from potential sellers. The offers or proposals you can make once you find the right property, together with motivated a seller, can be very profitable for you. The big money comes from doing your homework before ever making your offer. Always learn everything you can about the seller. Do it fast though, so you don't lose the deal.

LET'S GET STARTED—HERE'S WHAT TO DO FIRST

Let's look at a hypothetical situation, but one that's very common today. The property is totally trashed with broken windows and junk inside and out!

The county ownership records reveal that tax bills are being sent to out-of-town owners. We learn, after writing to them, that the owners are a middle-aged working couple with two children. The trashed house they own is a rental property. The last tenants left in the middle of the night without paying rent. A close friend who oversees the property is trying to get "fix-up bids" for them, but hasn't had much time lately, so the house just sits. Nothing is happening. The general area is residential. Houses are worth about $60,000 and 20% of them are rentals. The neighborhood is an older established tract approximately 25 years old. Rents average about $500 per month for a three-bedroom, one-bath house. We learn the mortgage balance is $29,500, with payments of $216.47 per month at 8% interest. We determine that fix-up work will cost $7,500 ($2,500 for material and $5,000 for labor). Labor generally runs about 70% of the total fix-up costs the way I do it.

IF THE NUMBERS WORK—WRITE YOUR OFFER IMMEDIATELY

Here's the proposal we will make: PRICE $54,000. I generally offer 10% under the average market price in the area for this type of a deal. There will be no cash down payment involved.

Instead, I will jump right in and provide all labor and materials to restore the house within 30 days. The seller will allow me $7,500 credit (down payment) for my personal efforts to restore the property. This deal will solve all his problems in one big swoop.

		Monthly Payments
Down payment—fix-up (labor and material)	$ 7,500	
Assume or take over the existing mortgage	29,500	$216.47
Owner to carry back private note for balance of sale price (10 year term) 8% rate (interest only payments)	17,000	113.33

I normally agree to pay closing costs—*THE REASON:* Sellers should not be saddled with out-of-pocket costs. Remember, there is NO CASH down payment to owner for escrow expenses. This small item will upset most sellers. I strongly advise you to pay the closing expenses, if possible. It's an excellent exchange for maximum owner concessions. Don't irritate the owner when you're asking him or her to carry a long-term note with very favorable terms (below market). It's not a smart move to try and get everything, *you should just be concerned about the most important things.*

Small Deals Build Large Bank Accounts

This deal is just a plain and simple everyday "Garden Variety" transaction. It's not very hard to do at all. If your offer is accepted, you will need $2,500 for material, plus closing costs. Use your credit card or split the deal with a friend who has the money. You can do the fix-up labor yourself.

I'll be the first to agree, this "two-bit" transaction will certainly not make you a millionaire; however, what is important—you now own a property! You've started. Starting is like making a touchdown in football, going on your first date or getting a jump-start when your car battery is dead. It provides the momentum you need to move forward. Now you are beyond the thinking stage—*you're in the action mode.*

When the dust settles, you'll have $10,000 equity and rental income of $500 per month. Your mortgage payment will be $329.80. After taxes and insurance you might even have enough left to make payments on your overloaded VISA card. When you've completed five or ten of these deals, you'll be on your way to an exciting future in real estate. After you have completed 15 or 20, you can start carrying a briefcase around, instead of a toilet plunger.

As you gain more experience and learn how to quickly "sniff out" the high potential PROFIT-MAKERS, this business will get much more exciting. It's sort of like being a private detective, but it pays much better and you'll seldom get shot at.

DOUBLE YOUR PLEASURE AND YOUR PROFITS

Larger properties have exactly the same problems and the same motivated owners as single houses. However, percentage-wise, the equity and profits are much greater. Your financial statement grows bigger and faster with multiple units. Also, don't forget groups of older houses and duplexes in run-down condition can be purchased for much larger discounts because fewer investors want them. It's a case of *LESS COMPETITION.* As problems grow worse, most owners become more motivated to sell. When this happens it's much easier for you to tighten up the deal. "Tightening up" means asking for more concessions. Here are several of my favorite TIGHTEN-UPPERS:

1. Deferred mortgage payments until property is fixed up or until all deadbeat tenants are evicted (generally no mortgage payments for six months).

2. Longer mortgage terms with less interest. Also, ask for delayed payments on owner carryback notes. Often you can get motivated owners to wait several years before payments start. This concession is a great way to improve operating cash flow when you first acquire a property.

3. Larger discounts on the purchase price. Especially when non-paying, deadbeat tenants occupy "junker-type" run-down properties. Let the bikers help you buy cheap!

THE KEY TO WEALTH IS SOLVING BIGGER PROBLEMS

This is the perfect combination for knowledgeable, problem-solving investors. Stated another way, owners will give maximum concessions to buyers who can both fix up junky, run-down properties and handle deadbeat tenants. Not everyone can, and even fewer are willing; therefore, there is a strong demand for talented problem-solvers. It's the same principal used by the famous oil well fire-fighter, Paul (Red) Adair. He asked for and got an astronomical sum of money, for capping gas fire blow-outs on oil wells, simply because he does it very well and because there are only a few others who even want the job to begin with.

You may recall it was Red Adair who was summoned to the North Sea to extinguish the raging fire and plug leaks at the world's largest oil field disaster—the Piper Alpha oil platform explosion—that killed 166 men. It was also Mr. Adair who was hired to cap the flaming oil wells in Iraq, set purposely on fire by Sadam Hussain during the Gulf War. Do you suppose the folks asking for help were very concerned about Adair's fee schedule?

Learning to be a skilled specialist, plus having very limited competition doing what you do best is a "sure-fire" path to *large paychecks* and *financial independence*.

FINDING THE RIGHT PROPERTIES

One of the methods for finding these properties is to read the classified "FOR SALE" ads. You can also look at ads for "LEASE WITH OPTION TO BUY." Remember what I told you earlier, sellers who advertise FOR LEASE may very well prefer to sell if given the opportunity. Ask them and find out. Your chances are better when no one appears to be taking care of the property.

Another method to find properties is to do drive-bys. Keep tabs on what's going on in your own town or buying area. Having a real estate contact (broker or sales person) can be a tremendous advantage if they will work with you and can understand that your cash to purchase is limited. To make this arrangement work, you must convince your agent that you will be a continuing customer (long-term benefits to him). Agents instinctively don't like buyers with no cash down payment. After all, where will the commission come from if the buyer offers "Nothing Down?" Sellers normally pay an agent's commission from sale proceeds put up by the buyer.

A GOOD REAL ESTATE AGENT PAYS OFF

Generally, if an agent can look beyond several "skinny deals" in the beginning and visualize much healthier commissions on future transactions, he or she will often go along with your initial investment strategies. When you find agents who will do this, make sure you treat them right. Never forget, you need good help to grow financially. In the beginning you will do everything yourself. However, as time passes and your successes add up, you'll need to build a trusted team to help you. A good real estate agent should normally be your first trusted helper when the time comes to start your team. However, as I said earlier, this chore is like everything else I recommend for "Do-It-Yourselfers." DO-IT-YOURSELF means exactly like it sounds. It means—YOU DO IT!

It's always been my feeling that your apprenticeship dues are never paid until you contribute your personal time and efforts. That's what I've done myself. That's what I preach, and of course, that's my advice to you. Don't expect your agent to do all the things I'm telling you about here. Remember, he's not on the job when you are first getting started.

I'll assume you're all excited about what I've told you so far. You've decided you're ready to start immediately. You want to hit the bricks running! You want to make money and you don't want to look back until you make it on your own! That's good—now let's review some starting plans.

HOW TO DEVELOP INSTANT EQUITY

Before I tell you how to proceed, let me first say a few words to help set the tone for any investment strategy you may wish to pursue. Ninety-five percent of all real estate is bought, sold and traded at retail prices. Real estate agents earn their living in the *RETAIL MARKETPLACE*, with very few exceptions.

If you race out to buy a property at the regular everyday retail price and somehow, some way, you manage to do it for no money down or a very small cash down payment, I can almost guarantee you'll never make a profit. Worse yet, it's quite likely you will end up having negative cash flow on the deal and your odds for going broke will never be better. My strongest recommendation to you is, DON'T DO THAT. Re-read this paragraph several times over in front of your bathroom mirror. Read it out loud and watch your lips as you say it. Above all, never forget it. *Paying too much for income properties will sink your investment ship faster than variable rate bankers in torpedo boats.*

Don't Fall in Love With Rental Houses

Many property owners "over-improve" the houses they live in. They add all sorts of extras (whistles and bells) that surrounding houses in the same neighborhood don't have—things like fancy kitchen gadgets, add-on rooms, finished garage interiors, costly backyard improvements, including "in-ground" swimming pools, are just a few that come to mind. There's an old saying in real estate circles—*"you can spend all the money you want on a $75,000 tract house, but you'll still have a $75,000 tract house when you're done."* That's pretty much a "real world" fact of life. You might be able to sell for a few more dollars than the neighbors, but rarely are you likely to get your full investment back. They call this "over-improving." It's done every day by countless thousands of homeowners. It's even done by rental property owners who haven't read my books about fixing rundown houses.

BEWARE OF PROPERTIES THAT DON'T PROVIDE CASH FLOW

After doing fix-up for so many years, I must make a confession! Fixing is much more satisfying and rewarding to me than buying the newer more expensive houses, although I do own a number of them now. Let me simply make this observation about newer houses. They are an excellent vehicle for reducing my excessive cash flow—and if you ever have a problem with too much cash stacking up in your safe deposit box, newer houses, with high monthly mortgage payments, will quickly eliminate the problem.

Someone recently said to me, "Newer properties have other good qualities, even if they are a bit short on cash flow! They provide a great sense of 'ownership pride'." I'm somewhat reluctant to buy that argument, but I will concede that money's not everything, so let's move on to something closer to my heart.

FIX-UP SKILLS EARN WAGES—KNOWLEDGE BUILDS FORTUNES

There are some folks who believe that fixing rundown houses is a job that only experienced carpenters or contractors can do. Nothing could be further from the truth. Almost anyone can do this job! In the final analysis, it matters very little who performs the physical "fix-up" work, so long as the right things get done and at the right cost!

Investors who do their own fix-up work will only enjoy a money saving advantage if they fix the RIGHT THINGS at the RIGHT TIME. Both are equally important! As you shall learn, KNOWLEDGE is what makes the big money. Swinging a hammer and swishing a paint brush will merely earn you average wages or save you a few bucks in the short run.

I am frequently asked two questions about fix-up. (I will answer these questions in the following paragraphs.) *FIRST,* does fixing up houses require any special licenses or professional skills? *SECOND,* do you feel it's best for new (start-out) investors to do their own fix-up work on the properties?

As a general rule, no licenses are required by owners who fix up houses for themselves. However, if you do it for someone else—for example, as an employee or independent contractor, that's different! It's very likely you will need a license to be perfectly legal. However, this publication is not exactly the best source of information on doing everything perfectly legal. After all, you've never seen a rich cop, have you?

Having fix-up skills can certainly be an advantage because it's one less thing you'll need to learn about But, having said that, let me tell you something you should immediately underline, and above all, never forget! It's very important information because once you clearly understand and accept it, you can direct your efforts where the real money is. **BIG PAYDAYS DO NOT COME FROM FIX-UP SKILLS. THEY COME FROM REAL ESTATE SKILLS AND SPECIALIZED "HOW TO" KNOWLEDGE.** When I first began fixing rundown houses, I had no idea how important this really was. It took me several years of hard work before I got the hang of it.

To illustrate my point, just think about all the thousands of licensed building contractors who can do almost anything to a building—they have plenty of skills! Now think about what most of them don't have plenty of!—if your answer is money, you're right! Believe me when I tell you fixing skills alone are not enough if you intend to make serious money in this business. Don't quickly decide that fixing rundown houses is not for you simply because you can't stop a toilet from running. Just let it run while we discuss some of the more serious stuff. You can always shut off the main until you call a plumber.

FIX-UP PROPERTIES REQUIRE MULTIPLE SKILLS

I have long held the notion that no one should ever hold the rank of journeyman real estate investor without first having completed an understudy assignment doing fix-up. My reason is simple. Fixing up houses and apartments gives the investor many "real world" experiences that he or she is not likely to get from any other kind of real estate investment.

Most fix-up jobs require a multitude of skills beginning with the purchase of a property cheap enough to allow the investor to earn a profit. Fix-up investors will be involved with employees and contractors. They must develop a specific fix-up plan to fit within a pre-determined budget. Finally, they must handle tenants for rent-up or negotiate a buyer into a profitable sale.

By the time you've completed a fix-up project or two, I can promise you this—You'll have enough "Real World" experience to deal with almost any real estate transaction that comes your way. Also, your confidence level will increase at least 100 fold.

BIG BUCKS COME FROM UNDERSTANDING ECONOMICS

One of the most important skills you will need to develop in order to be profitable is learning to fix up things that will pay you back, and not spending much effort on things that don't. Your goal should be to get the biggest bang for each fix-up buck you spend. *You want the highest possible returns for each of your investment dollars!*

My monthly newsletter, TRADE SECRETS, is an excellent tool for helping you improve your profit-making skills. Every month I write about specific methods and strategies I use to maximize my profits and rental income. Many newsletters are written about properties I currently own and have fixed up; therefore, the dollar numbers and values are actual, not hypothetical or Disneyland!

Remember, too, that I continue to learn and develop new improved ideas and strategies as I go along! We never stop learning in this business—but, the point is, you will benefit yourself a great deal by continuing to learn as you go along. One of the best ways to do that is to learn everything you can from a teacher who does the same things you wish to accomplish.

I realize this sounds like a late night TV commercial to sell my newsletters! But, since you've already made it to Chapter 6, I believe you have some serious interest in my investment knowledge! If I'm right, you'll never find a better bargain or more information for $48 a year, anywhere. To order TRADE SECRETS (Product #2130) fill out Order Form in back of book.

Judge a Property by the Benefits Provided

Folks who have little experience in the ADDING VALUE Business will often judge a property by how it looks, rather than by the benefits it will provide. You should stop and ask yourself every now and then, "Why am I doing real estate? Do I want properties that look good or do I want properties that pay good?" There must be no confusion about what your goals or purposes are if you wish to be successful.

One of my first recommendations, especially for new investors, is to actually sit down and plot out exactly how much money you'll start getting back every month, once you become the owner of an income property. Do this exercise before you buy, not afterward! People who purchase single family houses often skip over this drill. If they were to "plot out" negative cash flow month after month on a sheet of paper so they could sit and stare at the depressing negative numbers, I'm sure many would avoid the heartaches of doing "dummie deals."

REMODELING A HOUSE IS NOT THE SAME AS FIX-UP

There is no cheap method to change older houses into shiny new homes. Many would-be fixers attempt to accomplish this task only to find their bank account is gone before the house is finished. The big difference between what I do and what remodelers do has to do with cost. Ask yourself, "Is what I propose to do going to earn more rents?" If the answer is "No," maybe it's not worth doing.

Quite often remodelers will replace entire electrical systems with all new wiring. Sometimes they'll have the entire house insulated. They tear out old flooring and replace floor joists and supports. They replace all the windows with new double-insulated glass. It's common for contractors to jack up a house to level it. This means they must also fix all the cracks and often redo the exterior plaster. Before you start doing fix-up like this to rental houses, make sure you have a sizable bank account—you'll need it!

TAKE ADVANTAGE OF OLD FASHION CHARM

Don't attempt to make older houses the same as newer properties—it's not necessary! Many older houses are oozing with charm. Charm is worth big bucks if you take advantage of it. Cleaned up, older houses very often radiate loads of charm! They have that homey feeling, with high ceilings, plenty of woodwork, big porches, private space, old-style windows, separate cooling, independent heating units, storage sheds and detached garages. Many also have an abundance of mature shrubs and trees. These are all marketable items that add to the charm of older properties. Don't try to fiddle around with chopping up room sizes and doing full scale remodeling! It will lose you money. Naturally, a thorough cleanup and painting are money-makers. In time you'll discover, as I have, renters think of themselves as temporary, therefore, they don't dwell too much on design. Spend your money where it counts!

THE BIGGEST MISTAKE IS OVER-FIXING

Just about everything on a house can be fixed, no argument about it! The real question you must ask yourself is, "Will it be cost effective?" You must sit down and calculate how much it will cost. This information will help you decide how much work is too much and when it's best to simply pass over the deal and move along to the next one. A fix-up investor must be concerned with fixing for profits, *NOT JUST FIXING!* The big difference between making a profit or losing your shirt—let me repeat—are OVER-FIXING and FIXING THE WRONG THINGS.

In order to determine what to fix, you must first answer two simple financial questions. *FIRST,* what will it cost to complete the proposed fix-ups or repairs? *SECOND,* what will the fixed-up value be after the work is completed? Cost and value knowledge are critical whether you intend to keep the property for rental income or sell it quickly for turn-around profits.

SAVE 70% DOING YOUR OWN FIX-UP

Fix-up work has two parts that cost money. First, and most expensive, is the labor on average, labor will cost about 70¢ of every fix-up dollar you spend. Supplies and material will cost 30¢ of every dollar. Certainly these numbers can vary a little for jobs such as adding or overlaying exterior siding where the siding materials (4- x 8-foot sheets) are expensive and the labor involved is relatively quick and simple. However, the 70/30 split is well within the ballpark for estimating average fix-up jobs.

What all this means is that if you are able to purchase a fixer property for 30% under the estimated "fixed-up" market value—say for example, a $70,000 market value house for $49,000—then you estimated it would cost $10,000 to fix it up to it's full market value of $70,000, it will cost you somewhere around $3,000 for material and supplies. That means if you do all the labor yourself, you can expect to have about $18,000 equity in the property when the job is done. Doing fix-up or "adding value," as it's called, means you won't need to play the inflation game to make profits in this business.

Always Scrutinize Your Fix-Up Plan

All fix-up work should pass some financial scrutiny! Ask yourself, "Does it really need doing?" I believe most improvements should be justified on the basis of paying for themselves. I expect the payments to come from higher rents or bigger profits as my reward for doing the work. Fixing or changing things around purely on the basis of personal likes and dislikes will seldom provide a justifiable "mark-up" (profits). Those kinds of changes should be avoided! This happens to investors who quickly charge forward without a plan. It also happens to folks who fall in love with investment property. I advise you—

be very careful and avoid these common pitfalls. Remember, fixing up dumpy-rundown houses is not glamorous work; but if you do it right, it will pay you better than anything else I know of. *You'll earn $10 an hour painting and $200 an hour for executing a good plan.*

WORK THAT SHOWS FAST—EARNS FAST

Many fix-up jobs will return their costs very quickly. Leading the list of "quick returners" are painting (inside and out), general cleaning, yard and landscape work, fencing, carpeting, window coverings, and installing modern faucets, light fixtures and Formica countertops in both kitchens and bathrooms. *It's not uncommon to get three or four dollars back for each fix-up dollar you spend, and when you do, it will make you very rich.*

It's not by accident that I always begin my fix-up project in the front yard. I've seen professional appraisers value identical houses with a difference of as much as $10,000 because of plain old filth and junk on one property. In other words, a clean house is worth $10,000 more, simply because the owner hauls away the trash and keeps the house looking nice. Think about that for a minute! That's a lot of money for ordinary clean-up skills. Suppose it takes a whole week (40 hours) to haul away garbage and clean up a property—that's $250 an hour, or nearly as much as a brain surgeon gets on his days off!

CHAPTER 7

The Price is Determined by Income and Location

COMPLEX FORMULAS ARE NOT NECESSARY

*F*iguring out how much to pay for a property and how to estimate the profits does not require a complicated, "long-winded" set of financial calculations! If it did, I can assure you there's a whole bunch of millionaires out there who would have never made it. After many years of doing this work, I'm convinced that simple formulas are the best. But, before I tell you how and why this is true, I must warn you to be very skeptical of "High Tech" computer programs that tend to bowl you over with tons of elaborate financial computations. Hyped-up input data that shows rent increases year after year can make any property look like a gold mine! However, any landlord worth his salt will tell you that rents don't always go up and neither do values!

Another trap that nails inexperienced buyers is relying too much on financial information supplied by selling agents. *I have found the income data they provide is nearly always overly generous, while the expenses are usually understated!* Before you put too much faith in the income and expense data printed in the multiple listing book, let me remind you of the words you'll find printed somewhere on each of the pages—"THIS INFORMATION DEEMED RELIABLE BUT NOT GUARANTEED." You can get more assurance than that for a kidney transplant! Take my advice, pay very little attention to any financial numbers that can't be verified. If they can't be proven, they just aren't so (as we discussed in Chapter 5).

GROSS RENT MULTIPLIERS GOOD ENOUGH FOR FIX-UP

Gross Rent Multiplier (GRM) numbers are not the same for every location. For example, in Redding, California, the numbers vary from four on the low end, to ten for premium-grade buildings; the average is seven. Just 250 miles away, in the San Francisco Bay area, average apartments sell for 12 times their gross income. GRM numbers also vary up or down based on the condition of a property and its location within a town or buying area. If a building is managed poorly, the GRM will be lower, also. Remember, the GRM numbers are set by buyers. When there are many buyers wanting to purchase a very limited number of available properties, the GRM goes up because competition increases. Conversely, during a buyer's market, when a glut of buildings is available, the GRM goes down.

GRMs are not precision measurements—they would not work for brain surgery or splitting neutrons, but they work well enough for making money with fix-up real estate. You must, however, know how to use them properly. It goes without saying that any investment must ultimately be valued on the net income it produces—***not the gross***. However, as you will see, there is direct correlation between the gross income and what's left over. In order for me to establish values and estimate future profits, the first thing I must do is determine the range of multiplier numbers for my particular buying area. This is the essence of knowing how much I can pay, as well as estimating how much profit I can eventually earn. The following chart shows you the range of GRMs in my town—but, do remember, all locations are different. You must develop your own chart based upon where you invest!

GRM	DESCRIPTION OF PROPERTY	Rents $
10	Premium-grade, long-term tenants. Excellent owner management. Shows pride of owner-ship every day.	525
9	Good property—tenants drive nice cars. Management very good. Quality maintenance.	500
8	Solid rentals—most tenants have jobs. Man-agement satisfactory. Little deferred mainte-nance.	475
7	Decent condition—blue collar tenants. Man-agement okay. Average deferred maintenance.	450
6	Generally older buildings—trashy rundown. Marginal tenants. Poor management—needs work.	400
5	Older properties—ugly junky. Deadbeat ten-ants. Lousy management—heavy fix-up.	325
4	Older "falling down" construction—"pig-sty," Hell's Angels tenants. No management. Com-plete rebuild necessary.	250

DON'T COUNT PENNIES DOING FIX-UP

Most beginners figure too close when determining how much to pay and how much profit is involved. You can't count the pennies in this business! That's much too close for estimating. Instead, it's best to be very generous with expense guesti-mates and conservative with profit projections. Hardly anyone over-estimates fix-up costs! Besides, you'll always end up spending extra money to manage tenants. It's quite common to get stiffed by deadbeat tenants who don't pay rent after you become the owner. Deadbeats most always come with the purchase of low GRM properties. Still, all in all, low GRM prop-erties are by far the most profitable properties when you know how to fix the problems.

Big Difference in Values is Real Eye-Opener

Most folks are really surprised to learn how much difference in value there is between quality buildings and rundown properties. Playing with GRM numbers, figuring values will give you a good idea why fixers are such an attractive opportunity for profit-making. There is a tremendous mark-up in value between a five GRM property and one that sells for ten times the gross rents. I will tell you now, before you start trying to calculate the numbers, that fixing up or attempting to convert a five GRM property to a ten is generally out of the question. That's far too much change to expect. You simply can't make a silk purse out of a "sow's ear."

For one thing, the location of the property will limit you! For example, on the low income side of the tracks, it's quite reasonable that the highest GRM sale you could ever expect to find would be seven times gross rents—but that's where the biggest profits are made. Conversely, in the "Nob Hill" section—or the ritzy side—of town, you can't buy a piece of property for less than nine or ten times gross (even if it needs some fixing up). Furthermore, if you insist on acquiring properties in the more up-scale part of town, you'll certainly be pleased with the beautiful glossy pictures they make; however, you'll be a lot less happy with the money in your bank account! Folks who buy this type of property lust after prestige and those plaster of paris plaques you get from the Rotary Club!

CHEAPEST PROPERTY, WILL LIKELY PROVIDE BIGGEST PROFITS

It goes without saying, if you live in a "scumbag" apartment, your rent will be a lot cheaper than if you live on Snob Hill. Naturally, apartment buildings on Snob Hill are worth much more than the ones at scumbag Villa. Suppose there is a six-unit property for sale at both locations. Scumbag has a GRM value of five times gross and rents for $325 each. (Refer to the chart on Page 111 for these numbers). The indicated value would be six units times $325 rent, equals $1,950 per month, times 12 months equals $23,400 annual income. Multiply this amount times GRM 5 for a total of $117,000. Thus, each unit is worth $19,500. Doing exactly the same math for Snob Hill, which is valued at ten times gross, with rents of $525, you'll notice that each unit is worth $63,000.

Let me ask you this question: Which property do you think is more profitable?—the $19,500 apartments that earn $3,900 in annual rents or the $63,000 units that earn $6,300 annually? Looking at rent returns, you'll see it takes five years of $3,900 annual rents to pay back a $19,500 value. However, it takes ten years (twice as long) to collect enough rents to pay off the $63,000 units. Although it's very important, the "quickest rent pay back" is only one consideration. Which property owner would you imagine is likely to offer the best selling terms; which would require the smallest down payment and provide long-term seller financing? Which one would be the most motivated to sell? I'm sure you guessed right—it's the owner who is selling the property with the most potential for upgrading and improving the cash flow.

JAY'S PROFIT STRATEGY: UP THE RENT—INCREASE THE VALUE

My investment strategy is to fix-up rundown properties enough to improve the GRM by at least two full points and increase the rents by 40%, and that's it! If you do it right, you can almost double the value of a property without waiting around for regular appreciation or the passage of time! Naturally, appreciation increases value over a period of time because of inflation. To me, waiting is much too slow and its not a sure thing. My strategy is called **FORCED APPRECIATION**, because I force the property value to increase by improving the GRM.

By increasing the GRM value by two points, I can automatically raise my rental income by 40%. Both improvements result from four basic actions on my part.

1. **I clean up the ugly property to make it more attractive in appearance.**

2. **I evict the problem tenants along with the deadbeats.**

3. **I fix up deferred maintenance and make improvements to attract better quality tenants.**

4. **I provide the proper management to operate the property more efficiently.**

That's it! *See how easy it is?* In order to achieve maximum results with this strategy, it's necessary to purchase low GRM properties—maybe not the very bottom, but as near the bottom as your fix-up skills will allow. Naturally, your skills will improve with each new purchase and, eventually, you'll end up just like me. NO PROPERTY IS TOO BAD, NOR PROBLEM TOO TOUGH, IF THE PRICE AND TERMS ARE LOW ENOUGH. Read that again, it will help you make a ton of money, if you do it right! Also, notice I made it rhyme! That should help you remember.

HOW TO CALCULATE WHAT YOU SHOULD PAY

We'll use a six-unit apartment as our example to show how to calculate what you should pay. The property, however, can be two buildings, six detached houses, duplexes or any combination. One single family house is not well suited for the GRM value measurement. I'll use the numbers from the GRM chart on Page 111 to illustrate how to put big bucks in your pocket, once you get the hang of this.

Let's assume we find a rundown fixer property in Redding that can be purchased for five times the gross rents. In Redding, as you can see by referring to the chart on Page 111, apartments valued at five times gross are renting for $325 per month. That means the purchase price will be approximately $117,000 (six apartments times $325 rents equals $1,950 per month, times 12 months for a total of $23,400 annual rents; $23,400 times five GRMs for a $117,000 price) As the chart shows, you can expect the property to be older, ugly and junky with deadbeat tenants and lousy management. Heavy fix-up is needed—that's the challenge, **but it's also an opportunity.**

Looking at Profits and Cash Flow

Increasing the rents and improving the GRM work together like two singers harmonizing. If one improves, it helps the other, as well. For example, in my town you couldn't rent your apartment that's worth five times gross for $450 per month. The value wouldn't be there. That's why $325 is the most you can get. Conversely, apartments that rent for $450 per month are worth much more than five times gross value when they sell. To be exact, they're worth seven times gross rents like the chart shows.

Back to our example with the six-unit apartment. The fix-up is done; rents are now $450 and the building is worth seven times gross rents. *What's a small two point increase in the GRM worth?* I think you'll be pleasantly surprised to find out the new value is $226,000. Almost double the price we paid. Also, the rents have increased $9,000 annually, which should put us well into positive cash flow! These are the kinds of numbers that turn "poor fix-up investors" into wealthy tycoons in the shortest period of time.

STAY AWAY FROM WEIRD STUFF—STICK TO BASICS

Everyone who knows me or seeks my advice, understands rather quickly how strongly I feel about doing the basics. That means buying properties that will start returning profits before very long. I determine this by studying all the $ numbers before I make the purchase. Most of my long-hand studies are done on yellow legal pads. They usually take the form of a sketch with a bunch of $ numbers plotted in year by year. What I want to visualize clearly, in picture form, is how much money I'll be putting into the deal (down payment) to start! Next, how much more for fix-up and what will my monthly fixed costs of operations be when I'm done. After that, I calculate my expected income, month by month, usually for five or ten years, or sometimes to a predetermined sale date.

My yellow pad studies are purposely simple. I'm only interested in cash returns. The two questions I ask myself are:

Will I make profits on this deal if I spend X number of dollars? And secondly, when do I get to have those profits in my hand? That's what most investors need to know.

Do-it-Yourself Investing Does Not Require Genius

Too many people get all "honked-up" with complicated strategies. They seem to have the same mindset as body builders—Who insist: "Without pain, there will be no gain". If you want pain, I suggest you buy a $100,000 house with $850 mortgage payment and rent it out for $600 a month. If that don't hurt enough, keep doing it. For those who need higher levels of pain—Buy a motel or small business opportunity. That should be enough.

Keep Your Wealth Plans Simple

If you over-complicate your wealth-building plan, quite often you'll become distracted by information you don't really need to make money. For example, a friend of mine is a computer buff. He has drawers full of printouts that tell him everything you ever wanted to know about rental houses—Plus lots of other stuff I'm not smart enough to ask questions about. He calculates his rents per square foot, how much paint he'll use in the next 20 years and the number of qualified renters in the county where he lives. Still, he has a very serious problem!

His computer shows his net worth is nearly $500,000, but he's paying out $1150 every month in hard cash money so he can stay in the rental housing business. Even though he has a good paying sales job, I figure if he buys just two more rental properties using his current strategy, he'll be broke in a year or so.

LEARN WHERE YOUR PROFITS COME FROM

Please don't get the idea I'm suggesting that you shouldn't do a thorough study before you invest your time and money in real estate! What I am saying is, *keep it simple and keep it basic*. That means analyzing all monies going out and money coming back in. You don't need to know the IRR (*internal rate of return*) on a duplex to make a profit. You're far better off knowing if both toilets flush. Later on when the money rolls in, let your accountant explain about the IRR. Perhaps he might even agree to do your plumbing when he's not doing the books.

WITHOUT CASH FLOW TODAY - THERE MIGHT NOT BE TOMORROW

When I began buying investment properties, there was no question in my mind whatsoever about where I could find extra money if my properties didn't provide enough cash flow. The answer was quite clear to me from the day I started—I COULDN'T. Even though my cash down payments were small, it was all the money I had. I knew very well there was nothing left in my bank account to make up for monthly cash flow shortages. The only funds I would have to pay my mortgages and expenses would be the money I took in each month from my renters. Obviously, buying properties this close to the belt is both challenging and exciting. There is little room for making serious errors, as you might well imagine.

BUYING AT RIGHT PRICE IS 90% OF THE BATTLE

Baseball legend, Yogi Berra used to say: *"BASEBALL IS 90% PHYSICAL, BUT THE OTHER HALF IS MENTAL."* Real estate investing is something like that! Buying income properties at the *right price and terms* is about 90% of a successful transaction The other half is doing all the things you need to do to make the property a profitable investment.

Many do-it-yourself investors get started on the wrong foot, I think! They mistakenly spend too much time running around looking at various properties trying to make predictions about their future potential. What's much more important, in my judgment, especially for investors with limited bank accounts, is to spend more time and effort to determine how much money the property will start earning right now—that means as quickly after the purchase as possible.

Beware of Future That's Unpredictable at Best

What I'm saying here is, that for most of us, it makes more sense to buy cash flow properties with uncertain futures than to acquire high potential properties without enough cash flow to operate them right now. Future potential is great, don't get me wrong here—but, it's still "Pie in the Sky," in my view. The top three reasons for owning and operating income properties are INCOME, INCOME and INCOME—Everything else starts with number four.

You Can't Pay the Bills With Equity

Investors like myself seem to be afflicted with what I call, "Blind Faith" when it comes to buying income properties. We seem to view all investments as being profitable. If not today, then most certainly in the very near future. This strong affliction— "Blind Faith" can often distort our judgment and cause us great financial distress unless we learn to recognize the problem and make adjustments for it. I know enough about this business to tell you—There are many investment property owners fighting the wolves away every single month—Yet they still brag to their friends about huge equities and the future "killing" they'll make someday when they decide to sell. Obviously, love is not the only thing that blinds people.

True Values are Always Set in the Marketplace

Never forget that value is in the eye of the buyer, *NOT THE SELLER*. It's not the least bit uncommon for property owners to have an overly optimistic view of their own property values. Financial statements prepared for lenders quite often reflect very generous estimates of net worth. Seldom, however, are these same bloated estimates accepted by a harsher judging marketplace—And of course, the marketplace is exactly where true values will always be decided.

I have investor friends who can show me they're millionaires on overblown computer spreadsheets. What my friends can't show me is when they'll be receiving any money. They are equity rich, but don't have enough cash to buy lunch! Unfortunately, their situation is more common than not! It happens to many new investors before they learn to get a bit more serious about doing their financial homework. My property analysis form will go a long ways towards solving this problem. Read Chapter 5 again—it will help.

OPERATING YOUR OWN PROPERTY IS ALWAYS THE BEST TEACHER

Almost every investor I know has paid too much for income properties. It happens quite frequently when we first start out. There is almost no defense against paying "too much" as least once or twice. I've done it more times than I care to admit. However, in my case, buying fix-up properties allowed me to add value and improve the income stream more quickly than if I had overpaid for average non-fixer type properties. By fixing and adding value to properties, I've always been able to recover from my overpaying errors much faster. The best education in the world for understanding values and what the real expenses are, will come from buying and operating your own income properties. I'm not talking about a single house here. That's not quite enough education. Five or ten houses should certainly qualify. Small multi-unit properties provide an excellent experience for both managing people (tenants) and managing money (both income and expenses). When you pay the bills yourself, no one will ever again be able to fool you again about operating costs and how much money is left over when all the bills are paid.

KNOW WHERE PROFITS COME FROM BEFORE YOU BUY

So far I've tried to get you thinking about the main reason for investing in real estate to begin with! I have found that when

I'm forced to explain what I'm doing, i.e. provide reasons why I'm investing in a particular property, it makes me dig up all the reasons I can find to support my decision. That's good business! Remember what I said earlier about "Blind Faith"— Having faith is good as long as it's not used by lazy investors as a "cop-out" for not doing all the preliminary research, or detective work, as I often call it!

My good friend and skilled investor, John Schaub, says: "The day you buy an investment property, you should be able to sit down with your wife/husband or someone and explain, in detail, exactly how you will make your profit from start to finish." I find John's excellent advice is right on the money. I did not follow that same advice when I first started investing and I always had problems making my houses cash flow. My two biggest problems were overestimating net income and paying too much to begin with.

TWO NUMBERS MAKE ALL THE DIFFERENCE

I stumbled into fixers with only the limited knowledge that I could purchase them cheaper and pay less money down, compared with more desirable looking properties. I have since come to realize after years of experience that probably no better opportunity exists in real estate investing today than buying and fixing rundown rental properties. This can be especially true for ordinary folks, like beginners or career changers who have very limited resources to start with.

As I've already told you, fixers can be purchased for substantially less than comparably sized properties in good condition. After rehabilitation, however, fixers will normally rent for almost the same monthly rates as the better properties. By converting this information into numbers, it becomes easy to understand why fixer houses can help small-time investors move quickly to "big time" profits.

There are two key numbers I consider to be most important because they have a great deal to do with CASH FLOW ad PROFITABILITY—And by the way, income property is supposed to provide cash flow regardless of what your real estate professional tells you about the extra tax breaks you'll receive. As far as I'm concerned, tax benefits are merely a bonus for investing in real estate in the first place!

The *first number* I'm always concerned about is the UNIT COST. How much will it cost to purchase and rehabilitate each house or apartment? Let's say I purchase a four unit property for $72,000. That means I'll pay $18,000 per unit. Next, I estimate it will cost $6,000 each to rehabilitate the units. My total unit cost will be $24,000 when my fix-up work is completed.

The *second important number* has to do with rental income return. I call it **RENT-TO-VALUE RATIO**. If you don't remember some of the other things I tell you here, try not to forget about this part. It's very important to your financial future because it's the difference between having cash flow or not—Ultimately, success or failure. The **RENT-TO-VALUE RATIO** is a number that expresses the percentage of monthly rent as related to the unit cost or my assigned value to the property. For example—If I own an apartment that rents for $300 per month and the apartment has my assigned value or unit cost of $24,000, then the **RENT-TO-VALUE RATIO** would be expressed as 1.25%. To arrive at that number, you simply divide the rent of $300 by the unit cost of $24,000 ($300 ÷ $24,000 = 0.0125). Move the decimal 2 places right.

It is my goal to achieve a minimum of 1.25 ratio, but when you really get the hang of this stuff—2.0 and more is well within your reach. It's also lots more fun going to the bank.

LOCATION, LOCATION, LOCATION

When I first started investing in real estate, "Old-Timers" always used to tell me, "Son," they would say, "There's only three things you've got to think about when you buy real estate—*LOCATION, LOCATION AND LOCATION*. If you do that," they told me, "You'll end up filthy rich, someday."

I would agree with this early advice, except for one rather significant exception—just one little fly in the ointment! As I look back over 20 years, or so, the most important thing I've ever accomplished with my real estate holdings was to buy them so they would pay me money every month. The politically correct term for this is called POSITIVE CASH FLOW.

Income Properties are Supposed to Provide Income

You see, without money coming in every month, small-time investors, like myself, would not likely last long enough to reap the benefits of choice locations. Not only that, but early on I didn't have the financial horsepower (green stuff) to outbid my competition, even if choice locations had been my first priority. Don't get me wrong here, I'm not knocking good locations. I'm merely suggesting that you may need to broaden your vision when it comes to buying income properties! After all, income properties are supposed to produce income, as the name implies.

PREDICTING THE FUTURE IS ONLY A GUESS, AT BEST

When I look back at all the properties I've acquired over the years, it occurs to me that picking locations has not been my greatest strength. It's not that I don't have some good locations; I do, but to be perfectly honest, they weren't so hot when I bought the property. What happened in several cases was that new development sprung up around my older houses, creating much more desirable surroundings. There is simply no way I could have predicted this would happen 10 or 15 years ago, when I bought them.

On the flip side, I've purchased some dandy properties with future commercial zoning, which I thought would pay off "big-time," like slot machines, when future development came my way. The problem was—development came all right, but on the opposite side of town. My rental houses are still just sitting there doing the same thing they've been doing for years—producing cash flow! Not shopping center rents, mind you, but steady spendable cash every month. In the long term I've discovered that's really what makes you rich! Underline "steady cash flow every month"—*AND DON'T FORGET IT!*

Most Investors Can't Pick Over 50% Anyway

Most new investors spend far too much time trying to figure out the future of properties they are interested in buying. This is actually non-productive for novice buyers who have very limited knowledge about real estate growth and development to begin with! In my town, the city planners have giant maps with future projections for many years to come, and they're still bewildered when out-of-town developers select remote locations and re-zone to build huge shopping malls.

I'm convinced that most investors would be better off gathering information about the future from Shirley McLaine's home-study "Psychic" tapes! After all, she claims she was here a thousand years ago, which certainly exceeds the time on the job for most city planners I know. The truth is, I've lived in Redding most of my life, and I'd be hard pressed to tell you where the best or worst location will be ten years from today.

MIDDLE GROUND IS ALWAYS THE SAFEST BET

I have a few simple rules I follow regarding location, which have served me quite well for many years. I'll share them with you in the order of their importance to me.

My first rule about buying property anywhere is to ask myself "DO I THINK IT'S SAFE?" There is no such thing as a "deal I can't pass up," especially if I feel there's even a remote chance that I, or one of my handymen, could end up buried in the backyard some night! Life is tough enough for landlords without this added worry, and, besides, how could I ever spend my money if I'm buried?

Often seminar students quiz me about where to purchase ugly rundown houses for cheap prices, when I tell them that undesirable areas are off limits, and that the newer tract houses are most likely too expensive to ever produce cash flow! "Where on earth is there left for us to go?" they ask. *Here's the answer—so pay close attention.*

Most towns and cities, both large and small, have five separate, but distinguishable, sections where residential properties exist. There are obviously over-lapping and combination locations; but, for the most part, they're quite separated and if you think about it, you'll know exactly where I'm talking about in your town. HERE ARE THE FIVE LOCATIONS. You'll recognize them, I'm sure.

1. *SNOB HILL*—this is where the wealthiest folks in town reside.

2. *DOWNTOWN COMMERCIAL*—mostly businesses with acres of concrete and blacktop.

3. *OLDER RESIDENTIAL*—surrounding downtown area. Generally 50 years and older.

4. *DENSE SLUMS*—downtown or pocket areas, often older, plus newer projects, most notably "HUD Projects."

5. *SUBURBIA*—sprawling subdivisions, tract houses, mostly owner-occupied.

Let's take these five locations, one at a time, and discuss what's right and wrong with each in terms of investing.

SNOB HILL—This is the place to live once you make it! It's not the kind of real estate to make money with. If you're the user (owner) and you live in this location, it's most certainly prime residential property and will undoubtedly increase in value over time. If inflation breaks loose, obviously, this type of real estate will soar in value, which is exactly what happened years ago in Southern California. Many who bought homes to live in and raise their families watched the values increase 20 times or more during the life of their mortgage.

Still all said and done, modestly priced rental houses did about the same thing with respect to jumping values; but, of course, they earned their owners steadily-increasing rents as their values shot up. Naturally, the big difference between owning a home and owning rental houses is that tenants paid all the bills for the rentals. A home is an excellent investment for a family, but the houses you see on Snob Hill will seldom achieve cash flow, so long as there's a mortgage to pay. Even without one, the investment returns are not satisfactory for most investors.

DOWNTOWN COMMERCIAL—Cagey sellers will often attempt to market houses in the downtown areas using the proposition that zoning is commercial; therefore, the property is much more valuable because it has unlimited future potential! Do not fall for this! There's no such thing as unlimited potential and, whereas an older house in the downtown area may be an excellent rental property, it's still a fish in the wrong pond. Only speculators should buy for the future! *Investors should always buy properties and pay for them on the basis of their earning capacity right now!* I'm always happy to buy properties with commercial zoning, but I'll only buy it for the same price that I'd pay without the zoning.

OLDER RESIDENTIAL—This is the best location to buy for cash flow. This is the older residential sections surrounding the downtown area. Many of these older properties are functionally obsolete, with three bedrooms and one bath, or single-car garages. Today's homeowners are shopping for two bathrooms and double garages, therefore, investors won't find as much competition for these properties. Less competition means cheaper prices for investors, since the owners have fewer selling opportunities. Many long-term owners are in the retirement-age group, with mortgage-free properties. There are great possibilities for seller financing—A REAL PLUS!

Many of these older properties consist of several (multiple) units on one lot. Often there's a duplex in the back or a "granny" apartment above the garage. Renters are willing to pay top prices, because they're close to shopping, yet still able to enjoy the privacy of residential neighborhoods. There's also a bustling first-time homeowner market when these older properties are cleaned up, rejuvenated and sold to young couples or families. This is the area where rundown properties can be "fine-tuned" and rented for cash flow, or sold on easy terms to a first-time buyer.

DENSE SLUMS—I do not recommend investing in slum areas, even if the property is dirt cheap or half the regular price. As I mentioned earlier, the question I always ask myself about all properties—*"Is it safe to be there?"* I will not expose myself, or my workers, to unsafe conditions. Even if I were willing to work in slum locations, there's another equally serious question to answer—*"Who will be my customer?"* You must always buy properties that will appeal to your customers—renters or someone who will buy a home from you.

Can you picture the kind of person who chooses to live in the slums? In case your vision is somewhat blurred, let me just say—I don't think you'd want to rent to them! The bottom line is—when there's doubts about personal safety, coupled with very restricted marketing opportunities, you're simply taking on too much risk, in my opinion. There are always exceptions, of course, but not very many.

SUBURBIA—This is the location most investors choose when they decide to acquire residential properties. The giant subdivisions and sprawling tracts of houses most generally located a few miles from the city core, tied together by express lanes and freeways. Most subdivisions are occupied by middle-class America. It's where we live.

Buying middle-class houses in these locations makes perfect sense, but very little cash flow. The only opportunity to make any serious money is during inflationary periods or after holding the property 20 years or more until the mortgage is paid off. Few investors I know are able to quit their day job anytime soon when they invest in subdivision houses. *To me, building a dependable cash flow income is first prior-*

ity! Tract houses purchased at reasonable discount prices are an excellent long-term investment when properly managed. But, if you're in need of monthly income, this is not the investment to start with. My suggestion is to concentrate on older residential areas first. Build a solid income and then gradually transition out to the suburbs, if that's what suits you.

There is no ONE SIZE FITS ALL in the real estate investment business. You must choose how and where to invest, based on your skills and available funds.

Years ago, a highly popular book directed novice investors to buy real estate for little or no money down. This strategy will often work when employed by "ring-wise," more sophisticated investors; however, in the hands of the uneducated, it's almost a "sure-fire" path to disaster. The inherent problem is that financing 100% of anything will most likely cause the monthly mortgage payment to eat up all the income generated, sometimes even more. Investing is not much fun when you have to pay additional out-of-pocket money each month, just to keep the property operating!

INVESTING LONG-TERM FOR FUTURE GROWTH

Life is full of compromise and its the same way for investing! I'll be the first one to agree that beautiful properties in top locations will increase more in value over the long haul. The sacrifice, of course, is that these same properties will pay you nothing while you wait! The problem can be overcome by adopting a strategy to acquire two different kinds of properties. Why not start out with fixer-type properties that can provide a solid monthly income—you may call this Phase One!

Phase Two can be a gradual transition into nicer properties, with better long-term profit potential. However—do this only after the money to buy groceries has been well secured.

Each type of property has good and bad features. For example, small rental houses can earn twice as much net income as larger homes. Repairs and fix-up are much more economical in small, *NOT SO PERFECT* houses. Rough tenants can do far less damage in smaller units that are not equipped with modern amenities like many of the larger and newer houses.

As an investor you should weigh these considerations before you begin. That way you'll avoid spending precious time trudging down the wrong path before you realize it's the wrong direction! Take it from me, *cash flow is a long way ahead of whatever's in second place!*

CHAPTER 8

Finding the Right Properties and a Motivated Seller

TRADE SECRET subscriber, Robbie Taylor, visited Redding on July 23rd and the temperature was 106 degrees—it was probably the hottest day of the year. Naturally, our tour was scheduled in the afternoon of this hot day. Fix-up investors don't need to be the smartest folks around, but they must have endurance. After our tour, the only question Robbie had was, "How does one go about finding the kinds of properties we've been looking at today?" I promised to explain it to him in a note, then I realized I might as well tell everyone else while I was at it.

THE FOUR BASIC METHODS OF FINDING AND BUYING FIXERS

I like to keep things simple! Others may tend to embellish these four fundamental methods I'll discuss here, but my experience has been that these are solid methods that continually work. Naturally, you must make them work for you! Nothing will help if you simply sit back, watching TV, while you wait for others to make you rich. First, I'll tell you what these four methods are, then we'll dissect each one to learn what makes it work.

1. WATCHING THE DAILY CLASSIFIED ADS
2. USING A REAL ESTATE AGENT
3. WORKING THE MULTIPLE LISTING BOOK
4. INITIATE WRITTEN COLD CALLS

Before we tackle classified ads, let me simply make this statement. *Once you become proficient at doing this stuff, you'll begin to understand that there is no shortage of "Fix-Up Real Estate" anywhere! The problem is—the right fix-up properties are seldom located where the public can find them,* and most real estate agents don't know "good deal" fixer properties from over-priced junk! True bargains rarely reach the multiple listing books because they get sold by listing agents to a handful of "pet buyers" who can cut a deal before the agent is required to submit his listing to the multiple service. They're called "pocket listings" and listing agents see no point in splitting commissions when they have a "slam dunk" deal.

Watching the Daily Classified Ads

This is the method I have found least profitable when finding and buying fixers. Yet, I consider it worthwhile, because it doesn't take a lot of time and it keeps you tuned in to what's for sale in your buying area. I read the classifieds everyday, and have done so for years. I always read "INCOME PROPER-TIES FOR SALE," "INVESTMENT PROPERTIES" and "COMMER-CIAL REAL ESTATE." I also keep an eye on "HOUSES FOR SALE" because, occasionally, "mom and pop" owners will list their six little duplexes under that classification. Agents almost always list properties under the correct classified heading.

The location where you find the ad might tell you if you're dealing with a sophisticated seller or a do-it-yourself property owner who has decided to retire. Naturally, "mom and pop" deals can be very productive because, quite often, they will sell for less if they like you. Also, many are trying to sell on their own (for sale by owner) to avoid paying commissions! Perhaps more importantly, older retired sellers are much more likely to carry back a mortgage. It will pay you to watch classified ads regularly to catch an owner-seller bargain.

Another benefit of calling the interesting classified ads is that it helps you become more knowledgeable about *prices, gross multipliers* and what *terms and conditions* are important to sellers. Many times it turns out the property I call about is not what I'm looking for; however, during the conversation, the seller mentions other property that's not advertised in the paper, or that he knows someone who is wanting to sell another property. Telephone calling generates leads. One *good* lead can make you big bucks. I purchased my Haywood property (Chapter 2) from a classified ad. The property was comprised of 11 junky houses on a two-acre lot. With this one fix-up project alone, I earned $150,000 for just 13 months of part-time fixing! Classified ads have accounted for about 10% of all my deals—that's $400,000 worth!

Using a Real Estate Agent

This method for finding properties and making deals has been, by far, the most productive for me! At least 60% of my properties were acquired using a real estate agent to assist me. However, not just any agent will do. I've only had two since 1977—Merv and Fred. Merv passed away in 1985, and Fred is still working with me today.

You must work hard to find the right real estate agent. Agents won't hang around for long if you're just a "Looky-Loo"—that's a person who wants to see everything, but never seems to find anything that meets his approval. No agent can afford that nonsense! Fred works for me for one simple reason—it's profitable for him! He knows I can close deals fast if he does his job. Fred's job is to know exactly what I will buy. He doesn't call me about every property for sale in Redding. A good agent will immediately qualify the property to determine if it has potential. Fred knows I don't normally want deals where new bank financing is required. He also knows I want sellers who will carry paper and that I rate small multiple unit properties—like four to six houses on a single lot or a bunch of ugly rundown duplexes—at the top of my buying list. When he hears about these kinds of properties, he acts quickly!

Fred's job is to qualify the property. Is the seller in any sort of financial trouble? Does he live out of town? Are they managed for him? Are the tenants causing the owner problems? Fred tries to determine if there are reasons for the seller to be motivated, such as retirement, going broke, dying, etc. It's detective work like Colombo does on TV. He obtains a property profile. He asks for copies of notes (mortgages) on the property. He verifies rents, vacancies and liens. All this takes time and it costs Fred money. The only way he gets paid for his work is if I buy the property.

Fred doesn't make a lot of dry runs. He knows what I want and determines quickly if a property has the *right stuff.* How did Fred get so smart? How did he ever learn this detective business anyway? When Fred and I started, we spent a lot of time discussing what I wanted! Investors must be very clear about which properties they will buy when one comes along, otherwise agents will attempt to show them everything.

To start with, I would recommend that a new investor simply walk into five or six realty offices, sit down with an agent who is on the floor and discuss the kind of properties they would like to acquire. Some agents will jump through hoops to help you. Others will shine you on. Personalities will naturally play a role. You like some people—others, you don't.

Using this method, it shouldn't take too long to find one agent who seems more interested than others. He or she will call you more and bring you property profiles to look at! When this special interest develops, you might have found the right agent. If not, keep repeating the process until you do. Finding a good agent is not a lot different than finding a good wife or husband. It's simply a matter of "weeding out" until you find the right combination!

Working the Multiple Listing Book

Only licensed real estate agents can have multiple listing books! That's the reason I've always kept my copy out of sight. It seems like month old copies somehow find their way to my desk! It's perfectly okay with me if they're a month, two months, or even a year old. Here's why! First, understand that most income properties in the multiple listing book would have sold before they ever got listed if they were really bargains! Remember what I told you about pocket listings, they never get published! What you normally find in the multiple listing book is overpriced apartments and "dog properties" in the "Income Property" section.

The multiple listing service contract provides for selling agents to split commissions with the listing office when a sale takes place. Obviously, for a property owner, the multiple listing book provides the maximum exposure for his property. Every licensed agent is a potential seller. The value to me is keeping an eye on properties that stay in the book for long periods of time, say six months, even a year. Sellers who can't sell in that length of time often become extremely motivated!—offers, they never dreamed they'd accept when they listed the property, somehow become much more acceptable to them after a long dry spell!

I've had pretty good luck buying properties that nobody seems to want from old multiple book listings. On several deals, the listing had already expired and I was able to purchase directly from highly motivated sellers. Picture yourself in the seller's position—if you can't sell a property in six months with all the exposure of the multiple listing book, obviously, more drastic measures are called for—*usually, lower price and better terms*. I've done about 10 to 15% of my business as a result of tracking down multiple listings. It's well worth the efforts.

Initiate Written Cold Calls

Real estate investors often complain that their wealth-building plans are seriously hampered, because too few properties are available in the areas where they invest. You can't buy real estate and build much wealth if nobody will sell you a property, right? Obviously, it takes both buyers and sellers to complete transactions! So, the big question is, "Where do you find real estate sellers, when it seems like nothing is for sale?" Cold calling is a technique where you contact property owners who own the kind of properties you'd like to acquire and try to persuade them to sell.

One of the first questions I'm always asked is, "Why in the world would anyone consider selling their property to me merely because I contacted them and asked?" There are more reasons than you might imagine, but first, let me confess to you that I did not invent cold calling. Real estate agents have been doing it for years and with enough success to keep doing it. The problem that many real estate agents have with cold calling is that it's too much work! It's much easier when sellers walk through the door and hand them a listing. A big percentage of agents are willing to wait.

COLD CALLING IS A PROVEN TECHNIQUE

Naturally, there's a big difference between investors who operate properties for profits and real estate agents who sell them for commissions. Cold calling can be different, too! For example, agents will sit at the telephone for hours calling long lists of owners. Obviously, that's got to be very boring! The way I do cold calling is much different. To begin with, I write letters to property owners rather than make telephone calls! I also do some research before I make any contact with property owners. This allows me to customize my cold calls (letters) to fit the property and address any special circumstances pertaining to the owner.

Not Just Any Property Will Do

Not every rental property is a candidate for my "cold call" letters. To start with, my specialty is the fix-up real estate! I make money by acquiring properties I can "fix up." My strategy is to quickly increase the value. Average looking properties, without any visible signs of being rundown or neglect does not fit my profit plan, therefore, those properties are automatically eliminated from my cold calling list.

When I'm driving around, I always keep an eye out for interesting properties I would like to own—*assuming I could buy them for a reasonable price and terms*. I keep a little notebook in my car to write down addresses, number of units and the condition. I also jot down my own estimate of their current market value—and, the amount I would be willing to pay. I find it very helpful to draw a little sketch of the lot—then take a measurement from the nearest street intersection (cross street) to the lot. A sketch will help you locate the property on the county assessor's map, when you visit the courthouse. Assessor maps don't have street addresses, so the measurement will help you find the right location on a scaled map.

Information Hunting is Valuable and Free

All this information can be found at your local county courthouse or a title company, and is easy, if you happen to know someone who will help you. Assuming you don't know who owns the property, first take a look at the key location map. It will direct you to the proper assessor plat map where you'll find the parcel, or lot, you're looking for. Once you locate the parcel number, you can go to the name index file and find out who the owner is! Both the courthouse and title company will assist you with this chore.

My research consists, primarily, of finding out who owns the property and where they live. *It's generally the address where the tax bills are mailed.* I want to know how many mortgages or trust deeds are secured by the property and what their original amounts were. Also, the tax bill will show me what the "value for tax purposes" is. I like to know if my estimate of value is somewhere in line with the county appraiser's view. I always like to get an idea of how much mortgage debt is still owed because I'm looking for properties with a lot of equity. You can't negotiate a good discount purchase if the owner has very little equity to give. Also, I'm looking for the opportunity to create a long-term seller carryback mortgage, should I be successful purchasing the property. Obviously, this sets the stage for buying back the mortgage at a sizable discount sometime in the future.

FINDING PROFITABLE DEALS IS THE GOAL

The first thing to do is, *don't panic!* Good properties are always available, but it takes a little creative effort to find them. Finding the right properties—the kind that will produce *monthly cash flow* and *long-term profits*—is one of the most important skills you must develop to enjoy any success in this business. Remember, if this was too easy, everybody would be doing it and they'd all be rich! Finding properties that will earn reasonable profits is one of the biggest challenges for every investor, regardless of how many properties are for sale in the buying area.

You must always keep in mind, finding profitable deals is the goal, not finding lots of deals! You should pattern your buying strategy somewhat like the old Hills Brothers TV commercial. The buyer in the ad says, "90% of all the world's coffee beans are rejected by Hills Brothers. They're simply not good enough for Hills Brothers, so they're sold to the other guys." I would guess about the same percentage should apply to investors looking for the right properties, unless, of course, they're buying property as a hobby. Quality deals are the ones that make money. *Don't substitute your profits for volume. It makes little sense to hurry up and buy a loser.*

COLD CALL STRATEGY REQUIRES A PLAN

First, let me tell you what you *shouldn't* do—don't insult or offend the owner because his property looks like junk! Chances are, he already knows that, but even if he doesn't, you mustn't blame him. It is better to blame something else—such as the property managers or unruly tenants—for all the problems. That way, you can sympathize with the owner, instead of alienating him.

I also attempt to establish my credentials as an experienced property owner, but not as a "hot-shot real estate tycoon." Remember, owners of troubled properties, who might be willing to finance a sale, will not have much trust in a buyer with no experience. They can easily visualize the property coming back to them in even worse condition.

In a situation where the owner hasn't been around his property much, especially if he lives out of the area, I try to paint a picture of a property that is poorly managed and is in need of immediate attention. I often give the owner an idea of what it might cost to get the property up to snuff! Note that the letter on Page 138, refers to "a $20,000 lesson I will never forget." I let the owner know that I'm a local landlord and I buy fixer properties. I generally write something like the following is in my letters.

> "Each day when I drive by your place, I can't help noticing how many jacked-up cars the tenants have accumulated."

or

> "I can't help noticing the lawns have all died in the past few months."

or

> "I notice the houses need a lot of work and I'm wondering if you ever thought about selling the property."

Letter Suggests Solutions to Owner's Problems

You'll notice my reason for writing says nothing about buying his property. I try to keep my offer to purchase as an alternate solution to his problems. I even ask him to check me out. Drive by my houses and see what you think. This reinforces my credibility with him. It also proves I'm a real landlord like I said in my opening sentence. Out-of-town owners generally know about management problems and high expenses! My letter merely reminds them that it's true and owners are the only ones who can make a difference. I think every owner will agree with that!

I like the owner to start thinking about his monthly *net income*, especially out-of-town owners who receive a monthly check from a property manager for the amount that is left after fees and expenses are paid. When I tell him about earning more NET SPENDABLE INCOME, it generally grabs his attention. Again, I never tell him my intention is to purchase. I only ask that he keep me in mind if he should sell. I offer my financial statement and tell him how to save a big commission.

John P. & Mary Jones
1234 Easy Street
Lake Camanche, CA 95640

RE: 2020 West Street
AP# 07-360-14
Shasta County

Dear Mr. & Mrs. Jones:

I am a local property owner and landlord in Redding. I own several rental houses in the same neighborhood as your property on West Street. I drive by your property at least once every week on my way to check out my rentals.

Apparently, you're having the same kind of problems I have with tenants who own junk cars. Last week, I noticed a car motor hanging in the tree in front of your duplexes in the back! I would guess it belongs to that old rusty blue sedan with no wheels, parked between the two pink houses.

I realize it's difficult to manage properties from 200 miles away! I did it only briefly before moving to Redding, but never again. The management company I hired allowed the tenants to take over everything. I had junk cars, dead grass and you wouldn't believe what the houses looked like inside. It was a $20,000 lesson I will never forget! That's why I visit my properties at least once a week, now.

The reason I'm writing is because my company does fix-up and repairs for rental property owners. We just rehabed the three light blue houses on the n/w corner of West and 17th Avenue. Drive by the next time you're in town. I think you'll be impressed. Keep me in mind if you need work done.

When I spoke with your tenant, Judy, in the front house to find out who owned the property, she said the owners were retired, but she pays rent to the Ace Management Company on Hilltop Drive. I hope they are better than the managers I had, but by the looks of things, they don't seem much better! I found out the hard way. Owners are the only ones who really keep the property up and watch what the tenants do! I would never again own property in a town where I don't live.

By the way, if you should ever think about selling out, keep me in my mind! I'd be happy to show you my financial statement and several of my Redding properties, so you can see how I do business. Quite often, you can earn more spendable income from mortgage payments than from rents, and, of course, you don't have to worry about tenants and property managers when you sell.

Drop me a line or call if I can be of any service, and call me direct if you should ever decide to sell. You can save a bundle without paying a big commission!

Sincerely,

Jay P. DeCima
(916) 221-0123

When I discover the property owner lives out of town, and his property is not being maintained, I will try to emphasis things that indicate—unless changes are made immediately—his property value will decline and his income will suffer. These are very compelling reasons for an owner to take some action. Several times in my career, cold call letters have worked long after I've written them. In one case, an owner worked desperately to fix his problems. He hired three different managers in a 12-month period. Finally, sixteen months after I wrote him a letter, I bought his property for $41,000 less than my initial drive-by estimate of value. If your cold call letters don't work overnight, just be patient and don't give up!

How to Write a Cold Call Letter

The letter on Page 138 is an example of how I do cold-calling! Read it several times and try to picture yourself as John P. and Mary Jones. How would you feel if you owned the property on Easy Street and just received this letter from me? Would my letter start you thinking?

NEVER BLAME AN OWNER FOR THE CONDITION OF HIS PROPERTY

Never say anything about the condition of his property that puts the blame on the owner. Always put the blame on lousy tenants, bad luck, poor property managers or something other than the owner. The whole idea is to not alienate the owner! Instead you should sympathize with him, perhaps explaining that you have the same problems yourself, but you have more time to spend with your tenants. If any owner thinks that spending time with tenants is necessary to make his property get better, he may just give it to you! Of course, that's wishful thinking, but you get the idea, I think!

Your letter should not mention price, cash down payment or terms. It should state that you are a serious buyer who performs quickly if a sale could be worked out. You should also point out that, by dealing with you directly, he can save a large sales commission! If your letter is timed properly (obviously this is mostly luck and intuition), you can develop the selling idea in the mind of a leaning owner. "Leaning" means he has been thinking about selling, off and on, but has never gotten around to doing anything yet. Be sure to include your telephone number, so he can call when he gets your letter. If he calls, you're all set to grab your sketch of his property and follow along as he explains details over the phone.

Many owners have called and thanked me for my nice letter, then told me they weren't interested in selling right now. But, you can probably guess who they'll call first when they are ready to sell. Believe me, they'll never throw your letter away. One of my best properties was acquired exactly this way. The owner wasn't interested when he first read my letter; however, a month later he was and we closed the deal. Cold call letter writing puts you in a position to buy a property that no one, except you and the seller, knows about. *Finding a good deal that no one else knows anything about puts you in the driver's seat.* It's like going to an auction where you're the only bidder. Cold calls have accounted for about 20% of all my acquisitions!

FINDING SELLERS WHO TRULY NEED TO SELL

Contrary to what many novices think, the most difficult part of buying the right properties has nothing to do with contracts, paperwork, escrow instructions and presenting offers—*it's finding sellers who really and truly desire to sell their properties.* It's called discovery! You are looking for sellers who have very compelling reasons to sell. Sometimes, fear of losing the property in a foreclosure creates the right pressure. That's an obvious reason. Many other reasons are hidden, or not so visible, however. Always remember this—A REAL NEED SITUATION MUST EXIST FOR THE SELLER, OR YOU WILL HAVE GREAT DIFFICULTY MEETING YOUR PURCHASE OBJECTIVES—that is, buying the property with the kind of flexible terms you need to make the transaction profitable for you. Again, I will repeat myself here because it's so important.

You must locate sellers who, for whatever reason, have a real need or a very strong desire to sell. Nothing short of this will work—don't forget this part! Also, there is no point pursuing sellers who don't seem to have any urgency to sell—you'll be wasting your valuable time. In the long run, you'll discover your time is more valuable than money.

LEARN TO BE A HOUSE DETECTIVE

Initially, it's quite likely that you won't know the seller's real needs because he won't tell you about them. Eventually, as you gain more knowledge from practice and experience, your detective instincts will improve and you will learn how to dig out the information you need more easily. Knowing the real reason a seller wants to sell can be a tremendous advantage when negotiating the kinds of terms that you must have to be successful—that means profitable real estate that pays you back for all your efforts.

Motivated Sellers

Motivated sellers get "motivated" for many reasons. Here's a list of the most common reasons, all except one, which we'll tackle separately, in a minute.

1. PROPERTY OWNER LOSES HIS REGULAR 8 TO 5 JOB.
2. FAMILY PROBLEMS, MOSTLY DIVORCE OR DEATH.
3. POOR HEALTH, CAN'T WORK ON PROPERTY ANY LONGER.
4. CHANGE OF INVESTMENT GOALS. FOUND BETTER MOUSE-TRAP.
5. MOVING FROM AREA. JOB TRANSFERS MOST COMMON.
6. RETIREMENT TIME—READY TO HANG IT UP AND GO FISHING.

"FOR SALE" OFTEN MEANS "I NEED HELP"

I know from experience there are many owners of small rental properties—the kind I suggest you look for—who have their properties for sale, but in reality the "FOR SALE" sign really means "I NEED HELP." These sellers have a variety of problems, and many of them need some serious help. Their solution to getting help is to sell the property. Owners in this situation can provide big financial opportunities for investors who are willing to help solve their problems—namely, they can give you a great price with seller carryback financing.

"DON'T WANTERS"—MOST DON'T KNOW HOW TO MANAGE

Many real estate instructors refer to sellers who don't know how to manage their property as "DON'T WANTERS." The number one reason most sellers of small income properties are "DON'T WANTERS" is because they cannot manage their own properties. For the most part they can't stand the hassle of dealing with their tenants. The reason for this failure is that most of them have never taken the time to educate themselves in preparation for this most important task. They are victims of the "Columbus" technique. They don't understand what they're doing, and have little or no idea where they're going. They are simply drifting without direction. Since they have no plan and very little knowledge, they are unable to fix the things that are wrong. You won't find yourself in this situation if you do as I suggest and learn how to be a landlord (manage your properties), right from the first day you start investing.

Learning to Tell What's True and What's Not

When you read your local newspaper classifieds, you'll see all of the various reasons for selling over and over again—OWNER MOVING; POOR HEALTH FORCES SALE; GROUND FLOOR OPPORTUNITY; OWNER RETIRING; DIVORCE SPELLS MY LOSS, BUT YOUR GAIN. These are typical classified statements in the real estate "For Sale" sections of any regular newspaper. Obviously, the ads are written to persuade potential buyers that they (the owners) are motivated to sell! Some ads are true, most are not. It's nearly impossible to tell about true motivation from an ad. Still, it's a starting point.

You will need to make offers and have direct discussions with sellers before you can really determine their motivation level. One way to get a clear indication is to offer them about half of their asking price! If the seller doesn't throw you out of the room, perhaps he's really serious about selling.

MOTIVATED SELLERS SELDOM TELL THEIR SECRETS

Few sellers, motivated or not, will tell their secrets in classified advertising or any other sales pitch, for that matter. You must do some detective work! The fact is, a seller will do his darndest to hide this particular motivation—"HE'S MOTIVATED BECAUSE HE'S A FAILURE. CHANCES ARE GOOD, HE'S LOSING HIS SHIRT ON THE PROPERTY." If you can discover motivation of this kind, and if you have the knowledge and ability to fix the problem— that is, relieving the seller of the mess he's in—you're in an excellent position to be richly compensated for your skills. This situation will create "super paydays" for fix-up investors, believe me!

The Value of Long-Term Thinking

Often new investors are far too shortsighted when they begin looking for properties to acquire! Many lack what I call "long-term thinking." Folks without long-term thinking are really not investors. They're more like speculators looking for an easy path to wealth. Quite often they become easy prey for the "get rich quick" gurus whose techniques contain enough "hot air" to fill several Goodyear blimps. "Hot air" is not an ingredient you need for success; it only delays it!

An Investor With Vision and Long-Term Thinking

In 1946, investor and long-term thinker, William Zeckendorf, was offered a chance to purchase several east river slaughterhouses in New York. The property was owned by the

Swift/Wilson Meat Packing Company. The problem was that it had an exceptionally high asking price! Smith and Wilson were asking a lot more money than the adjacent properties could be purchased for. Six million dollars seemed an astronomical price for slaughterhouses back in those days.

Zeckendorf was a man with both vision and a long-term plan! He immediately recognized the only reason why adjacent properties were selling so cheap was because no one wanted to be neighbors with foul-smelling slaughterhouses. Conventional thinking real estate investors thought Zeckendorf had gone absolutely "bonkers" when they realized he had secretly signed options to purchase all the surrounding properties too! When the time was right, Zeckendorf sold the slaughterhouse, properties for construction of the world famous United Nations Buildings. His profit for having good vision was two million dollars. In addition, the surrounding properties, without the stinky slaughterhouses, earned Zeckendorf another three million dollars, before he finally finished his long-term plan for the east river properties.

VISION PLUS LONG-TERM THINKING WORTH BIG BUCKS

The moral to the story, and to Zeckendorf's success is quite simple!—*vision and long-term thinking are worth big bucks!* When I purchased the Hillcrest property years ago, I had a vision. I visualized that a small junky motor lodge, outdated and condemned by the city building department, could be resurrected and born again as 21 senior cottages. I would create a park-like setting, surrounded by three acres of lawns and lovely shade trees. When I became the owner of Hillcrest Cottages, it even smelled like Zeckendorf's properties; however, not from slaughterhouses, but from five collapsed septic tanks that had been "plugged up" and had long since quit functioning. I can understand why observers thought Zeckendorf had gone "bonkers," because that's exactly what my friends thought about me!

THE COURAGE TO LOOK WHERE OTHERS DON'T

People constantly ask me how I find these kinds of properties—"Diamonds in the Rough," some call them! The most common complaint I hear is, "There's no properties in my hometown like you write about!" With very few exceptions, I must disagree. The properties are there! You simply haven't found them yet, and there are several good reasons why you haven't. I've found most folks are not looking around nearly enough. Most investors I know do the traditional kind of search for properties—if they wish to be apartment owners, they tend to look at traditional apartment buildings; single house investors generally drive through the suburbs in search of property. Hardly anyone is willing to look at slaughterhouses or an old motor lodge. My suggestion is to broaden your vision a bit. After all, "non-traditional properties" are a lot more fun, and quite often, a lot more profitable, as well!

Breaking Ranks With the Typical Buyer

Often, you'll find the biggest rewards are found a little ways off the beaten path. Rich gold mines are seldom in plain view to all who would seek them! Think about investing a little differently. Isn't it the folks with vision who always seem to be the ones who make the most money? Followers are always lagging behind and are never quite sure what they're looking for! Most often, they show up too late after the competition has already "bid up" the prices. *My suggestion—lead, don't follow.*

Finding the right properties, with a high potential for profits and cash flow, is the first step for developing a successful investment plan. It's most important to break away from traditional thinking—stop following the average investor crowd. *Serious money is made by those who study the marketplace and develop the ability to spot bargains that others simply fail to see.* You don't need to buy slaughterhouses, but you do need to develop buying skills and open up your vision to different possibilities. Be especially "tuned in" if you run the numbers and a rundown property shows good cash flow!

BUYING PROPERTIES FROM THE BANK

Yellow Court did not look like the kind of property you see on the "Late Night" cable TV show "How To Get Rich Without Cash or Credit." There were no Rolls Royces, no testimonials from slick-looking pitchmen, and certainly no bikini-clad girls frolicking on the deck of an expensive speed boat.

Yellow Court was five ugly houses repossessed by an out-of-town bank for non-payment of the mortgages. When a bank forecloses on real estate, they call it "REO," or "REAL ESTATE OWNED." Banks don't like "real estate owned" because they're in the business of loaning out money for interest payments, not managing houses! When you learn how much money they lose on them, you'll clearly understand why REOs are poison to the banker's balance sheet and profitability.

Junky Houses are a Public Nuisance

In the case of Yellow Court, the bank had foreclosed on five junky houses that had been completely trashed by wave after wave of deadbeat tenants. The city had formally advised the bank that the houses had become a *public nuisance*. Finally, after continuous nonpayment of utilities, the city shut off all services and "red tagged" the houses. They were deemed unsafe for human occupancy, and the tenants moved out taking everything they could tear loose—medicine cabinets, plumbing fixtures, carpets, interior doors and even the kitchen sinks and toilets.

When I first heard about Yellow Court, the tenants had been gone for nearly eighteen months. All the houses had long since been "boarded up." The windows and doors were covered over with plywood that was bolted in place! There was nearly enough plywood and bolts to build another house. It's like you see on those "trashed out" HUD-foreclosed houses!—they look like "forts" under enemy attack!

BANKS GET VERY MOTIVATED WITH TAKE-BACK HOUSES

Banks not only despise REO properties because the mortgages are "non-performing," but also because they now own a liability, rather than an asset. They must set aside funds which could otherwise be loaned out in order to cover their loss. Also, they must now shoulder the liability costs involved when they become the owner. That means fire insurance and management fees! It costs them even more when no one lives there. After an extended period of ownership with continuous cash losses, even the richest banks will become extremely motivated to dump the property.

TIMING IS EVERYTHING FROM WINE-MAKING TO REAL ESTATE

Winemaker Paul Mason claims they'll sell no wine before its time. That claim is based on simple economics. Obviously, they can't sell unfermented grape juice or they would, believe me! Life is full of events that require perfect timing. Some timing must be more accurate or precise than others! For example, saving a drowning man in twelve minutes is critical! No need to try 30 minutes later because he'll only be blowing bubbles.

With real estate investing, you must have proper timing if you expect to make serious money. For example, if you decide to purchase a property simply because "you want to own it right now," you might be doing exactly the wrong thing unless, of course, you have a special pre-determined plan worked out in advance. ***Buying right requires a good plan***. Buying property and having a good plan go together like a pair of purple socks.

By the way, I make exactly this same argument when I discuss selling properties! You simply cannot make maximum profits unless you time your sales properly. *Buying low and selling high will only happen when you learn how to properly time both events!*

Yellow Court—Five Ugly Houses for Sale

I first learned about the five houses on Yellow Court when I saw the "Do Not Enter" notices posted on the buildings. The bank ran several classified ads in my local paper. No price was listed, so I immediately called the bank's REO department and asked how much they wanted. No prices had been set yet; however, they sent me their forms to submit a bid.

DETERMINING HOW MUCH TO PAY IS A MUST

I cannot over-emphasis the importance of knowing your particular marketplace! *You must know what properties are worth—either to sell or to rent out to tenants*. You will also need special knowledge about repairs and fix-up! This may sound a bit more difficult than it really is. However, estimating fix-up work on a house is really no more difficult than learning grocery prices at your local supermarket. Once you've done the first one and compared the actual expenses with your initial estimate, you'll catch on very quickly.

Bidding War Starts At Yellow Court

I always submit my bid to purchase REO properties written on plain paper or on the bank's forms. Verbal communications back and forth by telephone are often misunderstood, sometimes on purpose. They are also non-binding between the parties. If my bid gets accepted, I want to see a piece of paper with a bank officer's signature on it. That way all my time and hard work won't be lost in the event a higher bidder shows up, suddenly causing the bank to forget all their verbal promises to me. With fax machines, signatures are just as easy to get as verbal commitments—and much safer for investors!

Figuring Out My Offer to Purchase the Property

The five Yellow Court houses were three-bedroom, two-bath properties with approximately 1,350 square feet of living space. They each had double car garages and large back yards! *That's the good news*. The bad news was everything inside the houses was totally trashed! There were giant holes in the sheet rock walls, doors ripped off cabinets, broken windows and sliding glass doors, holes in the floors, leaking roofs and every hot water heater had been stolen.

There was great potential for over-paying, because every system needed major repairs or replacement. I knew full well if I paid too much, I would end up working my tail off several months for no profits! I had estimated it would take me and my crew the full summer (approximately three months) to rehab all five houses! That's a ton of labor and it still didn't include the licensed contractors I would need to re-roof the houses, install new gas lines and rebuild the electrical services. It was also necessary to install brand new fuse boxes, because the existing ones had been completely stripped by former occupants.

Over-paying can be a destroyer of dreams for wide-eyed, inexperienced investors. Take my advice seriously here—Work the numbers, backwards and forwards, then over and over! Always use Murphy's Laws of Estimating—*"IF SOMETHING ELSE CAN GO WRONG, IT WILL! If it can possibly cost more than you thought, it always will! If there's the slightest chance you missed something in your cost estimate, you did!"*

My Bid is in the Mail

I called in my numbers first, then I mailed the written offer! The REO manager did not like my numbers—I could hear it in his voice. Before I could even speak a word, he advised me the bank would need $42,000 per house just to come out even. "However," he said, "Any offer you make will receive our serious consideration!" I've always been fond of those words— "SERIOUS CONSIDERATION," so I told him my offer and terms.

Offer: $100,000 FULL PURCHASE PRICE for the five fixer houses. $20,000 CASH DOWN PAYMENT to be paid to seller upon acceptance of offer.

Terms:

1. BUYER TO PURCHASE PROPERTY IN "AS IS" CONDITION WITH NO CONTINGENCIES.

2. SELLER TO FINANCE BALANCE OF PURCHASE PRICE ($80,000 MORTGAGE). **TERMS** 10 YEARS WITH 10% INTEREST-ONLY PAYMENTS.

3. SELLER TO PROVIDE A SIX MONTH MORATORIUM ON THE MORTGAGE PAYMENTS TO ALLOW BUYER TIME TO REPAIR HOUSES AND BEGIN GENERATING INCOME.

4. BUYER WILL PROVIDE REQUIRED FINANCIAL STATEMENTS NECESSARY FOR SELLER TO APPROVE CARRYBACK MORTGAGE ($80,000) FOR BUYER.

I vividly recall that telephone conversation! There was a rather long pause on the line! I had to ask, "Are you still there?" Then came sort of a muffled whisper. "But that's only $20,000 per house!" "That's correct," I agreed. Just before he hung up, he said, "Jay, I'll need to get back to you on this. There could be some problems, but I will tell you right now, our bank never allows moratoriums! Also, if we provide carryback loans, they are always written at 12% interest for five years maximum!"

My Offer Was Respectfully Declined

Unknown to me, at the time, several local building contractors had also submitted bids to purchase Yellow Court! The bank knows competition makes for a better horse race! Looking back now, I imagine they probably mailed out information packages to everyone they knew. When my phone finally rang, it was the REO manager! He sounded a bit too gleeful. He went on to tell me that my offer had not been accepted. A higher purchase price, more favorable to the bank, had been accepted instead. He thanked me for my participation, then made some joke about maybe "next time." He said, "Good-bye," and that was it.

Three weeks passed and I never expected to hear about Yellow Court again! Here's a "street-wise" tip for all investors—never throw your notes or cost estimates away! I learned this lesson the hard way, several times. Keep files on every property you have an interest in! If it's worth your efforts the first time, chances are it will be the second or third time, too. Good notes last forever. Besides that, you will develop a good history about the property which allows you to act quickly if there should be another round of bidding. Dedicate a special file for this purpose, "PENDING PROPERTY DEALS." It will pay big dividends over time, believe me! Now back to our exciting Yellow Court episode!

THE REO MAN RINGS TWICE SOMETIMES

When I answered the phone, I recognized the REO manager's voice immediately. This time the gleefulness was missing! "Jay," he began, "Would you reconsider your Yellow Court offer?" I told him, "I most certainly would, with the same terms as I submitted before!"

He once again explained how the bank had this policy about no moratoriums and no carryback loans for more than five years. I said, "The deal can't work that way, I need more time!" He said, "Let me check with the loan committee and get right back to you on this."

It wasn't even two hours before my phone rang again! "We've just had a new policy revision for REOs," the manager said. "We can allow up to six months moratorium for mortgage payments, and we'll extend our financing for up to ten years."

I said, "Gee! That's wonderful! That's exactly how my offer is written; therefore, I accept. We've got ourselves a deal!"

About a year later, I learned the contractor who had bid higher than me lost out because he had lousy credit. The bank would not approve financing for him! My persistence had finally won out.

My cost estimate to fix Yellow Court was $12,500 for each house. Since there were five houses, the total estimated cost was $62,500. You can see how fix-up expenses quickly add up for trashed-out properties.

Where Does all the Fix-Up Money Come From

In this particular case, I was financially able to pay the bills as I did the work. This is not a job you should attempt without money. Obviously, this was not my first fix-up experience! By the time Yellow Court came along in my career, I was doing pretty well with my other properties. The point I'm making here is this—*you must purchase these properties and fix them up within your own financial limitations!* My first fix-up jobs were much smaller and took a whole lot less cash, believe me!

Some investors would be much better off fixing up one house and doing most fix-up work themselves. For average "run of the mill" fix-up jobs (not trashed like Yellow Court), the labor costs will run about 65% of the total job cost. That means if the total fix-up job costs $10,000, your hard costs for materials will be about $3,500. You'll save yourself $6,500 doing all the labor yourself. Everybody starts with small steps! If you don't trip, you'll soon be doing five-house jobs like Yellow Court, and I'm certain you'll have the money by then.

WHAT'S IT WORTH TO FIX YELLOW COURT

The job took all summer, like I thought! It was, indeed, a lot of work, but the houses turned out simply beautiful, I'm proud to say. I often take my "Fixer Camp" seminar students over to see them! I like to brag about the way the houses look today! The students are always interested to hear about the financial rewards at Yellow Court! Was it worth all that effort? I'll let you judge for yourself.

Remember the mortgage payment moratorium I worked so hard to get? Well, the truth is, I never needed the full six months they gave me. Instead, I completed all the work in half that time and refinanced the entire property (all five houses) with a local savings bank. My new appraisals averaged $62,500, per house.

I placed brand new $40,000 mortgages on each house. That was enough to pay off the existing mortgage debt and pay back all my fix-up money. There was even a few bucks left over for me. By the way, my actual expenses to fix each house was $15,000, and not the $12,500 as I had estimated! See how Murphy's Law works!

Here's a quick view of the financial picture at Yellow Court. You'll recall, I paid $100,000 total cost ($20,000 per house), with $20,000 cash down payment! My new financing paid off the $80,000 loan carried back by the REO seller.

BEGINNING FIX-UP COSTS—FIVE HOUSES

Purchase price for each house...............................$20,000

Cash down payment by Jay..4,000

Mortgage carried back by (REO) seller.....................16,000

Fix-up costs, paid out over three months................15,000
paid by Jay, (out-of-pocket)

COMPLETION SIX MONTHS LATER

New appraisal (value $62,500)
Income rents—$550 per month

New 30 yr. mortgage placed on each house...........$40,000

Mortgage payment $305 per month

Pay off REO loan...(16,000)

Pay back Jay for fix-up costs ....(15,000)

Return of down payment to Jay ....(4,000)

Cost of loan—escrow expenses.............................(1,250)

Balance of loan proceeds from escrow.....................3,750
left-over funds (returned to Jay's pocket)

The bottom line: I have all my money back, plus I get to keep $3,750 for each house (left-over loan proceeds), for a total of $18,750. My equity after the refinance was $22,500 per house, for a total of $112,500. The monthly rental income, less mortgage payment, was $245 per house, for a total of $1,225.

You won't need a computer to figure out the return on this investment! *I've got all my money back!*—my fix-up money, my down payment and all financing costs, plus I put $18,750 in my pocket. I no longer have one thin dime invested, and yet I end up with $112,500 equity and $1,225 per month cash flow. When you own properties like this, all you need to do is hang onto them, and start looking for more!

CHAPTER 9

Fixing Rundown Houses for Money

*F*ixing rundown houses for money is not the same thing as fixing rundown houses! Those two powerful words "FOR MONEY" make all the difference in the world. In fact, many people who are quite skilled at fixing houses are lost when it comes to making money doing it. A good example are the thousands of small-time, do-it-yourself building contractors who, for the most part, are excellent craftsmen! Yet, the vast majority of them are not very successful, if you measure them by their bank accounts. In short, they can do the fix-up work just fine, but that's not the same as making money doing it!

DON'T FIX THINGS THAT DON'T PAY YOU BACK

Many so-called "fix-up investors" haven't quite figured out the difference between RENOVATING, FIXING and REMODELING. Perhaps renovating and fixing are similar, like distant cousins; however, investors who remodel fixer-type properties can easily end up working their tails off for very little profits and sometimes big losses. The problem is, remodelers change too many things that don't need to be changed! For some reason, remodelers have a fierce passion to make things better from their personal point of view with almost total disregard for what it costs! They often rationalize by telling themselves— since I'm working here anyway, I might as well go ahead and do such and such because, eventually, it will need to be done. That's like saying, I might as well spend my whole paycheck tonight and get it over with, so I don't have to worry about spending it later on.

Remodelers Spend Too Much Money Fixing

In order to be a successful fix-up investor (that means making money at it), you must fix only things that need fixing. It's important to separate your "pet ideas" from proven "money-making" fix-up work! I'm not talking about repairs here—obviously, things that don't work must be fixed. I'm talking about things like moving walls around, adding rooms, making windows bigger, ripping out and installing new kitchen cabinets. Remodelers do these things, but it's not cost effective to do them in the fix-up business. *The important thing to remember is that almost everything fix-up investors do should add cash value to the property*—either for buyers or renters. Larger rooms and newer cabinets will seldom matter much when it comes to getting higher prices from your customer.

BIG BUCKS COME FROM UNDERSTANDING ECONOMICS

I am absolutely certain that Suzanne Brangham—house remodeler and author of a wonderful book on the same subject entitled "Housewise" (available through Harper & Roe Publishers, 10 East 53rd Street, New York, NY 10022)—is a better house fixer than I am. She is a very detail-oriented person and is obviously good at renovating houses. There is no question in my mind whatsoever that both men and women can do this work with equal ease. Like I said earlier, however, the fix-up work itself is only a fraction of the job—*making money while doing it is what really counts.*

I doubt that, by the time she adds in her own labor, Suzanne gets a bigger return for each fix-up dollar invested than I do. One of the most important skills you will need to develop in order to be profitable is learning to fix up things that pay—and don't waste time and effort on things that don't pay! Your goal should be to get the largest return for each fix-up dollar you spend. Try not to imitate the stock market dabblers—once they invest their money in stocks, they seem to constantly worry about the return *OF THEIR INVESTMENT*.

IT'S IMPORTANT TO KNOW WHERE PROFITS COME FROM

When I write about this subject, I often compare fixing up houses with playing golf. In golf there's an old saying *"DRIVING THE BALL IS FOR SHOW, BUT PUTTING IS FOR THE DOUGH."* There are literally thousands of strong young players who can hit the ball a mile and look very good doing it! Precision is not what's critical as long as the ball stays in bounds and ends up on the fairway! However, *PUTTING* the ball those last few feet to the flag stick is very critical. It takes all the precision and skills one can muster up to sink the ball in a very small cup. It's impossible to play winning golf unless you learn to do both parts—driving and putting—equally well!

The same can be said for fix-up investors—you won't make any serious money by simply fixing up houses without a financial plan! So, to repeat myself, lots of folks can do fix-up quite well; however, the serious money is made when you learn how to purchase houses with the right things wrong, then limit your fix-up costs to those things that will generate returns or profits from your customers!

INVESTMENTS MEASURED ON HOW WELL THEY PAY

Those who have little experience in the fix-up business often make the mistake of judging properties on how they look rather than on the benefits they will provide. (Obviously, I'm talking here about how the property looks when you buy it!) You must stop and ask yourself these questions every so often—WHAT AM I DOING IN THE REAL ESTATE BUSINESS? *DO I WANT PROPERTIES THAT LOOK GOOD OR DO I WANT PROPER-TIES THAT PAY GOOD?* There must be no confusion about what your goal or purpose is, if you wish to be financially successful doing fix ups!

One of my first recommendations, especially for newcomers, is to sit down and write out exactly how much money you'll start getting back every month once you become the owner of an income property. Do this exercise before you buy, not afterward. People who purchase single-family houses often skip this drill. If they were to "plot out" negative cash flows, month after month, on a sheet of paper, where they could stare at the depressing numbers, I'm sure many would avoid the heartaches of making "non-profitable deals."

One of the most common characteristics of house shoppers—both renters and buyers—is that they make their financial decisions based almost entirely on LOOKS and FIRST IMPRESSIONS. If you can understand and accept this predictable trait and incorporate it into your personal investment strategy, then what I'm about to show you can help earn thousands of dollars for your efforts.

Most Buyers and Renters Lack Vision

You must understand that most buyers and renters want clean, attractive houses and apartments to buy and live in. Most people aren't even remotely interested in hearing about what the property might be, or could be, in the future. They are only concerned with what it is right *now*! If you can grasp this concept, then my fix-up strategy will make perfect sense to you as we go along

WHAT YOU SEE COUNTS FOR EVERYTHING

When I was a small boy, my mother always told me that looks are not what really counts. "Beauty is only skin-deep," she would say, "It's what's on the inside that really counts." She was talking about people, of course. However, in the house fixing business, that advice doesn't apply and will keep you broke forever. People will never ask to see the inside if the outside is ugly. Instead, the skin-deep fixing strategy works extremely well for house fixers. The fact is, attractive *exterior* paint is one of the top moneymakers when it comes to making big profits in the fix-up business.

GOOD LOOKS is the key to *making sales* or *renting properties to tenants*. Houses and apartments are judged, like books,

BY THEIR COVER. Looks also have a great deal to do with *who* your customers will be. There's an old saying in the rental business, "The property attracts the tenant—UGLY HOUSES WILL ALWAYS ATTRACT UGLY TENANTS." When you purchase as many rundown houses as I have, you'll discover this old saying is at least 95% true. That's why house fixers, like myself, will generally end up replacing most existing tenants during the period of time when I'm fixing up the property. Tenants I inherit when acquiring property are most likely not the kind I want to keep when I'm finished with the job.

Busted Doorknobs and Running Toilets

My fix-up strategy is designed to accomplish two main objectives, and both are equally important! The first is the easiest part to understand because it's what I call the "practical fixup"—this is what must be done to rundown houses in order to make them habitable—and, for the most part, it's just plain common-sense fix-up. Almost every investor, experienced or not, will concede that you must fix things that are broken, worn out and missing—these are the basic necessities; they are not optional!

Things like cracked windows, broken faucets, torn linoleum, holes in walls, leaky roofs, interior painting, running toilets, busted doorknobs, hauling junk and replacing missing fixtures are, obviously, things that everyone should understand must be fixed. Unfortunately this part of fixing, while necessary, is not the part that attracts the big profits! If it was, a lot more fixers would have bigger bank accounts and nearly every small-time do-it-yourself building contractor would be a whole lot richer.

GOOD LOOKS IS WHAT ATTRACTS YOUR CUSTOMER

Fixing for looks is just as important as fixing the toilet or the roof! The reason is simple—people rent and buy houses based, to a large extent, on how the property looks.

It's important to understand that renting and buying decisions are made by potential customers, driving by your property in their cars, in a matter of seconds. *It's that first look and their first impression that counts the most*. If your property doesn't generate positive vibrations on that first drive-by, the new carpets and countertops inside don't count for anything.

Tenants and buyers, alike, will judge your house like they judge a book—BY IT'S COVER! Knowing this about your customers can help generate thousands of dollars for you if you'll apply it to your fix-up strategy. It's for this reason that I constantly instruct my readers to dedicate their main efforts to FIXING WHAT SHOWS. You must create good looks before you do anything else because that's where the fix-up battle is won or lost. More importantly—*IT'S WHERE THE MONEY'S AT!*

YOU MUST POSITION YOURSELF TO MAKE MONEY

When my eyesight was just a little bit sharper than it is today, I could shoot a fairly respectable game of billiards. If you know anything about shooting billiards, or pool, you probably already know that position is what makes a winning player.

Each time you shoot a ball in the pocket, you must concern yourself with where the cue ball (the one you shoot) ends up so you'll have a clear, unobstructed shot at the next ball. That's called getting in a good position! When you don't have a good position, chances are you won't make your next shot. If you don't, you lose your turn as shooter to your opponent. Obviously, when you miss your shot, you're no longer in control of the game.

Real estate investing is a lot like playing billiards. You can settle for buying properties with very limited potential (one shot deals) or you can position yourself to earn bigger profits by acquiring the kind of real estate that pays off in several ways. Obviously, your chances of being a winner are greatly increased when you make money a variety of ways with the same piece of real estate.

When I proposed this "pool hall analogy" at a recent fixer seminar, students immediately jumped on the idea that pool sticks (cues) might also provide new enforcement opportunities for landlords who are having difficulty collecting rents from slow-paying tenants. That idea, however, is not one of the advantages we'll be discussing here.

RUNDOWN PROPERTIES OFFER MOST POSSIBILITIES

Buying rundown or "under-performing" properties is the quickest way to build CASH FLOW and EQUITY. The main reasons for this are: (1) You can purchase at discount prices, and (2) The rundown properties are generally rented for less than their potential when fixed-up, which allows for rent increases, once the fix-up is completed. There is usually clear visible evidence of why properties are under-performing!

Poor management—usually less visible—is nearly always an underlying reason for poor performance. Fixer properties are easier to acquire because their owners must compete with sellers of well-maintained real estate. Naturally, the owners with ugly fixer properties must make selling concessions— such as *low down payments* and *seller carryback financing*—if they expect to sell.

As a buyer you can reap big benefits if you are willing and able to help solve distressed seller's problems. In the process, you can quickly build large equities, that might take years to acquire with non-fixer real estate. This technique is called "ADDING VALUE," and is especially suited for new investors who don't have much money to start with.

THE CONDITION OF THE PROPERTIES YOU SHOULD LOOK FOR

The kinds of properties you should look for will generally fall into one of the following categories:

1. **RUNDOWN AND UGLY**—*NO MAINTENANCE BEING DONE*

2. **BAD TENANTS**—*JUNK CARS, MOTORCYCLES AND LOTS OF VISITORS IN AND OUT*

3. **HALF EMPTY PROPERTIES,** *TALL WEEDS, GARBAGE ON GROUNDS, UNSIGHTLY*

4. **FINANCIAL PROBLEMS,** *FORECLOSURE, BANKRUPTCY AND BANK REPOSSESSIONS*

5. **PARTIALLY COMPLETED BUILDINGS,** *ACTIVITY STOPPED DUE TO LACK OF FUNDS, ETC.*

The reason you are looking for ugly, distressed and financially-bust type properties is because they represent serious problems for their owners. The idea is to use your time and personal efforts to fix their problems in lieu of paying normal cash down payments—*CORRECTED PROBLEMS HAVE CASH VALUES.* Often the price to correct a problem will amount to a much higher dollar value than the normal cash down payment might have been. But, your willingness and personal ability to fix problems for others can create a very profitable opportunity for you. Many investors, including myself, have used this wealth-building technique to quickly develop large real estate portfolios. It is the perfect solution for anyone who doesn't have much cash for a down payment, which includes almost everyone who comes to me for advice.

What Kind of Sellers Own Rundown Properties

The following list will give you some idea of individuals most likely to own distressed real estate and who might be willing to sell their property with good terms for the buyer (that's us).

1. Out-of-town owners
2. Owners suffering financial difficulties
3. Owners with personal/family problems (divorce, death, lifestyle change)
4. Owners who advertise "for sale by owner" (all newspapers)
5. Owners who have lost jobs
6. Elderly or disabled owners
7. Inherited property owners
8. Owners who advertise "lease option" or even "houses for rent"
9. Job transfers (moved and now they have two house payments)

Sellers Will Seldom Tell You the Whole Story

An important thing to remember is, people don't always say that they mean. For example, "LEASE WITH OPTION" could really mean the owner doesn't want to be bothered for an extended period of time. Most out-of-town owners don't like long-distance renting, especially with problem tenants, but they seldom come right out and say so. Owners who have inherited their properties often think rental houses are just too much hassle. After all, they weren't the ones who bought the property in the first place, and don't really want to be tied down by them.

Sellers with financial problems, or those with two mortgage payments every month, will seldom tell you they're hurting when, in fact, they may well be. You must learn to "dig out" for this kind of information so you can propose a deal that will create a high level of interest for a seller who needs help. The offer you make, once you learn everything about the property and exactly how motivated the seller is, can be structured to be very profitable for yourself. *The key here is to become a good property detective!* You cannot make good offers without first learning everything you can about the property and the seller.

CASH FLOW AND EQUITY OPENS MANY DOORS

If you follow my advice and make it your goal to go after CASH FLOW and EQUITY, as quickly as you can achieve them, you will set the stage for many other opportunities to come your way as you continue to invest and learn new techniques.

When you achieve cash flow early in your investment plan, you'll position yourself head and shoulders above most of your competitors—many of whom struggle along with negative income for so long they become discouraged and give up! Attitude plays as important role in becoming successful and wealthy. It's very difficult to hold your head up and look successful when you're losing money every month. In my opinion, it's far better to fix up rundown properties—that will generate monthly income—than to get bogged down with beautiful beach front condos—that cost you out-of-pocket money every month just to own them!

When you quickly build up equity by ADDING VALUE to properties, you beat the *appreciation game.* You won't need to wait around for values to go up like most investors do! You will always be in a position to trade up or sell for an immediate profits.

FIX-UP PROPERTIES USUALLY FINANCED BY THE SELLER

Nearly 75% of all my real estate acquisitions have been at least partially financed by sellers who are willing to carry back mortgages. This is the way most older properties are sold. The same thing applies to rundown (ugly) real estate. The reason is because banks and other financial institutions want nothing to do with this type of property! When they do agree to make loans on them, you will only get approximately 65% of the appraisal or the selling price (whichever is less).

If you happen to be "self-employed" and unable to provide a W-2 statement, when applying for a loan to buy investment real estate, you had better sneak up to the manager's desk before he or she knows what you want. Otherwise, they will quickly disappear until after you leave the building. Non-employed borrowers rank very low at most banks!

Without either of you realizing it, the banker has probably done you a big favor by refusing to give you a loan. The truth is, you are much better off not having a bank loan at this particular stage. You should go after *seller carryback financing*, often called a *purchase money loan.* That means, the property you purchase will be the only security for the loan. If you should flop as an investor, the seller can only take back the property. With a bank loan, it's quite possible you'll get stuck with a deficiency (out-of-pocket) judgment, if the re-processed property is sold for less than the loan balance. As you might guess, seller financing is far less risky for you.

Seller Carryback Mortgage Provides Safety Net

Positioning yourself to make money also means taking the least amount of risk you possibly can. *Seller financing is far less risky than a bank loan.* The reason is quite simple—sellers of rundown properties who are sick and tired of the property when you buy it from them, seldom want it back! They will generally do most anything to avoid a foreclosure that would give them the property back. It's comforting to know this when you own the property.

Some years back, the vacancy rate in a seller-financed apartment building I owned soared to 50%. My seller-financed mortgage payments were $1,800 per month. I wrote a "panic stricken" letter to the seller explaining my problems. I told him I could only pay $950 per month for the next full year. There would be no make-up payments, no penalties, no add-on to the principal—just reduced payments for a year.

The local economy had turned sour and it was the best I could do! I also advised him that if my proposal was not satisfactory, he could have the property back! Naturally, he wasn't thrilled with my letter, but he felt $950 a month for a year was better than owning the property again, so he agreed to the terms. Banks will seldom do this kind of compromising. Most often their by-laws and their institutional charter will not permit the option, even if the manager would go along with such a plan.

DIVERSIFICATION OKAY—BUT ONLY AFTER CASH FLOW

One major downfall with many investors is that they develop what I call "tunnel vision." They have, unknowingly, created blinders for themselves. They are more concerned about the *vehicle* than the *benefits*. If you do likewise, it's very difficult to achieve your goals in a reasonable time frame. That's the main reason investors go along for years buying properties but never achieving cash flow. They can only talk about imagined equities. I can tell you from my own experiences, if you own properties that, for the most part, fail to produce cash flow, you are not in the right position to create wealth.

Specialized Help is Available

Jay's ONE-ON-ONE COUNSELING service is "tailor-made" for beginning and accomplished investors, alike! It's really a *customized mini-seminar* designed specifically to fit your personal needs. In other words, we will teach you profit-making techniques in two full days of non-stop education. You simply tell us where you need the most help.

If your goal is to acquire **money-making houses**, this is your opportunity to actually see for yourself the kind of properties that will earn the *most cash flow—and the fastest*. Jay's "high-profit" counseling will easily make the cost of your trip to Redding worthwhile, many times over. Seeing Jay's houses for yourself will speed-up your learning process at least 100 times over.

If you have a specific area where you need the most help, this is where you'll find it—everything from *financing, creative selling, obtaining HUD fix-up grants, setting up small partnerships, fix-up that pays the most, etc.* You'll also get to visit Jay's home office management operation where he keeps track of nearly 200 rental houses and all the tenants. Learn how it works—plus, *we'll give you all the special forms to take back home.*

It's all here for you to see for yourself, and it's all part of Jay's ONE-ON-ONE COUNSELING service in Redding, California. See the Appendix Resources Section for additional information about how to register or call toll free, **1-800-722-2550**, to check available dates.

CHAPTER 10

Jay's Moneymaker "Foo-Foo" Fix-Up Strategy

WHAT YOU SEE IS THE FOO-FOO

*F*olks who attend my fixer camps and visit my rental houses are always fascinated by my "Foo-Foo" beautification techniques! "Foo-Foo" is the term I use for cosmetic, or gingerbread, fix-up.

The primary goal is to upgrade the looks. The term originated under the big-top in the golden days of the circus. Paint and makeup worn by the clowns was known as "Foo-Foo!" Webster's Dictionary defines cosmetic as *CORRECTING DEFECTS, OR TO MAKE BEAUTY.* In some cases, my techniques might stretch Webster's definition a bit far, but most folks agree—my "Foo-Foo" techniques really do make ugly houses more pleasing to look at.

Let's be realistic here—if you decide to own and operate older houses you will, obviously, need to compete with landlords who own bright and shiny new homes. You might be thinking to yourself, how in heaven's name can a **fixer house operator** be expected to compete with newer property owners? I'll tell you how—**YOU MUST ACCENTUATE THE POSITIVE**.

Older houses and apartments may lack many of the up-to-date, push button amenities found in newer properties, but there are more than enough positive values found in the oldies—things that many newer properties are without. Heading the list is what I often call "old-fashion" charm. Charm is worth big bucks to creative landlords and it can be developed inexpensively by skillful application of my "Foo-Foo" techniques. Charm is present, both inside and out; but *outside* is always the place to begin the fix-up task.

THE "FOO-FOO" COVER-UP STRATEGY EXPOSED

Many older houses have ugly siding that detracts from their street appearance. One economical technique to create more pleasing looks is to install new exterior wood siding panels (4 x 8 foot sheets) directly over the top of the existing siding. The new wood panels have a stylish design and, once they're painted, along with new cedar wood trim surrounding doors and windows, the house will take on an exciting new appearance. No need to do this treatment on all four sides! The **street view** is what's important here. You can patch up and paint the other three sides if possible.

I once fixed a small cottage-type house that was listing badly to the "port side." To my non-seafaring disciples, that means leaning to the left. After many years, several of the foundation piers had actually sunk in the ground a bit, causing the house to tilt! I didn't have much money invested, but when the foundation contractor told me his fix-up bid was $6,000, I decided on an alternate plan; YOU GUESSED IT—the siding technique!

I left the house leaning, but installed new siding panels in front, straight up an down (plumb) over the old exterior. Today the house appears nice and straight when you view it from the street. You only realize it's crooked if you spill a bag of marbles on the living-room floor and watch them race toward the sunken left wall. I might point out that not one tenant in this house has ever asked for a rent reduction because of living in a crooked house. My total fix-up cost to solve the looks problem was under $500.

Paint Changes Looks and Image Quickly

Exterior painting is among the highest ranking outdoor improvements because it can quickly, and inexpensively change the looks and property image. Similar to decorating a naked Christmas tree, outside painting always creates that first layer of excitement! We can add more bells as we go. For older houses, with less than perfect exteriors, a light base color—such as off-white or beige—is recommended. Use a darker contrasting color for the wood trim, fascia boards, window surround and porches. On stucco houses, use the base color and a darker trim color for the woodwork. Painting is a very effective "Foo-Foo" technique, because it covers up so many imperfections found in older properties. In terms of pay-back, a new paint job can add approximately 20% more value to a medium-price house and sometimes double the price of a cheap one.

"Foo-Foo" creates the right look and adds the extra sizzle to help older, inexpensive houses compete with comparable-size, newer properties! When you consider that I generally pay about half the price that would be paid for newer, glossier houses, then rent them for approximately the same monthly rates as the new houses, you get some pretty good proof that my "Foo-Foo" techniques are working—plus earning more income.

LAWNS, SHADE TREES AND PICKET FENCES ARE HOT

Many older houses and smaller apartment buildings are just oozing with **charm** and **homeyness**! The problem is, you must educate yourself to spot these things while searching through many dumps! *Identifying potential beauty before it actually exists can quickly move you ahead of the competition.* Most buyers are frightened off by dirt and deadbeats!

Be especially tuned in when it comes to providing yards. Everyone loves a yard—and a nice yard can be especially appealing to potential renters. The combination of nice green lawns, shade trees and a freshly painted white picket fence will keep your properties rented while others are left vacant.

Fix-up investors should always begin outside improvements before ever setting foot inside. The idea is to turn the property around (make it look better) as quickly as possible! Remember, people will judge the looks—*and people make decisions outside, not inside*. The very first thing I always do is trim trees and shrubs, cut the weeds (which I call grass), get the sprinklers turned on and revive dead yards. Watering should be done at the same time you're fixing the houses. The goal here is to have an attractive yard to offer by the time the house is fixed up and ready to market (*RENT OR SELL*).

THE FIX-UP REVOLUTION—MADE TO FIT AND READY TO USE

Newcomers might be slightly ahead of me on this stuff, but most old-timers will agree with me. Fixing houses today is a whole different ball game in terms of ready-made, easy-to-use materials and instructions on how to do things.

Bob, a friend of mine who is very close to being a total klutz, recently re-wired two add-on rooms in his house with just a little help from the clerk who sold him materials. Very knowledgeable, the clerk explained exactly how to install new wires and even drew a simple sketch showing my friend how to connect all the circuits together! The fact that Bob didn't burn down his house or electrocute himself sold me on the proposition that today's do-it-yourself movement has come a long way since the time I did my own plumbing work!

Hook-em Up and Glue-em Together

It seems like only yesterday that do-it-yourself owners had to learn some basic soldering skills if they intended to do their own plumbing chores. Today, unless you're messing around with copper supply lines, basic plumbing parts are all made from plastic! You simply cut them with a hacksaw and glue them together. All that maze of twisted pipes under the kitchen and bathroom sinks—it's all plastic now! And, the whole works comes in a packaged kit, with "how-to" pictures on the back. Every clerk at the do-it-yourself store is happy to explain what piece goes where if you get stuck or end up with parts left over.

The Do-it-Yourself Explosion

Giant handyman stores are in fierce competition for the do-it-yourself trade. They stock acres of every building product imaginable for doing your own home improvements—or for fixing anything that's broken! The whole idea is to get the new generation hooked on doing things for themselves. Don't read me wrong here, not every job is a candidate for do-it-yourself investors. As a general rule, I'd be very careful with jobs that concern safety—like gas piping and electrical upgrades. These jobs normally require building permits and because of the added liability, you may wish licensed contractors to do the work.

THE MAJORITY OF FIX-UP IS CREAM-PUFF

Drain lines are glued together and those hard-to-bend chrome supply lines under the sink can now be substituted. Flexible polyethylene lines of various lengths, with screw-on fittings at both ends, are the amateur plumber's dream come true! Personally, I never did master those flared ends with compression fittings. My plumbing jobs generally leaked for a week or so. Nothing serious, but with screw-on connectors I can plumb like a pro.

Not so long ago, installing plastic laminated countertops (FORMICA) could be a major undertaking. Plywood or pressboard panels and thin plastic sheeting had to be glued together, then held snugly in place with special wood clamps until it dried. Gobbs of glue or air bubbles could make the job look like a third grade ashtray project.

Today, you can buy countertops already made! They even come with 45 degree angle cuts so you can install professional looking corners. All you have to do is cut a hole for the sink and glue the counter top to the base cabinets, and you've taken a giant step toward modernizing any ugly kitchen or bathroom.

Doors and Windows Just Plop In

With ready-made, pre-hung doors, all you need to do is install them in the wall and "trim out" both sides with standard molding and your replacement job is done. Most doors are already factory drilled for lock sets (door-knobs).

Steel pre-hung exterior doors are my favorite, because they don't crack, split or swell up in wet weather like most wood doors. And, there's no drilling—all you do is nail the door and frame in place, install a key set and dead bolt, trim around the edges—inside and out—and you've got a professional-looking door.

Windows are the same. Aluminum sliders, both single-glazed and double-insulated, are ready-made to fit the same hole that you removed the old ones from. Replacement windows come in every shape and size you'll ever need.

ELECTRICAL GENIUS IS NOT REQUIRED

Modern electrical fixtures of all kinds are easy to install. Fans, blowers and a large variety of light fixtures take up whole departments in the giant handyman stores. There's no longer any need to tear the walls apart when you need to add an extra circuit or two. Decorative conduits can be installed to conceal the wiring along the baseboard without cutting into walls. Surface-mount boxes are available in various sizes for hooking up new fixtures. Remember, the old rigid conduits that used to require an electrician with a pipe-bender, forget that! Now there's flexible, aluminum conduit that goes around corners and anywhere else you wish to run a new electrical wiring. They even make it plastic covered (weather-proof) for outside uses. Electrical wires are not soldered and taped anymore. They're attached together with plastic threaded (screw-on) connectors. As long as you remember to turn-off the power at the main meter panel, you can tackle most of your basic electrical jobs.

It's quite easy to learn exactly what you need to know about installing ready-made fixtures and doing minor repairs. Most salesmen working at the handyman stores can actually draw you a sketch to show how your project should be done. There's also several good illustrated books—*like the "Time Life Series"*—that will show you every detail about any wiring job you may decide to undertake. Do remember however, turn off the main breaker switch or pull out the fuses before you start any electrical job. If you stick your fingers in a hot splice box and you hear a sizzling sound, stop immediately, pull your fingers back out and re-read this paragraph!

YOU WON'T GET HIGH ON PAINT FUMES TODAY

Not so long ago, painting was a stinky job. Oil-base paint, in a small room or closed-in area, brought tears to your eyes. Cleaning up the mess with solvents or thinner took almost as long as the painting job itself, and it smelled worse. Water-base latex paint put an end to these problems. Plus, you don't need to keep cans of thinner sitting around, because plain tap water cleans up the mess.

Painting is a major part of fixing up rundown properties and, fortunately, just about anyone can paint without much training or practice. Easy-to-use rollers, extension poles and airless sprayers can help you look like a pro with a couple simple lessons. Also—most physical therapists would agree— painting is one of the most complete body fitness exercises one can do. Almost all major muscle groups get an excellent workout when you paint. You might keep this in mind before you renew your membership at the local gym.

Cabinet-Makers are Gone With the Blacksmith

Fixing kitchen cabinets with damaged drawers and missing parts was once an expensive fix-up chore that required specialized skills. Today, you can purchase most any size cabinets you need—already built and ready to install in kitchens or bathrooms. They're available with finished surfaces or you can finish them yourself—and the best part is, almost anyone can install them with a few instructions.

Tub kits and one-piece shower enclosures, made from fiberglass and plastic sheets (skins), are available for any size bathroom. They are very easy to install if you simply follow the instructions that come in the box. I use plastic shower kits (plastic sheets) in my older houses where only a tub exists, and I always convert tubs to a tub-shower combination. The only tools you need for installation are a tape measure, a straight-edge marker and a pair of scissors. The waterproof plastic sheets overlap one another and are easily glued to any wall surface. By replacing the tub faucet with a shower diverter faucet and installing a short riser (chrome pipe mounted up the wall) and a shower head, you can have an inexpensive, but efficient, tub-shower combo!

THE HANDY-PERSON FIXER UNIVERSITY

If you're interested in learning how to install floors, hang doors and windows or plain old plumbing and wiring, then I'd recommend you attend a handyman college. No, I don't mean the regular college! Find out when and where you can attend the free clinics conducted by most of the big handyman stores.

The big handyman stores are after your business and nearly all are sponsoring evening or weekend clinics to show you how to use *their* materials. Taking advantage of one of these clinics, I recently learned some new tricks about finding and repairing leaks on my flat-roof houses. Naturally, in the process, I purchased several rolls of fiber tape and a couple buckets of high-price "goop" to fix my problems. You simply can't learn enough about fixing leaks on flat-roof houses short of just saying "No!" to the next seller who offers you one!

WINDOW COVERINGS ARE TOP SIZZLE ITEMS

Curtains, drapes, mini-blinds and decorative shades will make any house or apartment look more homey! Homeyness translates to about 10% more rent when you dress up your houses with window coverings. Also, attractive window coverings will make an empty house rent much faster than one with bare windows. These are good sound economic reasons for installing them in all your rentals.

Anyone can hang curtains, right? WRONG! If you don't believe me, chances are you haven't watched tenants install them! One of my tenant installed curtain rods with nails so long they went all the way through the wall and came out the other side. I've seen four foot drapes hanging on eight foot windows, curtains that were at least a foot too short, and traverse rods that won't open. Take my advice here—do this job yourself, while the house is empty, before your tenants ever show up!

If you use drapes and traverse rods—which I like to use for most larger front-room windows—be sure you purchase them from a supplier who specializes in selling to apartment owners. You'll find their prices are much cheaper, plus it provides for easy one-stop shopping, because they can supply drapes, rods and all the hardware to complete the job.

Rich Man-Poor Man Strategy

People often ask me if it's necessary to know everything about doing fix-up work to be a successful house fixer. The answer is "No, but any knowledge you have is very helpful, even if you don't plan to do the work yourself." The reason is because you need to know how much things cost—this includes the labor and the price of materials.

In my own case, when I first started out, I did everything I could (about 90%) because I didn't have the money to pay to have it done. Many investors do a portion of the fix-up—then contract with others to do the rest. Some investors lack handyman skills all together and hire everything out!

In the final analysis, it boils down to "money," no matter who does what! *Approximately 30% of every fix-up dollar goes to purchase materials, and the remaining 70% is the cost for labor*. Thus, you can, obviously, save $700 for each $1,000 you planned to spend by doing the labor yourself. Saving $700 is almost the same as earning that much! It makes a pretty strong case for learning how to do your own fixing, I'd say. But, even if you choose not to do any of the work yourself, the more fix-up knowledge you have, the more successful you will be.

FIX-UP SKILLS OFTEN WORTH MORE THAN CASH

It is not necessary to have a lot of money in order to make a lot of money. However, if you don't have money, you must have something to use as a substitute. Your fix-up skills and ability to find and acquire bargain properties—and, of course, manage them—are CASH-EQUIVALENT SKILLS. Often they're even more valuable than cash, because cash is much more common than fix-up skills. Don't forget this when you think about *partnerships* and *co-investors*, such as my 90/10 PLAN you will read about it in Chapter 17.

HOUSES WITH ALL THE RIGHT THINGS WRONG

Some folks say, "The real secret to making big money in the fix-up business is purchasing rundown properties with all the right things wrong." Indeed, there is a lot of truth to that; but, seldom have I found exactly the kind of conditions I'd like to have. The truth is the houses I find most profitable will generally require almost total reconditioning—notice I didn't say "remodeling." The following is a list of fix-ups and repairs I find most common for properties I acquire.

1. ***Paint exterior***—Includes siding or stucco repairs, and anything else that looks bad.

2. ***Fix yards or landscaping***. Front yards: Front street view is most important. Re-seed lawn areas, plant inexpensive shrubs and trees for improved looks. (Pyracanthas are one of the best shrubs for front yards, because they grow fast, they're inexpensive, they have thorns—which keep kids from pulling them up—and they are colorful.) Rebuild, and repair when possible, existing fences.
   Back yards: Cut and water grass, and that's it!

3. **Paint interior of houses** after tenants move out—not while the unit is occupied—using off-white semi-gloss latex paint.

4. *Replace carpets and linoleum*, only if required. If they can be salvaged, and will look good afterward, repair and clean the existing floor coverings. Do this step between tenants moving out and new ones moving in.

5. *All water valves, faucets, toilet assembly (guts), and shut-off valves*—also showers, tubs and sinks—must be fixed if not working properly. It is usually best to replace them if they are old and ugly, or in poor condition.

6. *Replace unattractive light fixtures* with new, but inexpensive (economy) ones. All switches must work easily, otherwise replace them.

7. **Repair obvious problems** like holes in walls, busted doors, malfunctioning door hardware, broken windows, etc. Exterior doors must have working lock sets—preferably with dead bolts. Re-key, as necessary, making sure to have a single key that works in all exterior doors—this way you won't have to make so many copies, plus you won't need to carry so many keys on your key ring.

8. *Replace old, ugly kitchen and bathroom countertops*, if needed, with new ready-to-use plastic tops (Formica).

9. *Providing window coverings is a big pay back item*, plus it saves you the heartache of having tenants hang ugly sheets and blankets over the windows, or split window moldings by using large nails to put up curtain rods. Most houses should have *full window treatment*—plastic mini-blinds or drapes in living rooms, bedrooms, dining or family rooms, and inexpensive, colorful curtains in kitchen and bathrooms. Always add shower curtains to make bathrooms look complete. Window coverings add that touch of class, plus they also hide minor flaws and imperfections found in older houses and apartments. If your money is running low, do the street side windows only. Let tenants

dress-up the back windows that no one can see from the street.

10. It goes without saying; however, I'll say it anyway—*heating units and coolers or air conditioners must work properly!* In my town the temperature gets over 100 degrees in the summer. I use evaporative coolers (swamp coolers) in most of the lower-priced, economy houses, and forced air heating with cooling in upper scale houses. I furnish kitchen ranges—either good re-conditioned or economy priced new ones. I don't usually furnish refrigerators in lower income rental properties, unless it's an upstairs apartment where it's difficult to move appliances in and out.

11. *Rotten wooden window frames that can't be fixed must be replaced*; however, most can be repaired and re-puttied without much cost or labor. A little paint will fix most anything on a house that looks ugly.

12. Adding frills is okay, as long as it is inexpensive to do so. **One of my favorite "frills" is ceiling fans**. One or two ceiling fans add a lot of class to older houses and it's usually money well-spent, because it attracts the attention of potential renters, who associate fans with higher-class rentals.

13. *Big ticket repairs—like repairing bad roofs, especially if they look ugly from the street-side view—are sometimes justified*, but can be made in several stages. For example, on pitched roof houses, it seldom happens that both sides leak at once—generally the south-side, or weather-side, leaks first, and the other side doesn't! Patch or repair only the half that leaks in the first phase of your fix-up project. The street-side of roofs that are old, weathered and ugly may not leak yet; however, ugly roofing equates to lower value because it distracts a great deal from the property looks. Replacing the street side is often justified for this reason. Do the back side later on Phase II. Remember, however,

you can't get higher rents for new roofs. The function of a roof is simply to keep the water out, and that's it!

14. Extra sizzle items that cost very few fix-up dollars are **exterior shutters**. Install them on front street-side windows. You can get inexpensive plastic shutters for various size windows in white, brown and black colors from most handyman do-it-yourself stores.

15. I always recommend building a three-foot-high, **white picket fence** around front yard areas! White picket fences show off the property—both to renters and potential buyers, alike! The fence creates a homey atmosphere, especially when you have a nice green lawn in the front yard.

    Fences in backyards are also hot seller items. They keep kids and dogs inside the property; thus, mothers like fences, and mothers make the decision to rent or buy. Need I say more! However, if money is tight, do front fences first, and save backyard fences until Phase II.

16. While you're working on the property, train yourself to water front yard areas while you are there. Always carry a hose and sprinkler with you at all times. Then when you arrive at the job, you can set up your sprinkler and water dry areas. **The idea is to make all front yards green!** Toss out some lawn seed anywhere you see bare ground. Lawn seed and weeds grow together to become a durable lawn when they're watered enough.

17. For multi-unit properties, *save most inside fix-up work until the units become vacant*. This will happen when you raise rents! As soon as the outside painting is done and the yards start looking nice, you should begin to increase the rents. The original tenants will begin moving out, one by one, and you can replace them with tenants willing to pay more rent for a better looking property. I call this process *"TENANT CYCLING."* Multi-unit fix-up jobs will normally take a year to 18 months when you do it this way.

18. *Exterior painting can make a tremendous difference in the looks of any property*. I recommend medium quality exterior latex paint. Use a light color for the main body of the house, with a darker trim color. With off-white or tan, almost any trim color goes well. The trim can be doors, window surrounding and the fascia or eave board. A two color combination is very attractive. Painting the exterior is the first thing you should do because it immediately changes the looks of a property from dirty and dingy to clean and attractive! This change needs to be done quickly so everyone—tenants, passer-bys, potential customers—gets used to seeing the "new look."

Many rehabers and most all contractors will save the painting until last. They do all inside repairs first, then paint the exterior as their last effort. To everyone who observes the property during fix-up, it continues to look unattractive regardless of how much work is being done inside—people can't see inside. *They judge the property by how it looks outside*. This sequence is very important—*DO THE OUTSIDE FIRST!*

19. Old dilapidated garages, carports and storage sheds can be made to look attractive by installing new light-weight metal, tilt-up garage doors and exterior paneling (wood or masonite). However, only do this on the front, or whatever side shows! Old-style, heavy wooden garage doors, with those hard-to-adjust springs and bent hardware, should be replaced. *The idea is to fix up the building enough to have a nice exterior appearance—that's the goal here!* Tenants like garages—not for their car, mind you— they love to store junk. The good news is that most renters are willing to pay $35-$40 more rent per month to have a garage. As a landlord, this is appealing for two reasons— first, it's more income, of course; and second, if they don't have a garage, junk will end up all over the yard instead. With garages you can usually convince tenants to keep their stuff hidden inside.

20. *The big difference between "successful fixers" and the others who will crash and burn, financially speaking, is learning where and how to end a fixer project!* Fixing older houses can cost you all the money you have today, plus every dime you might earn in the future. However, it doesn't have to be that way! Remember, it's not much fun fixing up junky houses and dealing with deadbeat tenants if you aren't getting paid well. Over-fixing is a common problem you must guard against. For me, the most difficult group of investors to teach this lesson to, is building contractors! Contractors are taught, from day one of their contractor training, that buildings must always be plumb and that dry-rot is worse than the AIDS virus. Technically, I suppose they are correct. However, no one says we're living in a perfect world! Perfection is okay for your hobby, but not this business, if you intend to make any serious money doing it.

Don't misunderstand me here, I'm not telling anyone to do shoddy work. *What I'm suggesting is that your work be creative and that it fit within the limits of your budget.* It can be done, believe me! One of the best examples I can think of are the highway builders—they construct six lane freeways, when every motorist can plainly see that 10 or 12 lanes are needed. What they are building is what they have the funds to build—and that's it. When more money becomes available (later on), they'll add more lanes. That's exactly the same strategy house fixers must adopt to be profitable!

HOW TO ESTIMATE WHAT THE JOB WILL COST

Fix-up work has two parts that cost money. *First and most expensive, is the labor*—on average labor costs 70¢ of each fix-up dollar; supplies and material cost the remaining 30¢. Certainly, these numbers can vary slightly for some jobs, such as installing or overlaying exterior walls with new siding panels. The materials (4 x 8 foot sheets) are expensive and the labor involved is relatively quick and simple. *However, a 70-30%*

split is well within the ballpark for estimating most fix-up chores.

What all this means is that if you were able to purchase a fixer property for 30% under the estimated "fixed-up" market value—for example, say a $70,000 market value house for $49,000—then you estimated it would cost $10,000 to fix it up to its full market value of $70,000, it will cost you somewhere around $3,000 for material and supplies. That means if you do all the labor yourself, you can expect to have an $18,000 equity in the property when the job is done. Doing fix-up or **"adding value,"** as it's called, means you don't need to play the inflation game to make profits in this business.

My rule of thumb is to plan on spending 10% of the amount I pay for the property for fix-up. For example, if I purchase a house for $75,000, I'll plan on $7,500 worth of fix-up. It can, however, be as much 20% for the "real uglies."

BIGGEST PAYDAY COMES FROM KNOWING WHERE TO KICK

Allow me to share the story of an electrician who repaired saw mill motors. He specialized in heavy-duty electrical motors and his job was to keep the big "high-speed" machines running smoothly. Large crews of mill workers and their families were very much dependent on the giant motors; when they didn't run, the crews were sent home. Obviously, their paychecks stopped at the same time the motors stopped.

One afternoon the repairman got an emergency call. The biggest motor at the mill had quit running! "Come quick," they pleaded, "We had to send all the men home. We're losing thousands of dollars worth of business. We need you right now!"

The repairman arrived shortly thereafter carrying only a small leather tool pouch. Very methodically, he opened the pouch and pulled out a weird-looking wrench; then he stood there a moment staring at the lifeless machine. When he finished looking, he began adjusting the bolts and turning a couple screws. In less than ten minutes he was done and yelled to his helper to turn on the switch. The helper did, but nothing happened! Once again the repairman paused and stared at the giant motor. Suddenly, with a renewed sense of direction, he walked around to the other side of the motor and gave it a swift kick with his "hard-toed" work boot! With a loud pop and a couple squeaks, the giant machine began to turn. Within seconds it was running at full speed again. Everyone smiled and the crew was called back to work.

Several days later, the repairman's bill arrived in the mail, and the mill owner, who had stood by and observed the complete repair job, stared at the bill in total disbelief! "$500 for only 10 minutes work," he screamed, "That's outrageous! I demand an itemized statement of the charges. I demand some explanation—*WHY SO MUCH?*"

Two days later, the amended bill arrived in the mail. It was itemized as requested:

Travel time .... $ 30.00
Adjustment to motor .... $ 20.00
Knowing where to kick........................ **$450.00**
Please remit promptly............................. $500.00
THANK YOU!

Almost anyone can eventually fix a leaky toilet, and most folks can do a half-way decent paint job, if they really try—but knowing how to turn these chores into big paydays is the real secret to fixing for profits. *UNDERSTANDING WHERE TO KICK IS WHAT PAYS THE BIG BUCKS!*

CHANGING THE LOOKS ADDS THE QUICKEST VALUE

It's not by accident that I always begin my fix-up projects in the front yard. *I've seen professional appraisers value identical houses as much as a $10,000 difference because of plain old filth and junk at one property*. In other words, a clean house can be worth $10,000 more simply because the owner hauls away the trash and keeps the house looking nice. Think about that for a minute! That's a lot of money for ordinary clean-up skills. Suppose it takes a whole week (40 hours) to haul away garbage and clean up a property. That's $250 an hour, or nearly as much as a brain surgeon earns on his day off!

Scrutinizing Your Fix-Up Plan

All fix-up work should pass some financial scrutiny!—ask yourself, *"Does it really need doing?"* I believe most improvements should be justified on the basis of paying for themselves. I expect the payments to come from **higher rents** or **bigger profits** as my reward for doing the work. Fixing or changing things around purely on the basis of personal likes and dislikes will seldom provide a worthwhile "mark-up" (profit). Those kinds of changes should be avoided. This happens to investors who charge forward without a plan. It also happens to folks who fall in love with investment property. I advise you to be very careful and avoid these common pitfalls. Remember, fixing up ugly-rundown houses is not glamorous work; but, if you do it right, it will pay you better than anything else I know of.

Fixing Average Rental Houses in Decent Locations

When considering location, let me just say, "Don't buy property in locations where you'll need an armored vehicle to drive through the neighborhood." It's not that you can't make money in a "combat zone," but houses like I'm recommending are not scarce in decent areas, once you develop your techniques for finding them. Bad areas are simply not worth all the hassle. Save your energy for painting.

SIZZLE FIX-UP OFFERS THE BIGGEST PROFITS

Let me take a moment to say that all fixer properties must be brought up to basic-minimum building code standards before you can expect them to generate income. *That's a must rule for all properties*.

When I talk about *fixing for dollars*, I'm primarily referring to what I call "sizzle" items—things like white picket fences, fresh paint, window coverings, ceiling fans, wall paper, new counter tops (Formica), attractive floor coverings, planters, shower curtains, decorative porches or entrance doors, trees and shrubs, green lawns, new faucets, modern toilets and new shower enclosures.

The reasons I call these "sizzle" items is because they are attractive and eye appealing, as well as useful. Sizzle items seldom have anything to do with code problems. For example, an old dingy carpet will pass a code inspection just the same as bright new carpets—trees and shrubs, or curtains or ceiling fans, have nothing to with codes or safety, but these items have a lot of customer appeal. This appeal translates to big dollars in the form of higher rents from tenants, and bigger profits when its time to sell.

PROVIDE YOUR CUSTOMERS WITH WHAT THEY CAN PAY FOR

Investors who intend to make serious money fixing up run-down houses must prevent themselves from "over-fixing" and fixing things that don't count for much! For example, don't waste time changing the wallpaper or redesigning the hallway. It's not cost effective! It's easy to fall into the *over-fixing trap* and it's about the quickest way I know of to loose money.

Fixers must learn to concentrate their time and money on things that clearly have *proven pay back values*. Improvements made on the basis of personal taste are, generally, not worthwhile. Don't ever forget that it's your customers (tenants or buyers) who you'll need to satisfy! It's not you!

"FOO-FOO" MAKES CHEAP HOUSES LOOK MORE EXPENSIVE

A difficult lesson for many investors to learn is that it's not wise to provide *too much* house for the money! You'll soon go broke if you keep doing this! Let me explain what I mean—

many tenants can afford to pay $450 a month for rent, but that's their limit! Your goal, as a profit-motivated investor servicing these tenants, should be to provide the very best $450 rental house you possibly can—and still earn a profit, of course!

If you over-improve and spend too much money fixing the property, chances are the deal won't be profitable! Plus, you might even end up spending out-of-pocket money every month to pay off your accumulated fix-up bills. I can tell you, from personal experience, that takes all the fun out of investing! It's my policy to give my tenants *everything they can pay for* with as many frills as I can. In other words, when I rent out my $450 houses, I want them to be the very best $450 houses available. Obviously, however, they won't be as good as my $600 houses—and they shouldn't be.

PROFITABLE FIXING BOILS DOWN TO "WHAT DOES IT COST?"

Since almost every problem can be patched up, repaired or replaced by skilled persons, it becomes necessary to further qualify fix-up work in terms of economics—"How much will it cost?" This information will help you decide how much work is too much and when it's best to simply pass over the deal and move along to the next one. The fix-up investor must be concerned with *fixing for profits*—not just fixing. This is a very important concept, one you must never forget. The two worst mistakes for beginning fix-up investors are OVER-FIXING and FIXING THE WRONG THINGS.

In order to determine what to fix, you must first answer two simple financial questions.

1. WHAT WILL IT COST TO COMPLETE THE ITEMS YOU PROPOSE TO FIX OR REPAIR?

2. WHAT WILL THE FIX-UP VALUE BE AFTER THE WORK IS COMPLETED?

Cost and value knowledge is critical whether you intend to keep the property for rental income or sell it quickly for turnaround profits.

CALCULATING THE FIX-UP RETURN

When you spend fix-up dollars to upgrade a rental property, you should, obviously, get your money back in the form of higher rents. My goal is to recover most fix-up costs from increased rents over a period of one to three years. "Why such a variation?," you ask. The reason is because many fix-up items also add more permanent value to the property. It takes a longer period to recover costs for permanent value items. For example, I often replace old, deteriorated wooden garage doors with new light-weight metal doors. The old doors are very heavy, hard to open and close, and quite often they are hazardous. The wood frames or jambs securing the lower end of heavy "coil-type" operating springs are often split out and rotten. Springs can easily break loose under tension causing them to fly wildly through the garage. Flying springs can seriously injure anyone who happens to be in the area when in happens.

Light-weight metal doors will cost several hundred dollars, installed. These doors will add much to the looks and have a tendency to increase the overall property value. However, they require a longer recovery period in terms of pay back from increased rental income. Perhaps a better way of explaining this would be to say, *I'm satisfied to recover a certain portion of the total garage door costs from increased rents and the balance from extra profits on the permanent value when I sell or trade the property.*

Remember, this example applies only to garage doors and similar replacements or repairs, that have longer term values. It does not apply to "short-life" repairs, like painting and adding curtains. Dual-glaze window replacements are in the same category as metal garage doors. It takes much longer to get fix-up or replacement money back, but, here again, energy efficient windows add much to the long-term values of any rental property.

There is another very important economic benefit that comes from installing modern insulated windows—which helps shorten the long-term cost recovery period. It cuts the cost of utility bills.

PASSING THE SAVINGS ON TO THE LANDLORD—THAT'S YOU!

Lower utility costs (heat and cooling) create an opportunity for owners to obtain higher rents immediately. Here's how it works—even though tenants are paying their own utilities through individual meters, most tenants try to budget a specific amount monthly for housing or shelter expenses. Tenants are concerned about their total monthly housing costs, not just the individual costs of rent, utilities and upkeep. Thus, new insulated windows can provide the opportunity for higher net rent profits. Let's say you own two identical apartments, and the rent for each is $300 per month. The only difference between the two apartments is that one has old wooden, single-hung, single-glazed windows—with loads of charm—and lots of air leaks; the other has new energy efficient, double-glazed windows—without the charm—but no air leaks. Let's assume the winter heating bills in the apartment with leaky windows average $100 a month, and the bills for the unit with energy efficient windows average only $50 per month. This provides the potential for being able to charge a higher rent for the apartment with new windows.

Tenants Understand Leaky Windows—High Utilities!

Let's say you are an apartment tenant who is accustomed to paying total housing expenses of $425.00 per month—$300 for rent, $100 for utilities, and $25 for upkeep. Let's also say you have an opportunity to reduce your expenses by $25.00 every month; would you do it? I'll assume the answer is "Yes" because all my tenants would. Also, would you be concerned over who gets what part of your total monthly housing budget, as long as you are saving $25 every month? Again, I'm sure I know your answer. Why should you care? After all, you're saving $25. Most tenants will agree to a $25 reduction with no questions asked! I think everyone is happy to save $25. The following chart shows the breakdown of expenses before and after installing new "air-tight" windows.

Tenant's Per Month Housing Costs	
Old windows (with charm)	New windows (without leaks)
Rent......................... $300	Rent........................$325
Utilities.................... 100	Utilities 50
Upkeep.................... 25	Upkeep 25
Total..................... $425	**Total......................$400**

As you can see, new windows will provide worthwhile benefits for everyone. As the above example shows, the landlord can get a $300 annual ($25 times 12 months) rent increase, and the tenant can save $300 annually, and the utility company will award you their prestigious "Blue Star" window sticker for energy conservation. Before we leave this subject, let me just say that tenants today are very much aware of ever-increasing utility costs, and they compare costs with their neighbors who live down the street. I will promise you this much, based on my long-time experiences as a landlord working with numerous tenants, you will have no problem with the dollar trade-offs I've used in this example. Once you convince your tenants that the improvements you plan will save them money, they will be 100% behind your project as long as they believe *they* will save money.

Your rental ads can read "CHEAP UTILITIES." Tenants know exactly what that means. As you acquire additional properties and rent to more tenants, at different locations, it becomes quite easy to develop a utility cost comparison chart showing the actual expense figures to back up your advertisement claims.

Work That Shows Fast—Earns Fast

Many fix-up jobs will pay for themselves very quickly. Leading the list of "quick returners" are painting (inside and out), general cleaning, yard and landscape work, fencing, carpets, window coverings, modern faucets, light fixtures and Formica countertops in kitchens and bathrooms. *It's not uncommon to get three or four dollars back for each fix-up dollar you spend on these items putting you in the high-profit mode!*

Garages are Quick Return Cash Flow Generators

Another quick return suggestion, which also features some extra rental income, is to look for properties with garages or carports when you're shopping. Garages and closed-in carports are worth $20 to $45 more rent per month to landlords. Even junker garages, with leaky roofs, are well worth their fix-up cost—within reason, or course. "Why garages?," you ask. Because most tenants want garages. They ask for them and are willing to pay extra for them. I don't mean that you should consider building garages where none exist. All I'm suggesting is that you look for them when you are buying. Properties with garages and carports, either attached or freestanding, will always have more long-term value. They will also help the property rent much faster when you advertise them. Simply stated, as a landlord, I like garages because **my customers like garages**. They are willing to pay more rent to get them. It's strictly a bottom line economic decision for me. *Remember, higher rents make wealthier landlords.*

Frankly, I've never understood why so many lower income tenants are willing to pay 10 or 15% more rent to have a garage. Seldom have I ever seen their cars inside. Apparently it's worth the extra rent to my tenants to have their own personal "on-site" storage unit. Most of my tenant's garages are filled to the rafters with cardboard boxes, boats, canning jars, weight-lifting equipment, hope chests, bicycles with missing parts and busted furniture. Generally, if an automobile is put in the garage, it is not running and never leaves until *I* haul it away, after the tenant moves out. Even considering my clean-up expenses, garages are still "money-makers" for landlords. Remember, higher rent is the goal—you can take my word for it that GARAGES PAY WELL, and understanding the reasons why don't count for much.

PAINTING IS A DRAG—BUT ALSO A TOP MONEY-MAKER

General clean-up and fresh new paint will return an owner's fix-up money about as fast as any other improvement I know of. It's also the most important job in terms of what new owner-investors should plan on doing first. The reason is quite simple. The general public (most people) associate value with looks. Therefore, an unattractive scum-bag property always looks worthless. The very same property "cleaned up," with sparkling new paint, suddenly looks like it has much greater value. If you truly understand what I'm telling you here, and if you learn to apply the profit-making strategies we're discussing, you're ready to start right now and I'm betting you'll become a very successful fix-up investor in a few short years. There is no limit to the financial rewards that can be yours if you'll apply my practical advice here—just remember—CREATING "GOOD LOOKS" IS THE LIFE BLOOD OF MY PERSONAL FIX-UP STRATEGY.

After doing fix-up for many years now, I must say that fixing is much more satisfying and rewarding to me than buying the newer, more expensive, sweet-smelling houses (even though I do own some now). Let me point out that newer houses provide an excellent vehicle to reduce excess cash flow. Also, if you have a problem with too much money stacking up in your safe deposit box, newer houses with high monthly mortgage payments, will quickly eliminate the problem.

It Pays to Invest Where the Money's At

The colorful, but most unlucky bank robber, Willie Sutton, was once asked, "Why do you keep robbing banks when you always get caught and you're tossed right back in jail?" Willie's reply was, "'Cause that's where all the money's at!"

People often ask me why I keep fixing rundown houses, when I, obviously, have enough money to buy the more attractive and cleaner properties. My answer is the same as Willie's, "CAUSE THAT'S WHERE ALL THE MONEY'S AT!"

ALWAYS GET A SECOND OPINION

My good friend and nationally syndicated real estate columnist, Bob Bruss, shares my love affair with fixer-upper houses. Many of Bob's excellent "How To" newsletters are specifically written about fixin' houses and I strongly recommend you read what he has to say. Bob shares his personal experiences buying, leasing and selling fixers. What makes the **ROBERT BRUSS NEWSLETTER** stand out is that it's packed full of wisdom and good solid advice that comes from a practicing fix-up investor—you get knowledge you can use.

I don't expect you to remember what I told you back on page 41, so I'll remind you. I said, "**KNOWLEDGE IS WHAT BUILDS FORTUNES**," and I still mean that right here, 150 pages later. Obviously, the kind of knowledge you need is somewhat limited, so you need to take advantage when you can find it. Here's what I recommend, while we're on the subject: write to **Robert J. Bruss**, 251 Park Road, Burlingame, CA 94010, for information, or call 800-736-1736 to order. You'll thank me for this suggestion, I promise.

CHAPTER 11

Secrecy, Plus Adding Value Creates Profits

In this chapter, I will show you that there are two sure fire ways to earn big profits investing in real estate. I'm repeating several things I told you in Chapter 3, but they are very important!

The first way is to acquire property for a price substantially below it's actual value—*30 to 50% below its potential market value*. To do this, you must find out about the sale before anyone else knows about it. It's nearly impossible to benefit from this method when other buyers know about the under-priced property. The reason for this is quite obvious! When others learn about a good deal, the increased competition will almost always drive the price up. Suddenly, the discount is not very big! If the property sells for 10 to 20% below its market value, the reduction is not very substantial, in my view. The deal might still be all right, but it's not the quickest path to the big profits I have in mind.

THE HOUSE DETECTIVE APPROACH

In order to make big profits using this method you must know about the availability of the property before anyone else does. More specifically, your offer to purchase the property should be the only offer! If others know about the deal, it's still possible to earn substantial profits; however, you must possess the ability to spot *HIDDEN PROFIT POTENTIAL*. That means you must have special knowledge and the skills to increase the property's income, or value, that other competing buyers don't have! These special skills can only be developed from experience. Obviously, new investors will need to own and operate several properties in order to get the kind of experience I'm talking about here. Generally, novice investors without much real estate experience will not be able to recognize the properties *gold mine potential*, even if they get close enough to fall down the shaft. *EDUCATION and EXPERIENCE* will take some time. Completing several actual deals will provide the necessary experience. I do not know of any shortcuts to develop these skills.

THE SECRET PATH TO THE GOLD MINE

Be sure you understand what I'm telling you here! Once you do, it will be less of a shock when I tell you that most HUD, FHA and VA foreclosures or take-back properties are *not* very good deals for investors—especially investors who have visions of becoming rich anytime soon. "How can that be?," you're probably thinking. The answer is really quite simple. Their normal selling prices are not discounted nearly enough to allow you to make anywhere near a *SUBSTANTIAL PROFIT*. You probably won't even be able to get cash flow if you rent the property. The truth is that sellers don't need to cut prices very much when the entire public is notified about the sale.

You need to understand the public is comprised of countless "dummies" willing to bid up the price! Well advertised properties are seldom sold at prices low enough to be considered substantially below the market.

Your only consolation, if you do decide to purchase a government-owned foreclosure, is that you'll also receive a heaping supply of top-grade plywood, as an added bonus. Both FHA and HUD use tons of plywood to board up all the windows, when they foreclose properties. However, the selling price rarely ends up less than 10 to 15% under true market value. To me, that's simply not enough discount for a cash flow investment.

First in Line With the Only Offer

The most important thing to remember here is that *THE MORE BIDDERS, OR POTENTIAL BUYERS, WHO KNOW ABOUT THE DEAL, THE LESS CHANCE OF YOU BEING ABLE TO SNAG A BARGAIN FOR YOURSELF.* If you can secretly find out about property that's for sale and get there with the only offer, it puts you in the driver's seat! Without any competition, your offer will be the only game in town. *You must learn how to be a "house detective" and find good deals before the competition arrives.* That's one of the true secrets to buying property substantially below market.

You'll soon discover, as I have, it's nearly impossible to buy properties cheap enough—or get real good terms when there are lots of eager bidders trying to buy the same property. *Acquiring cash flow properties should be as private as you can make it.* You can tell the whole world about the deal, after you own it!

GETTING STARTED ON THE RIGHT FOOT

The very first thing I do, when I hear that property I'm interested in becomes available, is to begin what I call *"detective work!"* Sometimes my broker, Fred, will perform this task—but it took four years with my coaching before he became "snoopy" (skilled) enough to suit my taste! Brokers and sales agents, typically, don't do the exhaustive research or "snooping around" I insist must be done.

The biggest difference between me and most sales agents is that they will accept the word of sellers as being mostly true! I accept it as being mostly exaggerated and often untrue! It's never considered true until it's proven true to me. I'm not trying to be overly critical, but you must never forget this important fact of investment life—*ONCE THE ESCROW CLOSES AND EVERYBODY GETS PAID, IT'S YOU, ALONE, BY YOURSELF, WHO MUST LIVE WITH THE DEAL YOU SIGNED!* If, somehow, you've failed to uncover the true property expenses and it turns out they're considerably higher than you were led to believe, guess what? It's you, alone, who is now stuck! That's the reason I learned to become a very snoopy house detective early in my investment career.

Don't Save the Seller, but Lose Yourself

In order to acquire properties at bargain prices or for prices that will allow you to make a reasonable profit, you must first determine if the property can actually be purchased for a reasonable price! I imagine you are thinking to yourself, "That doesn't seem so tough! Why not simply ask the seller. Certainly he can tell me in a minute." Friends, here's a bit of advice—*sellers may not know why they have problems, especially, if they paid too much themselves.*

Many times I've found properties for sale that are very desirable, but are seriously over-financed. Owners (sellers) who have over-financed properties are often extremely anxious to sell—many times for little or no cash down payments. The reason is because they need to stop their negative cash flow! You must be wary of these kinds of properties. An over-financed property can spell big trouble, no matter how small the down payment is. Once you determine how much you can afford to pay for a property and still make money for yourself, you should forget the whole idea if it looks like the selling price will exceed your estimate! Sellers with too much existing debt have very limited flexibility for negotiating the price downward. Conversely, **sellers with lots of equity have room to reduce their asking prices. Those are the sellers you're looking for.**

FORGET SELLERS WHO CAN'T MAKE YOU A GOOD DEAL

Quite often you'll find sellers who have over-financed their real estate by paying too much in the first place, or by adding additional loans during their ownership. The reason doesn't matter, the important thing is that these sellers are not in a position to make you a very good deal! The only way they could, would be to pay down the existing mortgage debt or to pay you to take their property! Obviously, these options are not very attractive for most sellers. The simple truth is, when too much money is already owed on a property, it will seldom be a good deal for you. Don't waste your time on deals that don't show you a clear plan for making reasonable profits! Before you invest one penny—do your homework and check things out!

Doing The Undercover Detective Work

The first thing I do is visit the title company where I do the most of my business. I request a parcel map, the tax bill and copies of all the deeds or mortgages recorded on the property. Many title companies provide this service free for repeat customers and real estate agents. Some folks refer to this information as a PROPERTY PROFILE. Obtaining the property profile documents is a good way to start a file on property you are interested in, because they will provide you with the following information:

1. *A parcel map*, which gives you lot size and often shows easements and right-of-ways, as well as location.

2. *The tax bill* showing who owns the property and where the current tax bill is being mailed each year. It will also show how much the taxes are now and how much the county assessor has appraised the value of land, improvements (buildings) and personal property for (the dollar amounts).

3. Additionally, with a parcel map and the tax index (owner names) you can find out who owns the properties around the parcel you're investigating.

4. Copies of *mortgages or trust deeds* showing the amount of debt against the property at the time of the last sale or transfer of title. They also show who owes the debt (trustor) and who receives payment (beneficiary). Any due-on-sale clauses will normally show up on trust deeds or

mortgages. Deeds and mortgages will show the chain of title with recording dates and notarized signatures.

5. *The documentary transfer tax*—which is a state tax on the sale of real property and is computed on the full selling price or selling price less the liens remaining at the time of sale. The amount of tax is generally stamped on the grant deed. The current tax rate in California is $1.10 per $1,000. Therefore, when you have a copy of the grant deed, you can easily determine what the current owner paid for the property, based on the tax amount. For example, if the documentary tax is $46.75, computed on the full value, then the purchase price would be $42,500 ($46.75 divided by $1.10, times $1,000 equals a value of $42,500). Sometimes, knowing what the owner paid can help you develop your own offer to purchase. Remember that sellers dislike offers for less than what they paid, no matter how motivated they are to sell. Always keep this in mind when negotiating.

THE MORE YOU KNOW THE BETTER YOU CAN NEGOTIATE

Generally, motivated sellers are not "stupid dummies" like many late night infomercials might lead you to believe. Most are just regular people who have gotten themselves in a financial bind. *Also high on the list of motivated sellers are folks who are just not good landlords, and dealing with tenants is driving them "bonkers."* I've purchased several excellent properties from sellers who were not willing to take me to the property for a showing, because their own renters frightened them! Knowing this kind of information is extremely helpful when I'm negotiating a deal.

OFFER RELIEF FROM PAIN IN LIEU OF CASH

When I check out a rundown property, I try to find out if the tenants are basically running the place. There's a good chance that poor management, or lack of landlording skill, is really the motivation or reason the property is for sell. If this is the case, sellers are often willing to give very favorable terms in exchange for IMMEDIATE "PAIN RELIEF."

Problem tenants, who are out of control, are worth big bucks to investors who know how to establish law and order! Many sellers who find themselves in this predicament are willing to forego traditional cash down payments in order to rid themselves of the unpleasant situation! I have long emphasized the value of solving these kinds of problems in my training materials—people pay handsome rewards to investors who can straighten out the mess.

BEWARE OF SELLERS WHO HAVE NO EQUITY

Properties that have been owned by the seller for a substantial period of time—*six to ten years, or more*—will offer far greater opportunities for negotiating the selling price downward. The reason for this is because the existing mortgage debt has most likely been paid down over the years. Always look for sellers who have a substantial amount of equity in the property, because its the equity they can discount.

For example, let's say the seller is asking $250,000 for his property. You can assume the existing mortgage of $195,000. The balance of the asking price ($55,000) is the seller's equity. You've already filled out your INCOME PROPERTY ANALYSIS FORM and concluded that $210,000 is the right price to pay. If the seller accepts your offer, he will receive $15,000 for his equity. He may not like the low price, but he still receives some money from the sale.

Now consider a similar situation where the seller had purchased the property just a year ago for the price of $235,000, with a down payment of $20,000. Obviously, he still owes almost $215,000 on the mortgage! Your chances of buying this property for the *price of $210,000* just flew out the window, because the seller would need to pay you $5,000 make the sale work. Even if this transaction would help the seller financially, it's almost too much of an obstacle for him to overcome emotionally. Sellers will seldom pay buyers to take their property. In short, this seller is not in a position to make you a good deal!

Verify Income Plus the Market Rents

I have found tenants who are paying $500 rent for a $400 house. When you see the type of tenants who are willing to do this, you'll often understand why. In one particular case, four occupants were listed on the rental agreement, but I counted 12 coming out one morning, when I was inspecting the property. Sometimes you can get an idea about who lives there by checking out the cars after dark. When I feel that negotiations to purchase a property are going my way, I spend more time driving by and observing what goes on at the property. The word that best describes how I conduct these observations is *"sneaky"*—I don't bother calling ahead for a formal visit. You can't find out what's really happening at the property unless you snoop around.

If you plan to rent your properties to average tenants at prevailing market rents, you must first know, or find out, exactly what those rents should be. You can do this by matching the rents in classified ads to comparable properties in the same neighborhood as your rentals (or soon to be yours). Some valuable information is learned by this exercise! First, you'll learn if the rents for the property you're negotiating to purchase are currently **too high** or **too low**. Hopefully, they're low so you can raise them. Perhaps, equally important, you will quickly discover how much income is the right amount for the property you're after! When you know exactly what the *true market income* should be for a particular property, it becomes a lot easier to figure out how much you can pay for it—and still make a profit for yourself!

The "Don't Bother Me" Landlord

Over the years, I've had good luck finding properties that were *UNDER-RENTED*. Here are two common situations to look for. First, *older owners of small income properties*—perhaps they own six or seven small houses or apartments—where the owner takes care of everything, including maintenance, rent collection and dealing with the tenants. Often, these owners will live in one of the houses on the same property. It's quite common, in this situation, for an owner to allow the rents to remain low, or lag behind the market value, hoping this will keep the tenants from asking for repairs or maintenance. These owners simply don't want to be bothered. Tenants understand full well when their rents are low, if they're smart—and they generally are about their own rents—they don't call the landlord with complaints. They simply let things be. The problem for these owners doesn't become apparent until they sell. Income property selling prices are based on the rental income the property generates. *Low rents means a lower price when it's time to sell.*

FREE AND CLEAR PROPERTIES—NO MORTGAGES

The second situation is somewhat like to the first. *It also involves tenants paying under-market rents.* This frequently occurs when owners have no mortgage payments on the property—it's free and clear (no loans). Quite often, the original owner will pass away, leaving the rental property to the kids. The rents on the property bring in a large net monthly income, because there is no mortgage payment to pay out! Typically, the mortgage on income properties will eat up 50 to 70% of all the rent monies coming in. It's easy for an complacent owner to get careless with property management when the net income is so high—vacancies are not nearly so critical, raising rents annually is not really necessary and routine maintenance expenses are easy to pay from the rental income.

Properties that require 50 to 70% of the rental income for the monthly mortgage payments must be tightly managed or there will be very little, or no, net income. Sometimes, these owners must dig into their wallets each month just to keep their rental operation afloat. Similar to the first situation, with under-rented units—properties generating less net income than they should—must also be sold for lower prices. Sophisticated buyers will expect to purchase them using the *CAPITAL-IZATION METHOD* to determine the value. The Cap Rate Valuation Method, as it's called, is based on the *NET INCOME ONLY*. If the net income is low due to poor landlording, the sale price will be lower as a result. As a buyer, you should look very hard for both *LOW RENTS* and *POOR MANAGEMENT* when trying to find the properties that will generate cash flow quickly—after you fix the problems.

MY FAVORITE PROFIT-MAKER IS ADDING VALUE

At the beginning of this chapter, I told you there are basically two ways to make big money in real estate. I'm about to introduce you to the second way, which happens to be my favorite. This method deserves most of the credit for increasing my personal wealth beyond anything I ever dreamed of when I started out. It's also the method that has made me a very successful real estate investor. It's given me an exciting career and made me financially independent. Naturally, you can understand why it's my favorite. I call this method THE ADDING VALUE STRATEGY. There are many things I like about this method; however, at the top of my list is that it gives me *PERSONAL CONTROL OVER MY INVESTMENTS.*

Investment Control Worth More Than Pretty Buildings

When I buy rundown properties I want to be in the driver's seat. You need to understand exactly how important this really is. For example: Quite often I find myself as the only buyer making an offer to purchase an ugly rundown property. No one else seems the least bit interested in the deal. Why is that, you ask? It's because most buyers judge the value of a property strictly by how it looks. For that reason it's very important to have vision and a plan that goes deeper than the appearance of a property!

Let me offer this advice—***the uglier the deal, the better your chances of buying at a substantial discount***. Equally important, it's very likely you'll be able to obtain softer and easier pay-back terms and seller carry-back mortgages. Obviously, I have the upper hand (control) when I can dictate my own price and terms, within reason, of course. This happens because I buy properties where there is very little competition from other buyers. Remember too, that most of the older rundown properties can be financed, all or partially, by the sellers. **Sellers can and will give excellent terms when they need to**. Banks will rarely do that.

SETTING UP "WORRY-FREE" INVESTMENTS

If I can structure my purchase and the terms so that the property will generate cash flow quickly, and if I can negotiate a mortgage that can be paid off over a long period of time, I'm *HOME FREE* in terms of maximum control. With monthly cash flow and plenty of time to pay off my mortgages—all I need to do is keep the property in good shape and provide proper management. Appreciation or a value increase is not needed to make my plan work; even in poor economic times, I will still have no money problems! My rents are already the lowest in town, so I'm never without waiting customers. I have no short-term balloon notes to be concerned about. Therefore, I'm not under any pressure to sell or refinance. Monthly cash flow gives me many choices—and about as close as I can get to total control. If I should decide to sell, it's on my terms and never the buyers. Can you begin to see the attraction here?

Cash Flow Gives Investors Many Options

If you can "set up" each investment property where you have the kind of control we're talking about here—your investment plan simply becomes a matter of moving forward, one property at a time, until you build up the wealth or income you want. Ugly, rundown properties don't need to be your game plan forever! My suggestion is to buy a few of these properties to start generating monthly cash flow. Later on, with money in your pocket, you can become more selective if you choose. Most people start out investing completely backwards. They get very excited and start by investing in the over-priced, "nicer-looking" properties (pride of ownership). They end up with negative cash flow, and are constantly worrying about short-term mortgages that will require refinancing. Quite often their investment plan depends solely on appreciation or a good economy in order to raise rents or sell for a profit. I can tell you right now, that's not investing—it's *speculating*, or *guessing* what the future might bring. My good friend, Jimmy Napier, says, **"Never let other people control your money."** To which I would add, "If other people control your investment properties, they are controlling your money."

ADDED VALUE STRATEGY BUILDS WEALTH FASTER

Investment real estate is bought and sold by the numbers. Simply put, **the more income a property generates, the more it's worth**. Several other factors, besides income, are considered; for example, single family houses being used as rentals will generally have additional market appeal, because they can be used as either *income generators* or *personal residences*. Home buyers will often pay more than investors, because they're not concerned about the income potential. They want a home to live in and raise their families.

Small groupings of houses and apartment buildings are primarily *income generators*. Investors who buy them are looking for income. The most common pricing method used in the real estate sales business to establish a selling price, or value, for income properties is called the *GROSS RENT MULTI-PLIER*. This method is not scientific, but it's easy to use. Everyone in the business knows exactly what it means. Here's the way it works—if my rental apartments, that are in average condition and are located in an average neighborhood, bring in $25,000 annual gross rents, chances are they will sell for about eight times gross rents (8 x $25,000 equals $200,000). See how simple it is! Remember, eight times GROSS RENTS is average for my town—it's different in other places, so you'll need to determine the multipliers for you own particular area.

THREE VALUE FACTORS WHEN CONSIDERING INVESTMENT

Gross rent multipliers are also influenced by *RISK*, *CONDITION* and *LOCATION* of a property. For example, in my town, a low gross rent multiplier of *five times gross* would most likely mean the property is in bad condition (trashy and rundown) and is located in an older, less desirable section of town.

Be careful when you think about the area! Don't get this confused or mixed up with your personal preferences. Quite often good solid "money maker" rental units can be found behind a soup factory or in a mixed use commercial area. Obviously, you wouldn't normally consider purchasing your home there; however, *income real estate is different!* Also, keep in mind that one day, re-zoning or higher use of the property might land you "a super jackpot bonus." Neiman Marcus may want your lot to build a high-rise department store and parking lot.

GOOD LOCATION EQUALS HIGHER RENTS AND TOP MULTIPLIER

Property in my area selling for a high GROSS RENT MULTI-PLIER—for instance, *ten time gross*—would most likely be top quality units located in the best rental neighborhood. Convenience to schools, stores and parks add to the value of location. Also, a multiple property, with six or eight older detached houses or a small apartment, surrounded by owner-occupied homes, would have excellent renter appeal. Older rental properties are often grandfathered in when new subdivisions spring up around them. Renters will pay more to live in neighborhoods where most people own their homes because generally it's much quieter and cleaner than the so-called rental neighborhoods.

Properties That Work Best for Adding Value

Looking back over many years of investing, my biggest and quickest profits have come from the small multiple-unit-type properties. A typical example was my Hillcrest Cottages which I discussed briefly in Chapter 1. Hillcrest was a 50-year-old motor lodge. I fixed it up and converted it into 21 senior rental apartments. Basically, I paid no money down when I purchased the property for $234,000. (See details in Chapter 14.) I sold it less than two years later for $435,000. Another group of eleven rundown houses on Haywood Avenue earned me $150,000 for one summer's work; when I sold it, I received a "whopping" $50,000 cash down payment, Eight small single-family houses on my Hamilton Avenue property were bringing in $1,680 total monthly rents when I purchased them. Just five year later, my rents had jumped to $3,800 per month. When your rents increase and your biggest single cost or expense item—*the mortgage payment*—remains the same, it certainly doesn't take a rocket scientist to figure out that you're on the RIGHT TRACK to something big!

EARN BIG PROFITS INCREASING THE GRM

A "two-point spread" or difference using football jargon is not very significant. It means that both teams are almost evenly matched, as far as being able to win the game. Odds-makers generally give the hometown team a two or three point win advantage, simply because they are playing on the home town field and have local fan support. If a real estate investor can achieve a two-point upgrade or improvement in the GRM of his income property, it's absolutely amazing how much money we're talking about! *Two small points can make a very profitable difference*. Just two points are enough to provide an overwhelming victory in the game of *FIXING FOR CASH FLOW*. Let me show you how the "fix-up" game is played.

Buyers like me are constantly searching (*doing house detective work*) for properties that are ugly, rundown and poorly managed. These are the kinds of properties where the buying competition is minimal, because most investors simply don't want to bother with them. Deadbeat tenants living on the property help me the most! That's because I'm one of those rare buyers who gets very excited when I see a group of "biker types" overhauling their motorcycles in the living room. I know that in this situation my purchase offer, no matter how weak it might be, will receive serious consideration. Sellers are serious when they are truly motivated! *BAD TENANTS ARE THE BEST MOTIVATORS I KNOW OF.*

A TWO-POINT GAIN NEARLY DOUBLES THE VALUE

Let's say I've learned about the sale of ten junky little rental houses on a half-acre lot near the edge of town, suffering from poor management, and rundown conditions, which always means *low rents* and *deadbeat tenants*. Let's assume my offer to purchase the property at 5.5 times the gross rents of $325 each, or $3,250 per month totals, is acceptable. That means I will buy the property for $214,500, or $21,400 per unit (12 times $3,250 per month equals $39,000 annually; $39,000 times 5.5 GRM for a $214,500 purchase price.)

Let's also assume that I can fix-up the property and upgrade the tenants for about $3,000 per unit, including all the clean-up work and materials. I estimate it will take 12 to 18 months to do this job myself and to turn the property around. *Property turn-around* means getting a better class of tenants and establishing good property management policies.

It's not my intent here to discuss how much the down payment should be, or where to find fix-up money! Rather, let me simply say with regard to both that down payments are substantially less and terms are substantially better with low GRM deals. The following chart will give you an idea of the terms and conditions you might expect for various 10 unit properties with GRMs ranging from 4.5 to 9.5 times gross.

Rent multiplier	Condition	Monthly	Annual income	Estimated selling price	Probable terms required to purchase property
9.5 x gross	Excellent top grade	$500	$60,000	$570,000	$75,000 to $120,000 cash, no trades—firm price
8.5 x gross	Very clean, good shape	$475	$57,000	$484,500	$45,000 to $60.000 cash, trades probably not accepted
7.5 x gross	Medium	$440	$52,800	$396,000	$35,000 to $50,000 cash, possible trades for strong buyer
6.5 x gross	Dirty, needs some work	$375	$45,000	$292,500	$20,000 cash will work, trades acceptable—car okay
5.5 x gross	Junky, ugly, filthy	$325	$39,000	$214,500	Low down $10,000, maybe less—trades-boats-cars-airplane—anything of value
4.5 x gross	Falling down, pig-sty	$285	$34,200	$153,900	-0- down is good possibility, best terms available

Do-it-yourself investors should keep in mind that labor amounts to approximately 70% of the total fix-up cost. In this example we said fix-up costs would be $3,000 per house—or a total cost of $30,000 (10 x $3,000 = $30,000.) That leaves material costs of $9,000, which must be paid as you do the work and purchase materials. I've found credit cards are ideal for spreading out the expenses, just be sure you can pay the monthly bills if you decide to use them.

Looking Good Eighteen Months From Now

Finally, when the job is done, everything looks "spic and span," with nice green lawns (revived weeds) and white picket fences. The "jacked-up" cars are gone now and regular paying tenants occupy your units. Everyone agrees—the property looks great now. It's clean and the location is perfect for renters who need to be near shopping and schools. The nearby pizza parlor is an added bonus. Rents are $440 per month and worth every nickel. Your personal fix-up efforts have **added value** to make it happen.

INCREASE THE RENTS—DOUBLE YOUR EQUITY

It's my experience that keeping good income properties long-term is always best. That's my personal opinion; however, I do realize peanut butter sandwiches and spaghetti dinners get very monotonous after awhile. Also, your credit cards could be extremely close to a "plastic meltdown." Perhaps, you've made the decision that it's best to sell this one, even though the rents have increased by $1,150 per month. No one will ever believe your amazing cash flow, even at cocktail parties your friends will think the drinks are talking. Also, your real estate agent is now running around bragging that he can get you 7.5 times gross rents, if you choose to sell!

I think now is the right time to take a serious look at exactly how much *added value* you've given the property by improving the GRM by just two points. I think it will amaze you! The total rents are now $440 times ten for a total of $4,400 per month, times 12 months equals $52,800 annually. Now take the 7.5 GRM times $52,800 for a total value of $396,000. Remember how much you paid for the property just 18 months ago? (It was $214,500.) Wow! That's a $181,500 increase! Each *one point* improvement in the GRM has been worth $90,750. Now, I'm sure you'll agree that *two points* in the fix-up business is a very significant number, indeed!

Once you're off and running, you must discipline yourself to follow through with what you've started. I'm not trying to tell you this is easy, because it's not! It's hard work and you must learn a few new tricks along the way; but, what is important here is that once you get the hang of adding value, you'll find there's no quicker way to increase your cash flow and build substantial equity. Obviously, banking will be a lot more fun when you do this several times.

OVER-PAYING—THE DEADLIEST INVESTOR SIN

Overpaying for income properties is probably the biggest single mistake an investor can make! Especially investors who lack sufficient knowledge to recuperate from such an error! It's not just new investors or first-time buyers who make this mistake. Many buyers with years of experience fall victim to overpaying when they purchase properties in unfamiliar territory. *The old adage about buying wholesale and selling at retail is absolutely essential if you intend to make any serious money in real estate.*

Here's the bottom line—you can't expect to make money in real estate unless you learn to pick out BARGAINS and buy them at WHOLESALE PRICES. You must lock in profits and future equity build-up at the time you buy. ***Learning to master this technique is not optional***.

Let's suppose you're just starting out like many others who seek my advice. You don't have much money, knowledge or experience. However, you are sold on the idea that investing in real estate will make you wealthy, and you are very excited.

You can't wait to get your feet wet, so you "jump the gun." You race out the door and find a seller who will accept no down payment or very little cash. However, in your haste, you forget what I told you about the price—you pay the seller retail—that's not what you should do—*that's what dummies do!*

If you do start out by paying retail, chances are you'll never make a profit on the deal. To make matters worse, you'll most likely have a *negative* cash flow every month; plus, if you're married, it will certainly put a strain on that relationship. ***The worst mistake income property buyers make, in my judgment, is buying income property that won't produce income***. Overpaying is like a fatal disease for many investors; unlike other fatal diseases, science is not working on a cure for this one. So, you must protect yourself by following the SIMPLE RULES I'm suggesting.

CHAPTER 12

Where Do All the Profits Come From

Buying real estate is easy—becoming financially indepen-dent, doing it is another story! To begin with, many investors simply don't understand where profits come from or what they must do to produce them. Many are of the opinion that proper timing is the most important ingredient in making real estate millionaires. Others think patience is more important for success—simply buy choice properties in an appreciating market and sit back and wait. "You can't go wrong," they say!

PLAYING THE APPRECIATION GAME

The biggest question for real estate investors who bet on appreciation is—*WHAT IF THERE IS NONE?* Worse yet, what happens if property values drop? Real estate is a *cyclical business*; prices and values go up and down! Betting on short-range appreciation is like shooting craps—you can sometimes double your money overnight, but you can also lose the ranch just as quickly.

Reaping the benefits of appreciation, without having to worry about timing, is a much safer strategy. *Real estate will almost always increase in value over time*. For example, consider a three-bedroom, single-bath house in Sacramento, California that cost $20,000 brand new in 1968. It reached a high value mark of $181,000 in 1992, and then dropped to $160,000 by the end of 1994. For short-term-profit investors, most of the '90s have been declining or standing-still years. However, over the long haul, the Sacramento house has appreciated nearly $5,400 each year since it was built.

Appreciation Should be the Bonus

Appreciation should be a wealth-builder's "helper," not the whole program. If you view it as a bonus, or the gravy, you don't have to worry about short-term up-and-down cycles. Instead, you can concentrate on how to make much bigger profits in the long-term.

To begin with, you need to understand that most wealthy real estate investors buy more properties than they sell. Selling properties, especially when you're just starting out, is like digging a big deep hole and then filling it back in. The reason is because the minute you sell a property, your investment stops earning you money. Worse yet, any gains or profits you might make are immediately taxable. Obviously, paying taxes and building wealth don't go too well together.

THE MAGIC OF COMPOUNDING

Compounding real estate equities works the same way as your savings account, only better. That's because you can use something called *leverage* to speed up the money-making process. We'll talk more about leverage later, but first, I want you to fully grasp just how powerful "compounding" really is!

Compounding means earning interest on both the principal amount and on the accruing interest. As it keeps adding onto itself over time, the results are simply astonishing! Most wealth-builders are amazed when they discover how quickly money grows and multiplies. The following example serves to illustrate the power of compounding.

Let's say you deposit $1,000 into your bank account every month for 20 years, never drawing any money out, and you earn 12% interest on all your deposits, plus the accumulating interest—**THAT'S ENOUGH TO MAKE YOU A MILLIONAIRE** (well, almost anyway). You will end up with **$989,255** in your bank account at the end of 20 years.

Saving $1,000 a month might sound like a ton of money when you're just barely making ends meet. But, $12,000 a year is not really a huge amount anymore. Many folks have mortgage payments twice that amount and vehicle expenses that equal it.

HOW CAN JUST 20 YEARS OF INVESTING $12,000 A YEAR MAKE ANYONE A MILLIONAIRE? After all, 20 years times $12,000 is only $240,000. Where does the rest of the money come from? It comes from the 12% interest compounding— and, all you need to do is make your $1,000 deposits and leave the money in the bank to continue earning interest. In 20 years the interest alone adds up to $749,255. It goes without saying, compounding is very powerful, and it will do the same thing with $1,000 worth of real estate, only better!

Killing the Golden Goose

The magic of compounding comes from leaving your investment alone! In other words, if it's money in the bank, leave it there and don't touch it! If it happens to be four rental houses, instead, don't sell them for a short-term gain. I must warn you right now—the short-term profits you earn will never be enough to make up what you'll lose in the long run if you break up your compounding cycle.

Disturbing the cycle in the early years will cost you dearly. I'll show you why. At the end of the second year in our 20 year investment plan, the accumulated amount of principal and interest earnings would be approximately $27,000. If you withdrew $10,000 of that amount, it would result in a $230,000 loss by the end of the 20 years. Your accumulated total would only be $757,860, because you withdrew $10,000 in the second year of the plan.

In later years, as compounding builds, the account and the dollar amounts get larger, and $10,000 will be much less significant in terms of slowing down the earning power. For example, in the 17th year, the account balance would be $661,300, and will grow to $757,800 in 18th year. If you were to borrow $10,000 at this point, it would hardly be missed! That's because the annual earnings have nearly reached $100,000. During the 20th year, the account will earn a total of $123,000, which is more than half the total amount you actually invested during the entire 20 years.

Buying properties to fix up and sell for quick turn-over profits might seem like a great idea to some. But, it doesn't help you build the lasting kind of wealth most investors are seeking. The reason for this is because selling does not allow long-term compounding to work at full strength! When you constantly buy and sell properties for short-term gains, you wind up losing a ton of money in the long run, as our example of withdrawing $10,000 in the second year shows. Don't forget that a small $10,000 withdrawal costs $230,000, in the long run.

NOT JUST ANY REAL ESTATE WILL WORK

In order to maximize compounding, it's necessary for you to select the right vehicle to get you there! Not just any properties will do. ***There are four basic ingredients that produce profits***.

1. CASH FLOW EARNINGS

2. TAX SHELTER BENEFITS

3. EQUITY BUILD-UP

4. APPRECIATION/INFLATION

Keep each one in mind as you search for the right investment!

Cash Flow Earning

Cash flow is the most important benefit for any investment. Without it, you'll eventually be forced to sell the property, or worse yet, you'll lose it! Cash flow gives you the freedom to keep your property through ups and downs in real estate cycles, without being forced to sell at the wrong time. Selling in a low cycle can easily cost you $30,000 on a $100,000 investment. Believe me, you don't want many sales like that!

As a general rule, when real estate is hot (at the top of a cycle), it's not at all difficult to sell property for 115% of the normal price! However, when the cycle hits bottom, 85% sales are more common. On a $100,000 deal—earning an additional 30% profit is certainly a worthwhile objective. Cash flow earnings of 12% is a very reasonable expectation for investors who buy rundown properties and add value to them. Don't forget what 12% compounding does for your bank account.

Tax Shelter Benefits

Depreciation is the magic expense item that causes your property to show a loss for tax purposes, but still generate positive cash flow! The reason for this is that the IRS allows investors to deduct—as an expense item—a certain percentage for things that wear out, like buildings, coolers, refrigerators and carpets. This expense is really a "phantom" expense, because you don't need to write a check to pay it, like replacing the toilet.

The cash benefit comes from two sources. First, the depreciation expense will shelter the property income from taxes—for example, if the rental property generates $2,000 positive income, before deducting a $4,000 depreciation expense, the property will show a $2,000 loss for tax reporting purposes. Second, the same $2,000 loss against the property can be used to offset, or eliminate, taxes on $2,000 worth of income from owner's salary or from another positive income source.

Equity Build-Up

Equity build-up adds to your wealth on a monthly basis each time you pay the mortgage payment. Although very small at first, a certain percentage of your mortgage payment goes to reduce the principal balance. With each principal reduction payment, you own a little bit more of the property—or, stated another way, *THE BANK OWNS LESS, SO YOU OWN MORE.* Typically, on a 25 year mortgage, with an interest rate of 9% and a 20% cash down payment, equity build-up amounts to something like 3 or 4%, annually. As you can see, however, there are many variables. Obviously, in those cases where only interest is being paid, equity build-up is not occurring from a mortgage pay down.

In certain situations where you assume or take over a mortgage with only half of the original payments left before the mortgage is fully amortized, principal reductions will be substantially higher. But, regardless of whether they're higher or lower, the important thing to keep in mind is that *your tenants are paying off the property for you with their monthly rent checks*. They buy the property and you own it. You can't beat that!

Appreciation or Inflation

Appreciation or inflation comes in *two different flavors!* First, there's the kind that comes from natural causes. Everyone gets a taste automatically if they happen to own properties in an area where appreciation or inflation is happening. Sometimes it's as little as 2% a year, but I've seen inflation jump to 25% or 30% for a short period of time. I've also watched properties nearly double in price in just two years. The problem with natural inflation is that there's no guarantee it will happen! And of course, that's why I recommend you treat it like a bonus or an extra benefit. It's not wise to build your investment plan on the assumption that it will appreciate. Just be ready to reap the profit when it does.

I FORCE THEM TO APPRECIATE

Forced appreciation is my specialty—and, I can count on it to work, because I have complete control. For example, when I buy rundown houses and fix them up, I am forcing them to appreciate. The higher value comes from being able to attract better tenants who, in turn, are willing to pay higher rents! I can do this because I intentionally upgrade the properties from the rundown condition they're in when I buy them. It's simply a matter of being able to provide a better product to my customers (renters). On several occasions I've increased the property values by nearly 100% in just eighteen months of ownership. You can do the same thing for yourself—and when 100% starts compounding, it won't take very long to add a couple extra zeros to your net worth. You'll also find that smiling at your tenants becomes a little easier!

Leverage Lets You Soar With the Eagles

High leverage can make you richer faster than any investment tool I know of. The idea is to safely borrow as much money as you can to put with your own down payment (if you have one) to purchase income-producing properties. For example, 90% leverage is where I purchase a $100,000 apartment using $10,000 of my own cash for a down payment and sign a promissory note or mortgage for $90,000 back to the seller. If the property earns $10,000 annual rents that means the return on my cash down payment is 100%! The problem is, that can be good or bad! If the expenses are $4,000 and the mortgage payments are $7,000, my 100% return doesn't mean much! I'll still be losing my shirt.

Leverage is a double-edge sword! Safe leverage is the kind you want! In this particular case, if I can increase rents to $12,000, or negotiate a mortgage that would cost only $5,000 annually, I would then earn $1,000 on my $10,000 investment. A 10% cash flow, using 90% leverage is a very respectable return, especially for apprentice investors.

Not Everything can be Measured in Dollars

A rich man once said: "Money is not the most important thing in the world—but, it's still a long ways ahead of whatever's in second place!" I certainly won't disagree that money is very important, but it is not the only measurement of an investor's success. There's are many very attractive benefits that come with an investor's lifestyle. Most folks, however, never get to experience them, so let me share a few with you.

To start with, I don't need to operate my real estate business on a rigid time schedule, like most of my friends who work regular nine to five jobs! Since I've been there, I know the routine. My schedule is far more flexible and if you try it, I'll guarantee you'll like my way better. Also, I'm not stuck in a dead-end job going nowhere fast or working for a "nit-wit" boss who should be my helper. I'm also not part of that morning madness on the freeways, packed with hundreds of cars moving about half the speed I normally walk. That alone makes it worth dealing with my tenants—who generally act more civilized than early morning freeway drivers.

AN INVESTMENT SHOULD WORK FOR YOU

It's difficult to give each of my lifestyle benefits a specific dollar value. However, each one provides a good measure of comfort to me. In his excellent newsletter, the "MONEY-MAKER REPORT," my good friend, Jimmy Napier, often writes about his personal business philosophy. In so many words, Jimmy says that his investment goal has always been to get his properties *WORKING FOR HIM* so he has something that earns income besides doing the work himself. He is referring to his sources of income—renting houses, buying notes and even leasing out his bus for an escort travel service. Jimmy and I may communicate things slightly different because we live in different parts of the country, but I still like what he says. It's excellent advice and that's always been my goal too!

BRAIN COMPOUNDING CAN INCREASE YOUR WEALTH

When I started investing in "fixer-upper" houses, my simple plan was to buy them cheap (I had very limited funds), fix them up and rent them out. Selling was never my first choice for making money! I always felt that having a continuous income from my investments was the best way to go. At that early stage, I was not yet aware of all the various strategies I could use to make extra money with my houses. *Renting* or *selling* them was all I knew about! As I went about fixing houses, I began reading all the books I could find about real estate investing. I soon learned new and different ideas about how to invest. You must continue your financial education as you go along—otherwise you'll get stale!

Reader's Digest (May 1973) published an article, by Edwin Diamond, entitled "Can Exercise Improve Your Brain Power?" In it, Diamond said that through selected mental exercise, like reading, you can actually increase the capacity of your brain to make it function better. Mental exercise, such as reading real estate investment books and learning at seminars, can actually cause your thoughts and ideas to compound much like the compounding of money!

That's exactly what happened to me! Brand new ideas and methods of acquiring real estate filled my thoughts. A good example was when I first came up with the idea of "LEMON-ADING"—buying properties by using a combination of cash and personal property I no longer had any use for (details in Chapter 23). It works extremely well when you find sellers who don't want their rundown real estate anymore and are willing to trade. This technique allows a small amount of cash to go much further, like hamburger helper. It can double or triple your buying power!

Before I began reading and educating myself (brain compounding) I had no idea I could present lemonade offers and get them be accepted! I never realized I could buy back my own mortgages at big discounts. When I first began fixing houses, these techniques simple never occurred to me. When I look back today, I realize *BRAIN COMPOUNDING* has been a major contributor to my accumulation of wealth.

DON'T WALK AWAY FROM YOUR GOLD MINE

I'm aware that most new investors tend to follow along the same path I did! First, they learn how to acquire income-producing real estate, and, of course, that's a good start. However, what most of them *DON'T DO* is follow through and expand their money-making opportunities! Many small-time investors miss out on real profits, similar to inexperienced gold miners who, after easily finding shiny nuggets on the surface, leave the real fortune hidden by only a few inches of sand. Brain compounding happens when you continue to seek more knowledge. It will make your investment houses produce far greater yields than you can ever imagine, believe me!

ADDING NEW PROFIT BULBS ON MY MONEY TREES

I often refer to my investment houses as "MONEY TREES." I call each of my various profit-making strategies "PROFIT BULBS." In the beginning, I had just my bare tree with two lonely bulbs—namely, profits from my rents and, occasionally, profits from a sale. Those were the only two profit strategies I knew anything about. That was back before I knew anything about *BRAIN COMPOUNDING,* when my knowledge was very limited, and before I started reading and attending seminars.

Over the years, I've managed to decorate my "money trees" with many more "profit bulbs!" Obviously, the more bulbs I add to each tree, the more it glitters like gold. As we discuss these various profit bulbs, keep in mind that my basic tree has never changed. It's the same old tree (fixer houses) I started with, but now, with the addition of many new bulbs over the years, my gold mine has become much more productive.

Profit Bulbs Provide the Golden Glitter

My first *PROFIT BULB* has always been my rental income. Although the IRS refers to rents as passive income, I don't, and neither will anyone else who manages tenants and collects the rents. Still, it's my best source of continuous income every month. Once you develop your landlording skills, you can easily net 5 to 10% of the gross rents, even with leveraged properties. I once owned 216 houses with average rents of $385 per month. You can see that monthly income adds up rather quickly, even with very modest rents. *It's also important to remember that when you're doing this correctly, tenants are actually paying off your houses for you!* Realizing this has always made me feel much better at the end of a hard day fixing toilets.

FIXER JAY'S FAVORITE PROFIT BULBS

SELLING PROFITS is certainly a bulb with a whole lot of glitter. However, as I said earlier, I still favor keeping most properties and allowing them to earn continuous monthly income for me. About the only time I'll consider plucking a rental bulb from my tree is when a buyer gets very motivated to purchase my property, regardless of the price. Naturally, I can deal with a few less bulbs on my money tree when there's serious profits involved. This will most likely happen during a hot seller's market or when I'm weeding out properties that I don't think will meet my long-range profit expectations.

SELLER FINANCING is a profit bulb that will earn you a lot of money. It's available when you sell properties and carry back the financing yourself. Interest income is *EASY MONEY* and costs very little effort, except walking down to the mailbox to pick up the check. You can design your carry-back mortgages to be long-term when you're ready to travel around the world or retire. Also, when you're nervous about a buyer, or the down payment is a little on the thin side, you should insist on additional collateral to protect yourself (see Chapter 14 for the details).

BUYING BACK YOUR OWN MORTGAGE DEBT is a very high-profit bulb. When I first started buying rundown houses, I had no idea that property owners could do this! I was in shock (but happy) when I asked for and received a $29,000 discount for simply paying off my mortgage seven years earlier than I had originally promised to do. That's over $4,000 a year for doing nothing—you can't make money much easier than buying back your own mortgages at steep discounts.

HALF SALES are another excellent profit bulb on my money tree. Rather than selling a good income-producing property—giving up all the depreciation and effectively selling the goose that lays golden eggs—why not just sell half the property? That's enough to sustantially improve your cash flow! The basic strategy here is to purchase a rundown property with low rents for $100,000, and fix it up until it's worth $200,000, with higher rents, to justify the increased value. Then, sell half the *FIXED-UP* property to a passive-type investor for little or no cash down payment. This makes a hard-to-refuse proposition.

By doing this, I'll wind up with a ***mortgage receivable*** for the half I sell! This means I have mortgage payments coming in every month. Also, I'll be able to increase my personal income by collecting a ***monthly management fee*** and ***payments for maintenance and repair labor*** on the half I don't own. This is one of the slickest strategies I know of for making negative cash flow properties produce positive income. It's an excellent strategy for investors who can do the kind of fix-up (adding value) we discussed in Chapter 11. You can read more about HALF SALES in Chapter 16.

CREATING NOTES FOR DOWN PAYMENTS—this profit bulb can make your money tree glitter even brighter when you already own properties with equity, but lack sufficient cash to keep buying new ones! You can simply type up your own promissory note or mortgage—and use it for your down payment—simple, huh!

This strategy allows you to *select the note value*, the *amount of your monthly payments* and the *length (term) of the mortgage*. "How can I do that?," you're probably wondering—because you are the preparer and it's your typewriter—that's why.

Many times after fixing up rundown properties, I've increased the values from 50 to 100%. That means I've created ***borrowing equity***. This new equity will be security for the promissory notes I create. For example—suppose I purchase a property for $50,000 and increase its value to $100,000. I've created at least $30,000 worth of new borrowing equity. I will still maintain a 20% loan-to-value safety margin, like the banks do! This technique works extremely well when combined with *lemonading*—you'll recall that's where my offer includes some cash (the sugar) and something other than cash (the lemons). This combination is often irresistible to a motivated seller who can't find anyone else to purchase his ugly "scum-bag" property.

LEASING HOUSES WITH AN OPTION TO PURCHASE is another of my favorite profit bulbs. I use this technique to increase the monthly cash flow for my larger, more expensive houses. The smaller houses I own have excellent rent returns without offering leases or the rights to purchase. For example, I can rent smaller two-bedroom houses, with 600 square feet, for about $425 a month—that's 70¢ per square foot. My nicer houses with three bedrooms, two baths, double garages, and approximately 1,300 square feet of living space, will normally rent for $650. That's only 50¢ per square foot and not quite enough income for the exposure, in my opinion!

The lease option remedy allows me to collect higher monthly rents ($800 to $850 range) in exchange for giving potential buyers (optionees) a higher than normal rent credit back toward the purchase price if they exercise the option. My leases are for three-year terms and my selling prices are generally somewhat higher than appraisals to begin with. *I have never yet met a house buyer who can tell whether the value is $95,000 or $105,000.* I'm sure I don't need to tell you which price my option contract will specify.

BUYING DISCOUNTED NOTES OR MORTGAGES is another high profit bulb on my money tree, one that I accidentally discovered along the way! Buying discounted notes or mortgages can be a potential double-barreled opportunity for hands-on fix-up investors like myself. The reason is because I would always prefer to own *THE PROPERTY* that secures the note or mortgage rather than owning the mortgage. If I should take the property back for non-payment, I'll also get the rental income and tax shelter, plus a 40% discount the day I take over! You couldn't possibly design a better purchase deal than that!

I have **two questions** I must answer for myself when a note or mortgage buying opportunity is available to me.

1. Would I be happy to own the collateral, meaning the property which secures the note?

2. Is there an adequate equity cushion—that's the difference between the property value and whatever is still owed against it (total mortgage debt)?

If both answers are "Yes," I'll move forward with the deal! Obviously, there's a good opportunity for me to earn even bigger profits should the note default and I end up foreclosing like we discussed above. Again—If my answer to both questions is yes, I'm always happy to purchase the note.

My debt limit rule for income-producing properties is 65% *LOAN-TO-VALUE.* That means, if there are three existing notes or mortgages on a $100,000 property and they all add up to no more than $65,000, I'm generally willing to purchase any one of the notes at a reasonably discounted price. As you can see, buying discount notes or mortgages is really TWO PROFIT BULBS IN ONE—to start with, if everything goes well, I'll receive high yield note payments; and if it doesn't go well, I'm more than happy to take over a $100,000 income-producing property for 65% or less of it's full market value.

Every fix-up investor should explore this profit bulb opportunity very closely, so they can fully understand what I'm saying here. This bulb can truly add "extra padding" to a skinny bank account—believe me!

CHAPTER 13

Negotiating Deals That Earn Big Profits

*I*nvestment real estate might be created equal, but I can assure you it doesn't stay that way! It's easy for anyone to see this by simply taking a 30-minute drive around the area where you live. You'll see luxury apartments, new construction, older houses, junky properties and ugly pig-sties. I'll guarantee you this mixture exists in nearly every city or town across the country. If a 30-minute drive is not enough, take an hour, you'll find them all!

AN AMAZING DISCOVERY

Sometimes I ride around with my seminar students to offer advice and to, more or less, prove what I just said about the investment mixture. At the conclusion of these trips, I ask each student which of the properties we looked at would they be interested in buying for their own personal investment? Almost without fail, they pick the wrong properties! The reason is that their decision is nearly always based on how the property looks. This basic human characteristic is what gives me a distinct advantage over my competition when it comes to making big money with older, unattractive properties, known as "FIXER-UPPERS." Let me show how my buying advantage works.

To begin with, the folks who sell about 95% of all rental properties have their own numbering system when it comes to evaluating what income-producing properties are worth. It's far from scientific, nevertheless, it's the measurement that almost all real estate agents use. It's called the GROSS RENT MULTIPLIER (GRM) or sometimes *NET RENT MULTIPLIER*. We discussed this in Chapter 11.

Property value is easy to calculate using GRM. It simply says that if an income property generates "X" number of dollars, annually, you can multiply those dollars by a number called the GRM—which is based on the LOCATION and CONDITION of the property—and you've established its value. For example, let's say the GRM is seven and the gross annual rents generated are $20,000. The indicated value would then be $140,000 (seven times $20,000). As you can probably guess, a calculation like this is not exactly suitable for brain surgery, but I can assure you, it's good enough for making big bucks with investment real estate.

To give you a better understanding of how GRMs work, I've prepared several *MAKE-BELIEVE,* "FOR SALE" ads, similar to those you'll find in most newspaper classified sections. These properties range from a low multiplier of 5 times GROSS RENTS to a high of 10 times GROSS.

DESCRIPTION OF PROPERTY FOR SALE	GRM
Pride of ownership "PREMO." One of a kind. Garages with openers, no deferred maintenance, waiting list of qualified tenants, best location, highest rents in town.	Sale Price: 10 x gross
Newer units. Landscaping, pool, easy drive to stores, schools, shopping. Shows good upkeep, covered parking, top rents for area. Low vacancy.	Sale Price: 9 x gross
Solid rental units. Good location, near freeway and bus lines. Fenced lawn area for privacy. Numbered parking. Competitive rents.	Sale Price: 8 x gross
Established neighborhood, 24 year old property, shows well, located near jobs at factory and county hospital. Always full, special rates to couples. Owner may finance part.	Sale Price: 7 x gross
Exterior needs paint, but structurally sound. All units upgraded, new carpets, painted within past three years. Some deferred maintenance, otherwise perfect for wise investor who can spot potential savings.	Sale Price: 6.5 x gross
Dirty, but great potential for high profits. Retiring owner says rents should be raised substantially. Rare opportunity for astute buyer to do minor repairs and maintenance and enjoy big rewards. Good terms with owner financing available.	Sale Price: 6 x gross
"Handyman special"—bring your paintbrush and hammer. Low down payment for investor with vision. Low rents should be raised. Owner will finance. This property won't last long.	Sale Price: 5.5 x gross
Dirty, but cheap—great investment with big upside potential. Rents have not been raised in four years. Land value will triple when city extends sewer line. No legitimate offer refused.	Sale Price: 5 x gross

STRIPPING THE PUFFERY FROM "FOR SALE" ADS

It will pay you to familiarize yourself with the kind of jargon real estate agents use in their classified advertising. As you read the property descriptions (typical sale ads) you will notice that saying GOOD THINGS about properties become more and more difficult as we descend from a 10 GRM to a 5. As with all advertising, a certain amount of "puffery" can be expected. However, once you get proficient at reading between the lines, you'll become quite good at guessing the condition of a property and how weak the seller might be. Naturally, you are looking for a motivated seller who will give you liberal terms and who is in a position to carry most of the financing (seller financing).

In case you're wondering what I'm leading to here—let me re-focus on PROFIT-MAKING and CONTROL. In order to make profits from selling, you must first buy properties for less than their potential market value. I'm not talking about waiting for appreciation here! I'm talking about buying property at *wholesale prices* that will insure a built-in profit when it comes time to sell them (assuming you wish to do that).

In addition to buying properties wholesale, you must acquire them with financing that will allow you the freedom to sell quickly when the opportunity arises. Quite often bank mortgages that come with newer properties are not assumable to anyone without the bank's permission. **This restriction limits your control** and it's one of the primary reasons why owner financing is generally much better for the do-it-yourself wealth builder.

If you acquire income properties for the purpose of immediate cash flow from rents—and, long-term appreciation for later on, the sale terms are extremely important! First of all, you must have **long-term mortgages**. If you don't, you may be forced to refinance at a time when no one will loan you money, or, just as bad, when interest rates are very high. Either way, there's too much risk involved. Short-term loans (less than ten years) cause you to sacrifice too much personal control, in my view. An unsympathetic mortgage holder can take your property away from you, if you can't pay, no matter how much, labor and money you've spent fixing it up. Underline this paragraph and re-read it often so you don't become a victim!

An important part of my profit-making strategy is finding sellers who are capable of giving me the kind of terms I need, so I'll have maximum control after I acquire the property.

WHAT IS NEEDED FOR MAXIMUM CONTROL

What I mean by "maximum control" is that I'll be in a position to sell the property anytime I choose without having the mortgage interfere with my sale. I don't want a mortgage that requires a pay-off if I sell the property! Also, I don't want a short-term mortgage (less than ten years), because potential buyers don't like "short-fuse" financing. By short-fuse I mean that if they buy my property, they won't have much time before a balloon payment is due or a refinance is necessary!

These kinds of terms are too restrictive, because they limit my selling opportunities; plus, they also hamper any ability to negotiate the highest selling price. Stated another way, *IF I CAN SELL MY PROPERTY AND OFFER THE BUYER ATTRACTIVE TERMS, IT'S WORTH AT LEAST 20% MORE TO ME AT THE BARGAINING TABLE.*

Another term or concession I must have when I purchase property is a mortgage payment I can live with! A good rule of thumb is that my mortgage payments should never exceed 60% of the total income the property earns on the day I close the deal. As I already mentioned, I don't want mortgages with less than ten years before they're all due and payable. I always ask for 20 years when I begin negotiating.

When you can acquire properties with long-term mortgages and low payments, you are in a position to offer the same easy terms if you decide to sell. If you don't sell, low mortgage payments will help you generate cash flow every month. Either way, you're a winner!

Long-term mortgages are especially important to fix-up investors, like myself, who earn their biggest profits by adding value to the real estate. It could be very tempting for a ring-wise seller who has carried back a short five-year mortgage to quickly foreclose on a much improved property, if you are unable to pay off his mortgage that soon! **Never count on being able to refinance when you need to!** Instead, negotiate longer-term mortgages to begin with. Concentrate on those sellers who can give you the terms you'll need, so you'll always have maximum control. ***Don't forget this part***.

WHAT THE CLASSIFIED ADS ARE TELLING YOU

Classified advertising is the marketplace for everything, and, of course, that includes most all real estate that's offered for sale. In my seminars, I teach students that classified ads are one of the four primary sources I use for finding bargain prop-erties. I also point out that you usually can't tell if the ads are placed by sophisticated sellers or by "Mom and Pop" types who are wishing to sell out and retire. The key, I've found, is figuring out which ads are really saying "PLEASE, I NEED HELP."

I will shorten this discussion by telling you it won't be the same ads that are "pitching" the high gross multiplier proper-ties. Those ads are written about strong properties, by "hard-nose" sellers! Take a moment and re-read the first two ads, describing properties selling for 9 and 10 times gross, on Page 226. There is no sign of any kind of help needed! Statements like—"NO DEFERRED MAINTENANCE," "WAITING LIST OF TEN-ANTS," "HIGHEST RENTS," "SHOWS GOOD UPKEEP," "TOP RENTS" and "LOW VACANCY" are not terms that conjure up thoughts of weakness. Don't be afraid to call and check them out just to keep things honest. But don't be offended if the sellers shine you on! Sellers of premium-grade properties don't need to negotiate much. Generally, it's their way, *OR NO WAY AT ALL*.

I Like "For Sale" Ads That Admit Problems

Take a look at the ads describing 5, 6 and 6.5 times gross multiplier properties, on Page 226. These ads are much more

humble! They say things like "DIRTY," "CHEAP," "NO LEGITI-
MATE OFFER REFUSED," "HANDYMAN SPECIAL," "LOW RENTS,"
"OWNER FINANCING," "NEEDS PAINT" and "WISE INVESTOR
WHO CAN SPOT POTENTIAL." These are all the kinds of words
and statements that suggest the seller may need some serious
help!

Again, let me remind you, many ad writers speak with
forked tongues. Still, it doesn't hurt to call the ads that sound
like the ones in my examples! In the long run, you will find that
ads that admit something is wrong, will produce much more
favorable results for your calling efforts.

I don't suppose I need to tell you that my initial offer on a 5
times GROSS property will be 4 times GROSS or less. I always
like to see how sellers react before I give them my best offers.
Naturally, I recommend you do the same.

YOU MUST LEARN THE FACTS BEFORE MAKING OFFER

There are many books written on the subject of negotiating!
Some are good, others are nothing more than ego, combined
with "hot air," especially those that suggest various forms of
intimidation! Any method that "talks down" to sellers or, in
some way, leads them to think they are dealing with "Mr.
Slick" is not good business. Negotiating, while important, is not
what makes you rich in real estate. It's much more beneficial
to establish your creditability with a seller, than to attempt to
talk him out of his shorts.

Worthwhile Benefits are What You Need

There are several good financial reasons to establish creditability with a seller. However, none are more important than
persuading him to finance the sale, or at least finance a large
part of it. That's called owner financing or a purchase money
mortgage. Owner financing is a benefit that can make real
estate investors a whole lot richer in a shorter period of time.
Before we discuss the kind of negotiating to accomplish this,
let me first share something Dale Carnegie said, which puts
the right "spin" on the buyer-seller relationship, I think!

"THERE IS ONLY ONE WAY TO GET ANYBODY TO DO ANY-
THING—THAT IS BY MAKING THE OTHER PERSON WANT
TO DO IT. THERE IS NO OTHER WAY."

REAL PROFITS DON'T COME FROM PLAYING GAMES

Negotiating is sometimes nothing more than game playing! It's done by appliance dealers, horse traders and real estate agents. It's most often like a ritual, and sometimes it has very little effectiveness.

Here's how it typically works. Let's say you ask your real estate agent how much the price is for a particular property, and he tells you the seller wants $100,000. You ask, "What do you think he'll take?" Your agent says, "Let's offer $85,000 and see how he responds." The seller counters with $95,000, to which you quickly bounce back with $92,500. The seller says, "Okay—where do I sign?" When it happens fast like this, you always think you paid too much! The real question is—did you? Negotiating isn't worth all the effort, unless you first know exactly what you need, otherwise it's only game-playing.

Develop the Right Approach

One of the best methods for negotiating is what I call the **"Columbo Technique."** Lieutenant Columbo is the cigar smoking detective on TV with the wrinkled raincoat. He doesn't appear smart enough to ever solve a homicide case, yet he always does!

If you get the opportunity, watch Columbo solve a mystery. Observe how he does it. Let me offer you a suggestion of things to watch for! First of all, you should observe that Columbo is never intimidating or threatening. He never appears to be competing with anyone, yet in his own special way he quietly and forcefully moves directly toward his goal, which of course, is solving the homicide.

DETECTIVE STORY OFFERS EXCELLENT TRAINING

Real estate investors can negotiate in a very similar fashion, with equally successful results! You'll notice that Columbo always has a specific reason for everything he does. His questions are always supported by clues or information he develops. His attitude is courteous and often seems apologetic when he needs to ask the tough personal questions. He always gathers his information in such a manner that he doesn't intimidate possible suspects. In fact, most of them even go out of their way to cooperate with him because he's courteous and seems genuinely concerned about their needs. Notice how all the villains in Columbo episodes are steadily pressured into accepting surrender without putting up much of a fight. (I've never seen Columbo use a gun.) The reason is because the Lieutenant develops the facts, confronts his suspect with hard, indisputable evidence, which makes a conclusion very obvious. He does it all without allowing his personal emotions to interfere!

Good Reasons for Having Two Ears and Only One Mouth

A major part of Columbo's success in solving mysteries comes from his listening. Columbo asks short questions, then waits for long answers or explanations. That's a very effective method for learning about other people and it's just as true in real estate negotiating as it is solving homicide cases. You're probably wondering what all this Columbo business has to do with negotiating to buy real estate? "Has Jay been watching too much TV?," you are probably asking yourself, "His channel must be stuck!" Allow me to show you the connection. I think you'll get the picture.

Practicing Columbo's Winning Ways Can Help You

First, before you start arbitrarily changing or negotiating terms and/or conditions, make sure you can show the reasons why. If you can't, they're probably not valid; worst of all, proceeding without being able to show why, can do severe damage to your credibility. You don't need that, believe me!

Second, it's very important to listen to the other side. You can answer yes or no if you train yourself, but listening to others will provide you a wealth of knowledge and information that will help you structure offers or counter-offers. People love to talk! If you're a good listener, you'll be very popular with most folks. Columbo is very good at this, and suspects often tell him enough to hang themselves. This makes his job much easier, wouldn't you say?

Third, never get emotional. Don't be critical and, above all, never "talk down" to anyone. If you humiliate, embarrass or ridicule, you'll lose all chances for negotiating a winning deal. Even sellers who are about to lose their shirt won't do business with someone who intimidates or tries to overpower them. **Courteousness** and **understanding** are two of the most powerful tools in your "negotiating kit." Use them generously, they'll pay big dividends. Once again, Columbo is a master at this!

WINNING OVER THE SELLER LEADS TO WINNING NEGOTIATIONS

Consider the following exchange between a seller (owner) and a potential buyer, who were previously unknown to each other.

Seller: *Mr. Buyer, why are you offering me 25% less than my asking price?*

Buyer: *Well sir, the reason is that I know sellers always "mark-up" their sale price because they expect offers to be less! Obviously, you can still accept my offer because I think it's much closer to the real value of your property.*

If you own a nice piece of property (your opinion) and you have worked very hard to fix it up and make it look attractive. Then some buyer, a total stranger to you, tells you your property is worth 25% less than your asking price. What would you immediately think of him as a person? Chances are, your answer is not printable in this book! No one likes to be told their property is worth less than they think. That's human nature. It hurts your negotiations—even if it's the truth!

Columbo never tells his suspects anything that would alienate their relationship or stop the flow of information between them. In the end, they always end up giving him all he needs for a conviction. He wins without shouting matches, embarrassment to others or making enemies along the way. These are excellent skills for any negotiator.

AVOID GAME-PLAYING IF YOU WANT REAL BENEFITS

Arbitrarily cutting the asking price is "game-playing," especially for the reason given above! Obviously, if the price is truly "too high," it does need cutting. However, you should have a justifiable reason to cut it. The point is—if you have done your homework, and if you can reasonable demonstrate the reason why—your lower purchase price offer has merit; otherwise, it probably does not!

When your offer is considerably less than the asking price, and you can't explain the reason, it becomes very difficult to keep negotiations on a productive level. It's sort of like saying to the seller, *"You really don't know what your property is worth, but I do!"* If you mean it that way, you need to be in a position to tell the seller why.

I like to consider the terms I'm negotiating as if each one is a bullet in my gun, and I only have so many shots. If I shoot blindly at everything—that is, I attempt to change everything in the contract to suit myself, like asking for a lower selling price, smaller payments, reduced interest rate, longer length of mortgage term, less cash down payment and even a different closing date—I should ask myself this question, "Am I shooting at the right targets, or am I just wasting my shots, hoping to hit everything I can?"

DON'T WIN THE NEGOTIATIONS AND LOSE THE PROPERTY

I have known buyers who have done quite well at persuading sellers to reduce their asking price, only to find out later that the price was still too high! If negotiating is to serve a valid purpose, that purpose must be to make the deal workable for everyone. It should not be a contest to determine who can have his way the most, otherwise you'll end up like the little boy scout who cut the rope three times—*and it was still too short!*

What Really Counts for Every Investment

Here's some advice you should pay particular attention to. It's worth big bucks to every "small-time" real estate investor. In fact, if you do this right, you can quickly move up the financial ladder from *"small-time investor"* to *"big-time, cash-flow tycoon!"*

The three most important items you'll ever negotiate for are: **(1) the right purchase price, (2) the right amount of debt service** (monthly mortgage payment), and **(3) the length of time to pay the debt**. You can give everything else away if you have to, but you must get these three items within an acceptable range.

Obviously, financial circumstances for every buyer will be somewhat different, so there can be a little variation in the numbers. I'll show you how this variation might work for you as we discuss gathering income and expense data.

NEGOTIATING IS MUCH EASIER IF YOU KNOW THE FACTS

My chief negotiating tool is my "INCOME PROPERTY ANALYSIS FORM." What it does is develop a financial picture of the property I wish to purchase. It's a simple form that shows the financial situation of the property and approximately how much it will cost to operate the property. It also helps eliminate much of the emotion from negotiations (Chapter 5 explains this).

Often, a seller will argue that he only pays "X" number of dollars each month for expenses. But, you are certain that his numbers are short or his memory is fuzzy, a common seller affliction! When the analysis form is filled out properly, with the seller's assistance, it's difficult to argue the facts. Generally, the form has a tendency to jog a forgetful seller's memory.

IT'S ALWAYS BEST TO LET SELLER PARTICIPATE

You may wish to watch a couple of "Columbo re-runs" in order to get the hang of this. Pay particular attention to how the Lieutenant uses a "non-aggressive," "low-key" approach to uncover the facts. He's never pushy, nor does he ever accuse anyone of not telling the truth! Above all, he works hard not to alienate his suspects. Don't misunderstand here, I'm not saying sellers are suspects. However, just as Columbo does, you must verify all information. *Don't forget, the truth is always verifiable—loose words may not be!*

I always ask the seller to help me gather the information I need to fill in the blanks on my INCOME PROPERTY ANALYSIS FORM. After all, the seller should know better than anyone, since he operates the property and pays the bills. My experience has been that sellers will generally provide accurate information about the gross income; but, beyond that, they begin to develop forgetfulness. For example, on Line #2 of the form, hardly any seller I've ever purchased property from will admit to having vacancies. When they do, they claim they are rare.

This usually changes after I ask to see their Tax Form 1040 (Schedule E) for past year. Hardly any property owners tell the IRS the same dollar numbers they give to me. It's quite similar to comparing the automobile mileage a taxpayer reports on his tax return, with the mileage information he gives to his auto insurance company. Strange as it might seem, many folks apparently drive thousands of miles further according to their tax returns! This same thing happens with the *income property* numbers, I'm afraid.

No One Reports "Too Much" Income To IRS

It's quite common for investment property owners to be overly optimistic about their net rental income numbers. For example, the Property Analysis Form (Fixer Jay's Six Ugly Houses), on Page 242 shows a total gross income of $2,100 per month. If the seller's tax return, Schedule E, shows property rents of only $20,000 for the taxable year, my question is, "What happened to the $5,200?" My guess is vacancies and credit losses (non-paying deadbeat tenants). Obviously, transitioning properties might have gone through rent increases since the last tax filing; however, that's fairly easy to determine. The important point I'm making here is that investors should never allow less than 5% for vacancy loss or less than 5% for uncollected rents ("deadbeats and skips"). With run-down income properties, the percentages are often much higher.

VERIFICATION OF THE ACTUAL EXPENSES

Real estate agents seldom understand what it actually costs to operate income properties. If you ask them for a percentage figure, most will give you a number far below what the expenses will actually turn out to be! There are two basic reasons for this. The first, and probably the biggest, is that agents have little or no experience with paying the bills, so they don't really know. Obviously, they must rely on what sellers tell them! The second reason is that, if they did know the actual expenses and told their prospective clients, they might never sell another income property. From an agent's standpoint, knowing less serves them better!

If you doubt this, thumb through any multiple listing book and read the information about commercial or investment real estate for sale. You'll find the income numbers generally complete, but expenses are different. Most often, listing agents don't fill in the blanks. They either don't know or don't want to know! Keeping the expense information hidden or vague makes the property more attractive to naive buyers.

I have found what works best is to have the seller provide evidence (proof) of all expenses. The checkbook register or expense journals will generally provide this information. Expenses that need to be verified are costs for utilities, management fees (if done by someone other than the owner), repairs, maintenance and property insurance. I also want to see the county tax statements. In California, taxes are adjusted to approximately 1% of the sale price when a property is transferred (sold).

NO ONE MANAGES PROPERTY AND DOES THE REPAIRS FOR FREE

I've had property sellers tell me there are no management fees or maintenance and repair expenses on this property. "I do it all myself," they say. My typical response is, "Will you continue doing it, *for the same price*, after I become the new owner?" I've never had a single taker yet! Then I ask, "How much would you charge me to manage the property, if I buy it from you?" No one has ever told me less than 10%. Most won't take the job at any price. After a short discussion, they agree that my 5% management fee allowance is, indeed, very reasonable. After all, who will do it for less?

No one has ever developed a method for doing repairs and maintenance without spending money! When things break, it costs dollars to fix them. Painting, roof leaks, yard work, carpet cleaning and patching walls are just a few of the maintenance and repair activities necessary to keep income properties doing what they're supposed to do—*generating income!* At the very least, you'll spend ten cents out of every rent dollar you collect for maintenance on the property.

Don't include capital items like replacing carpets, coolers and roofs, as that's not maintenance or repair. Things like fixing broken door knobs, cracked toilets and broken windows will cost you approximately 5%—they are repair expenses.

Remember, these are the numbers I use on my INCOME PROPERTY ANALYSIS FORM. I consider these bare minimums. I rarely have any difficulty convincing sellers. Sometimes their own expense records, Tax Form 1040 Schedule E, will show higher numbers than my estimates.

THE MOST CONVINCING BUYER'S TOOL

The object of working closely (negotiating) with a seller using a tool like my INCOME PROPERTY ANALYSIS FORM is to develop hard evidence, or the written proof, of income and expense data. By assembling all the financial numbers on a single, one-page form, in an easy to read format, you have a very convincing tool for the serious negotiation of the three most important buyer objectives—*the right price, the right amount of debt service,* and *the length of time to pay the debt.*

Line 14 of the form on Page 242, shows the actual dollar amount (positive or negative) left each month to satisfy any new debt obligation. The example shows two existing mortgage payments totaling $795 per month (line 13A). The total mortgage balance is $100,000. Line 14 indicates that I would have a positive amount of $240 to pay any additional costs for the property. That means, if I purchase the property, I would only have $240 each month to pay the seller, assuming he will carry the financing. Figuring 10% interest, that means I can only finance $28,800 of new debt (interest only).

TYPICAL NEGOTIATION WORKS LIKE THIS

In my town, owners will expect to sell an older income property, like the example on Page 242—Six Ugly Houses, for about seven times the gross income ($2,100 per month times 12 months equals $25,200, times seven for an estimated value of $176,400). Generally, they won't sell for much less unless circumstances are unusual.

Let's say I'm willing to give the seller a note or mortgage with payments of $240 per month (all that's left over); however, at 10% interest, the note could only be for $28,800 (10% times $28,800 equals 2,880, divided by 12 equals $240). If I agreed to that and if the owner insists on a seven-times-gross selling price ($176,400), I would need to come up with a $47,600 cash down payment ($100,000 existing mortgage debt, plus $28,800 new seller carry-back note, plus $47,600 cash down equals $176,400). A 10% down payment is generally the acceptable amount for older income properties. It's also about all I am willing to pay down for ugly fix-up properties!

In this case, I might offer $175,000, with $20,000 cash down and ask the seller to carry back a note for $55,000, payable at $250 per month (interest only) for a term of ten years minimum (try for 20 years). If the seller says $250 per month is not enough of a payment for his $55,000 note, I would ask him, "Where do you expect me to find the extra money? After all, I'll be paying you $10 more than the property is earning right now! Are you asking me to reach into my pocket every month to pay you more money than the property earns when I've just paid you $20,000 cash for the privilege of owning the property?" *That's a hard question for sellers to answer*! They look at the property analysis form, then they stare at the $240 on line #14! They might mumble something like, "Maybe you won't have too many vacancies or credit losses. That would give you 10% more income. Perhaps you would consider giving up your 5% management fee. That's $105 per month, or maybe even 15% for maintenance and repairs! That really sounds too high!"

The real value of my form is that when you and the seller have taken the extra time to agree on expenses and to fill out each line together as a joint effort! It's extremely difficult for the seller to go back and ask to "undo" the results. The seller doesn't like the low payments on his carry-back financing, but he can accept them much easier when the form shows that $240 is all that's available to pay him. Don't talk about interest rates, discuss only how much money is available to pay him. Most sellers are willing to take back low interest notes (mortgages) as opposed to reducing their selling price. I've found that trying to get a lower sales price and offering only a 10% down payment is very difficult. The risk of losing a good income property is much greater cutting the price than it is negotiating the seller into a "soggy note" or mortgage *(soggy means low interest rate or weak terms)*.

SUCCESSFUL NEGOTIATIONS PUTS MONEY IN YOUR POCKET

If you are a *do-it-yourself operator* (you manage and do maintenance and repairs), and if you can negotiate a deal similar to what we've been discussing, you'll be in great shape! You'll own a property that has cash flow when you buy it or shortly thereafter.

Your profits will be enhanced by rent increases, and by minimizing vacancies and rent losses. Don't forget the 15% maintenance and repair allowance ($315 total in our example). 30% of that will go for buying materials, 70%, or $220 per month, is labor—*that's you!* When and if you sell, the soggy seller carry-back mortgage you have negotiated will allow you to sell for a premium price. Good financing is worth extra bucks because you can sell with good terms to the next guy.

ADDITIONAL HELP IS AVAILABLE

It is impossible to learn everything about this business by reading a single book, or even ten books for that matter! Your brain simply cannot process or retain all the information so quickly. Learning takes time and, if you're anything like me, you'll need to hear things and read about them several times before they really sink in! Your greatest benefits will come from continuous learning as you go along.

Don't wait until you learn everything about investing before you actually invest, because you never will! Instead, learn enough to start, then do it; but, keep your education active at the same time. A soldier always learns to shoot faster and straighter when he's in a foxhole with the enemy shooting back. The same idea works for real estate investors.

My home study training course, "FIXING RUNDOWN HOUSES AND SMALL APARTMENTS," can help "speed up" your learning efforts. You get my *360 page fixer book*, plus *eight information-packed audio tapes* with eight full hours of valuable lessons about making more money. My audio tapes with "how to" instructions will speed up your learning, because you can play them in the car, while you fix your hair or shave, even while cooking dinner. You'll be pleasantly surprised how fast you learn from listening several times! See the Appendix (Resources) section of this book for ordering details about this training product (#2100).

INCOME PROPERTY ANALYSIS FORM

Property Name _"Fixer Jay Property/Six Ugly Houses_   Date _Today_

LINE NO.	INCOME DATA (MONTHLY)	PER MONTH
1	Total Gross Income (Present)	$ 2,100
2	Vacancy Allowance Min. 5% LN-1 Attach copy of 1040 Schedule E or provide past 12 months income statement for verification	$ 105
3	Uncollectable or Credit Losses (rents due but not collected)	$ 150
4	Net Rental Income	$ 1,890
	EXPENSE DATA (MONTHLY)	
5	Taxes, Real Property	$ 175
6	Insurance	$ 100
7	Management, Allow Min. 5%	$ 105
8	Maintenance	$ 210
9	Repairs	$ 105
10	Utilities Paid by Owner (Monthly)	$ 160

Elec	$	25
Water	$	80
Sewer	$	-0-
Gas	$	-0-
Garbage	$	55
Cable TV	$	-0-
Totals =	$(	160)

11	Total Expenses	$ 855
12	OPERATING INCOME (LN 4 - LN 11)	$ 1,035

Existing Mortgage Debt

1st Bal Due	$75,600	Payments	(Montly)	Due Mo/Yr
2nd Bal Due	$ 24,400	Payments $	600	Fully amortized
3rd Bal Due		Payments $	195	6.5 yrs-then all due
4th Bal Due		Payments $		
5th Bal Due		Payments $		
13 Totals	$100,000	(13A) $	795	

14	MONTHLY CASH FLOW AVAILABLE	$
	(LN-12—13A) (Pos or Neg)	$240+

NOTE: Line 14 shows available funds to service new mortgage debt from operation of property.

REMARKS: All lines must be completed for proper analysis. Enter the actual amount on each line or Ø.

CHAPTER 14

Fixing Million Dollar Problems

CREATING EQUITY WITH VERY LITTLE CASH

*F*or most start-out investors, cash is scarce. That was my situation when I started out—I had big ideas, but very little cash. I soon discovered that my big ideas would only help me if I rolled up my sleeves and went to work! When you don't have much money to invest, you must supply something else that will persuade sellers to accept your offers.

As you will learn, solving a distressed seller's problem can have as much value as a cash down payment once you understand how it works. I decided to specialize in rundown fixer properties, because they would be easier to acquire with my limited resources, and I found that rundown properties quite often had owners who were in financial trouble. This combination is easy to find, because they often go together.

Understanding How Sellers Think

You need to realize that sellers who own average looking properties, that are being decently managed, will seldom have much motivation to sell for less than an average price and terms. *Investors in this market will find very tough competition!* That's because everyone wants to purchase nice looking properties with good tenants. If you don't have enough cash for an average down payment, chances are you won't have much luck investing in this type of property. *Don't forget this, because learning to understand how sellers think is much more impor-tant than learning to write great offers.*

My investment strategy is to create equity quickly with the smallest amount of cash down payment—sometimes none at all, as was the case at Hillcrest cottages. My plan is sim-ple—I purchase ugly rundown properties from highly moti-vated sellers and turn them into beautiful profit-makers! The financial rewards for doing this right will far exceed the profits you might earn from ordinary deals—plus, they will come to you much faster. You don't need to be a financial genius to make this strategy work either! What you must possess is a strong desire to succeed, a willingness to work hard and a firm commitment to keep learning new skills as you go along. You'll notice, I didn't mention money! As your skills develop, you'll need less and less money—that is, *less of your own money.*

HILLCREST COTTAGES—A MILLION DOLLAR PROBLEM

The Hillcrest cottage property was a perfect example of creat-ing instant equity using **personal skills** rather than cash. Hillcrest was a ugly rundown piece of property consisting of 21 cottage-type dwellings, that was situated on a hillside lot in my hometown. No other property I've ever owned, before or after, was uglier than Hillcrest, and none have ever had so many things wrong! Even the sellers of Hillcrest were all messed up.

The owners of Hillcrest, Pete and Mike (not real names), shared unequal ownership interests in the property. Pete—who owned 65% and lived on the property—had the most to say, because he had the majority ownership. This was a prob-lem for both owners, because Pete was a very poor landlord—he simply could not handle tenants and ran them off faster than he could find new ones. The results showed, and the rental income was not enough to pay the mortgage payments and expenses. Hillcrest was in serious financial trouble when I made the offer to buy it. Two lenders had already started fore-closure proceedings, even though neither really wanted the property back.

KNOWING THE REAL REASON FOR SELLING IS BIG ADVANTAGE

One of the most urgent reasons for selling investment property is when the income drops below the mortgage payments and expenses. The owner is faced with covering expenses out of his own pocket or allowing the lender to foreclose the mortgage and take it back. In the Hillcrest situation, Mike had a well-paying job, while Pete was living off the property. Mike was, naturally, unwilling to spend more of his hard-earned money to compensate for Pete's poor management.

Being aware of the financial problems of a seller is very important information when you're trying to figure out his motivation level! I always do as much probing as I can before I write offers. *Remember, the higher the motivation level, the less cash you're likely to need to buy the property.*

HILLCREST PURCHASE REQUIRED ZERO CASH DOWN

The offer I made to purchase Hillcrest provided for trading the sellers a three-bedroom house that I already owned, instead of a cash down payment! Naturally, they would have preferred cash; however, they didn't have a lot of time to negotiate due to the pending foreclosure, and my offer was the only one they had. Let me make a point here: If you can figure out the seller's urgency—like the property being in foreclosure, the seller in need of quick cash, job relocation or pending divorce—and then time your offer to relieve the seller's problem at the very last moment, you'll get offers accepted that wouldn't normally stand a chance. My Hillcrest offer was a good example of *perfect timing*.

Another threat facing the owners was the City Abatement Committee! Pete had repeatedly told city building officials that he would repair numerous building code violations on the property, some of which were hazardous conditions. The city had inspected the units earlier when several non-paying tenants filed complaints in an attempt to avoid paying rents. The Abatement Committee "red-tagged" the property and had even scheduled it for demolition, if repairs were not completed by a specific date. Obviously, this created a very serious urgency for the owners.

KEY INGREDIENTS FOR A SUPER DEAL

Let me pause for a moment so we can review why Hillcrest had the potential to be a **"super money-maker"** deal for me.

1. The property was **rundown** and *unattractive*, thereby eliminating most competition as prospective buyers.

2. There was an *abundance of handyman fix-up work*, which I could do myself—or knew how to have it done.

3. I learned early on that the owners were in *serious financial trouble*.

4. It was an excellent opportunity to *create immediate equity* from fix-up rehab.

5. It was a partnership property where unequal partners were not in agreement. In fact, *they were fighting partners!*

6. There were *low rent collections* because of deadbeat tenants and poor management.

7. The mortgage payments were delinquent, and *property was in foreclosure*.

8. There were *problems with a government agency*—the City Abatement Committee was demanding fix-up.

9. There were *"red-tag" building code violations* that had to be fixed in a short time period, which created *super urgency*.

10. The property owners had a *high level of motivation*, without many options.

Seldom do you find so many things wrong with both the property and the owners as was the case at Hillcrest. I had a feeling that almost any purchase offer I made, within reason of course, would receive serious consideration by both owners.

My Cash-Less Offer to Purchase

Remember what I told you earlier—I had big ideas but very little cash. I also said, "You must be ready to roll up your sleeves and go to work." That's exactly what I meant, and I was ready! So I prepared the following offer to purchase Hillcrest.

I proposed a property trade for the down payment, using the equity in a property I already owned. Naturally, when you think the sellers are *extremely* motivated, you can adjust your trade equity upward until you're challenged. Don't forget, there is never any harm is asking for the moon!

The property I was offering in trade was an older three-bedroom house on a large commercially-zoned lot in a major growth area. I had owned the property for a little more than a year and had it rented for about the same amount as the mortgage payment. I had originally planned to eventually convert the house to a small commercial venture, like an insurance agency or a real estate office; however, I had never gotten past the idea stage. I had purchased the property for $80,000, with a $20,000 cash down payment. Sometimes at my seminars people ask, "Where did you get the $20,000 cash?" It came from a $50,000 refinance loan on my Pine Street property. Details of that transaction can be found in my book entitled, FIXING RUNDOWN HOUSES AND SMALL APARTMENTS, Product #2105. See Appendix (Resources).

After a short negotiating session, Pete and Mike decided Hillcrest was worth no less than $234,000. They had been asking $295,000, until I came along! But now, they needed a very quick sale, so they agreed to my property trade proposal. I had not yet told them what my property was worth. Here's how I determined my sale price.

THE HILLCREST COTTAGES TRANSACTION

I agreed to pay $234,000 for Hillcrest. I would assume three existing mortgages totaling $143,000. It wasn't hard to figure their equity was $91,000 ($234,000 minus $143,000 equals $91,000). Knowing the amount of their equity made it quite easy for me come up with a fair market value for the property I was using to trade in lieu of a down payment. I simply added $91,000 to the existing mortgage balance of $58,000 and came up with a selling price of $149,000 ($91,000 plus $58,000 equals $149,000). See how easy appraisals can be!

Folks often ask me why the Hillcrest owners accepted my trade property at such a highly inflated price? They are quick to point out that buying a property for $80,000, and then selling it just 14 months later for $149,000, seems like a rather hefty mark-up! However, the $69,000 mark-up, hefty or not, was not even slightly challenged by either Pete or Mike. In fact, they were more anxious than I was to sign the deal.

The valuable lesson I learned from Hillcrest was that *distressed sellers are not nearly as concerned about how much money I make as they are about getting relief from their problems*. Solving problems and providing relief to sellers pays very big profits in the fix-up business. That's why I advise all beginners to become PROBLEM SOLVERS FIRST, and deal makers second. As I told you earlier, the Hillcrest cottage property, along with five small rental houses on Hamilton Avenue, would eventually be sold as a package to one buyer! This sale alone would earn me over a million dollars in profits, because I was able to solve the seller's problems. Problem solvers get paid big BUCKS for their skills—believe me!

FIXING UP HILLCREST COTTAGES

Fixing up Hillcrest was quite a chore, but, because of my large down payment (the $91,000 house trade), I had enough equity to borrow all the fix-up money. A local real estate lender appraised the property and loaned me $51,000 to do the fix-up work. Since I did most of the fix-up myself, except for hiring an electrical contractor, I ended up with about $20,000 left over. On average, it costs about 30¢ of every fix-up dollar for materials and supplies. The remaining 70¢ is spent for labor, *but when you do the work yourself, you get to keep this money for yourself.*

I worked for nearly two years getting Hillcrest fixed up and running smoothly. I filled in the ugly swimming pool and completely replaced most of the electrical wiring! I repaired sagging carports, built new floors and replaced nearly half the plumbing fixtures. I also spent considerable time watering and restoring almost two acres of lawns, trees and shrubs. Naturally, I painted the property inside and out. I built a new white picket fence around the front lawn area. Finally, the work was done and it was a beautiful property to look at. I had absolutely no trouble renting the cottages to good paying tenants! The fact is, I started keeping a waiting list. Needless to say, when I filled the place up, I had excellent monthly cash flow. Even with high interest payments on my fix-up loan, I was still making a nice profit every month.

Hillcrest was located along a major thoroughfare, where hundreds of cars passed by daily. People would slow down and stare at the property! I could see they were very impressed! Sometimes they would walk up to me and tell me what a wonderful job I had done. They couldn't believe how such a filthy, rundown property inhabited by winos and deadbeats could change so quickly. Of course, the answer was a bit more obvious to me! I spent at least three or four days every week working on the property. Later on, after I sold Hillcrest, I estimated that if I had worked 40 hours a week for two years, my earnings would have been something in the neighborhood of $300 an hour. I can tell you, I've worked at many jobs during my lifetime that paid a whole lot less!

Selling the Fixed-Up Hillcrest Cottages

Almost two years after I acquired Hillcrest, I sold the property to a local physician who was looking for investment real estate. He was also seeking a way to shelter his regular income from taxes. In order to create the right size tax deduction, the doctor needed to acquire depreciable property worth about $600,000. He wanted to use maximum leverage—like a very small cash down payment! We both put our heads together and I agreed to sell Hillcrest cottages, along with five small rental houses, to the doctor, all on a single contract, with no money down!

That was just what the doctor ordered! He agreed to my selling price of $594,000. You'll notice that my selling price closely resembles the amount of real estate the doctor told me he needed. I might point out, I've always prided myself on being able to accommodate a buyer's circumstances! This sale was one of several I designed to fit the special needs of my buyer. Obviously, it worked out very well for me, too!

REMOVING THE RISK FROM A "NO DOWN" SALE

People often ask me, "Isn't it a bit risky to sell a property for no cash down payment?" The answer is, "Yes, of course, unless you can offset the risk by using some other means." In this case, I asked for and received additional collateral!

I said, "Look doctor, I normally insist on a minimum 10% cash down payment! However, I am willing to take a promissory note and deed of trust for $60,000 (approximately 10% of $594,000) on another property you own in lieu of a cash down payment. I'm not asking for any payments on the security— and, it won't cost you a dime as long as you make the $6,000 monthly payments on the $594,000 Hillcrest contract."

I also told him that after he paid five years worth of payments (60 times $6,000, for a total of $360,000), I would release (reconvey) the $60,000 collateral note. This simple arrangement protects me, while, at the same time, it allows the buyer to use his existing equity to purchase more real estate without spending his cash.

Naturally, if he failed to make payments on the $594,000 contract, as promised, during the first five years, I would not only take back (foreclose) the Hillcrest property and five houses, but also the $60,000 worth of additional collateral property! I call this my "TWO-FOR-ONE" CONTRACT! I'll sell him one property, but, if he doesn't pay me as agreed, I'll take two properties back. After payments have been made for five years, I feel safe enough to release the additional security! By the end of 60 payments, a big percentage of the risk is gone. It's kind of like the first five years of marriage. Chances are— If you make it that far, you'll probably stick it out until the end!

Cash is Not the Only Way to Sell

There are several schools of thought about how to make money when buying and selling real estate. Some will say that selling for all cash is the only way to make a real sale. Cash is much more powerful than waiting for your money over time, they claim. I can't quarrel with the notion that all cash is powerful. However, I'm also aware of how small banks become big rich banks! They do it by charging rent for the money they loan out. Banks would not be nearly so rich if they simply got their loan principal back with a reasonable fee tacked on. *Collecting interest payments on loans is how banks earn their biggest profits!*

For example, a $100,000 loan amortized over 30 years, at 10% interest, will cost the borrower 360 payments of $877.58 each. It will cost $315,928.80 to pay the loan back. Which would you rather receive if you were a banker—the interest charges on the loan or the principal balance? Obviously, the interest payments add up to $215,928.80, which is more than twice the amount of loan principal. *Interest payments are what make bankers rich, and they'll do the same for you when you sell properties to qualified buyers and finance the deals yourself.*

When you become the banker (lender), you'll not only benefit when you sell your property for a hefty profit—but, you'll keep right on making even bigger profits from all the interest you collect.

It's important to understand exactly how you can earn your profits! Let's say I purchase a rundown house for

$50,000 and spend an additional $10,000 fixing it up. Then I sell the house for $80,000 cash. My gross profit is $20,000. In order to earn the $20,000 profit, I had to work hard and take some risks. Now, let's suppose I finance the $80,000 sale over 30 years at 10% interest. That will give me 360 payments of $702.06 each, for a total of $252,741.60. When you subtract out the $80,000 selling price, there's still another $172,746.60 I'll be collecting! That's the interest—and it's all profit!

Interest profits are easy money! They don't require hard work and there's hardly any risk involved! I think you can understand why bankers never get their hands dirty, but still end up filthy rich, just the same.

EXTRA PROFITS WITH WRAP-AROUND FINANCING

I sold Hillcrest Cottages and the five small rental houses to the doctor using an "All Inclusive" note and deed of trust. In California, trust deeds are used; in most other states mortgages are used the same way. This type of financing is called wrap-around because a new all-inclusive note simply wraps around or includes all other financing on the property. In this case I had a total of nine promissory notes on both properties when I made the sale! The total amount I owed on these notes was $333,055, which, of course, included the money I borrowed to fix up Hillcrest Cottages.

My selling price was $594,000, and since there was no down payment, the all-inclusive note was for the total amount of the sale. My profit was $260,945, which will magically grow to more than a million dollars, because I agreed to accept monthly payments, including interest, on the full value of my wrap-around note. I earned a full 12% interest on my profit portion of the note and an additional 2 to 5% interest on the underlying notes. This bonus interest is called "the spread." It's the difference between the wrap-around note rate (12%) and the lower interest rates—7 to 10%—of the underlying notes, in this case.

Interest income is very powerful. I will show you just how powerful in a minute, but, first, let me tell you some of the other advantages for sellers who carry back wrap-around notes when they sell.

BE AWARE OF LOAN-OVER-BASIS TRAP

First, you can minimize the income taxes you must pay when you sell! The IRS allows you to pay your taxes as you receive your mortgage payments. Only a portion of each installment payment is taxable in the year you receive it. Naturally a cash down payment will increase your taxes. However, there is one nasty tax trap you need to be aware of—it's called LOAN OVER BASIS. Pay close attention here so you can avoid a future heartache!

In simple terms, basis is what you paid for a property minus the amount of depreciation you've deducted on your income taxes. For example, if you paid $100,000 for a property and have deducted five years depreciation at $10,000 per year, your basis would be $50,000. If you sell the property for $160,000—your profit is $110,000. It's not $60,000 as many surprised investors find out the hard way!

LOAN-OVER-BASIS generally happens when you sell a property and allow the buyer to assume or take over your existing mortgages or loans.

The IRS calls this debt relief, which, of course, is exactly what it is! If you owe mortgage payments and then suddenly you don't owe them anymore—that's relief! Debt relief or allowing a seller to take over your mortgages when you sell can be very hazardous to your wealth. Let me show you why.

In the example above, let's say I paid $10,000 cash down and the seller carried back a $90,000 mortgage. After five years my mortgage balance is $85,000. However, I just recently borrowed $15,000, which was secured by the property, to make improvements so I can sell the property and earn a healthy profit. A buyer shows up who is willing to pay

$160,000, which is exactly what I'm asking! He agrees to give me $30,000 cash down and assume both of my existing mortgages, totaling $100,000. I have agreed to carry back a promissory note for the balance of $30,000. This sounds like a sale made in heaven until it's tax time, when loan over basis will "fry me"

WRAP-AROUND FINANCING WILL AVOID BIG TAX BILL

On the day I close escrow on this property, I'll owe taxes on the amount of money I receive ($30,000). I'll also owe taxes on money I don't receive ($50,000)—that's the amount of loan over basis (basis = $50,000, loans = $100,000, excess = $50,000). If my tax bracket is 35%, I'll get to keep $2,000 of the $30,000 cash down payment, assuming it's not used up for closing costs.

Cash down payment received............................	$ 30,000
Total amount of loan over basis.........................	50,000
Taxable income in year of sale	$ 80,000
Tax bracket...	35%
Payable immediately to the IRS	$ 28,000

Wrap-around financing will eliminate loan over basis, if done correctly. ***Here's what you must do***.

1. With wraparounds, the seller must remain liable for paying the existing mortgages (loans) on the property. With a wrap-around mortgage, the buyer pays the seller a monthly payment large enough to pay the existing underlying mortgages, plus an additional amount for the seller's equity or profit.

2. The buyer must not assume any mortgages or loans. A new wrap-around mortgage is created to facilitate the sale.

3. The seller must continue to remain totally liable for the existing underlying mortgages. The buyer should never

send mortgage payments directly to underlying note holders for convenience or any other reason!

4. The sellers' payments to the underlying mortgages cannot be dependent on the seller receiving the buyer's payment.

By following these steps correctly, you can save a lot of money. In this example, a wrap-around mortgage could save the seller enough money to buy another property.

WRAP-AROUND MORTGAGES GIVES SELLER BETTER CONTROL

Another important benefit for sellers who carry back financing using wrap-around mortgages is they will have FAR MORE CONTROL! On my Hillcrest sale there were nine separate underlying mortgages, which I was responsible for paying! Every month, when the doctor sent me a $6,000 mortgage payment, I would send payments to each of the nine underlying mortgage holders. My net income (what was left after the nine mortgage payments) was $1,950 each month, when I first started out.

Knowing that all the underlying mortgages are paid each month provides added security for sellers! Particularly those who sell for small cash down payments and finance the biggest share of their profits.

It can be a costly lesson or *"Very Expensive Seminar,"* as my good friend John Schaub puts it, when you sell a property and allow the buyer to assume or *take over* the mortgages, and then, later on, he stops paying them without your knowledge! You should always make sure you are notified immediately if a mortgage payment default occurs on a property you've sold and carried back part of the financing! This can happen, and often does, even while you are receiving your mortgage payments every month. The title company will show you how to record a notice so you'll be automatically notified if a default occurs. Make sure you ask them to do this.

I've sort of strayed a bit to help you understand about a serious pitfall of regular seller carry-back financing. However, it can be a very sad day, indeed, if you are ever forced to take your property back (foreclose) and then suddenly discover the buyer has not been paying the senior mortgage payments for six months or so. If that happens, you'll be in a real fix. *With wrap-around financing, this can't happen, because you are paying all the mortgages yourself.*

Besides earning extra interest with wraparounds on the spread, and saving money where loan over basis is a problem, there is yet another pleasant surprise in store—it happens when the underlying mortgages are paid off, and you get to keep the whole payment for yourself.

When you sell a property and wrap the existing mortgages like I did at Hillcrest, it's quite likely that the underlying mortgages will amortize (pay-off) long before the last payment is made on the wrap-around. For example, suppose I have a $100,000 wrap-around mortgage with $1,200 monthly payments coming in for the next 20 years. The total payment is not all mine to keep, because there's an underlying mortgage with a balance of $50,000 owing. The payments on this underlying mortgage are $500 per month, but the mortgage will completely amortize (pay off) in 13 years.

Keeping the Whole Enchilada for Yourself

In this example, I'll receive $1,200 payments every month; but, for the first 13 years, $500 of each one goes toward payments on the underlying mortgage. However, during the last seven years the full $1,200 payments will be all mine to keep! You just can't help but like wraparounds when they make your cash flow jump by 70%.

To give you a idea of how the cash flow jumps on my Hillcrest mortgage—eight of the underlying mortgages will be paid off about half-way through the wrap-around term. The one remaining mortgage, with a monthly payment of $438, will be paid off just before I receive my final payment on the wrap-around. That means my cash flow will jump from approximately $2,000 a month to $5,562, for the last 13 years and my net income, after paying off all the underlying mortgages, except one, will be more than $850,000.

BENEFITS OF SELLER CARRY-BACK FINANCING

People often ask me why I'm so eager to accept carry-back mortgages when I sell my property. There are several reasons why. First, I'm very fond of having the steady monthly income, which carry-back mortgages provide. Also, I can avoid losing a big chunk of my profits right away by using the installment sale method for reporting income taxes. But, perhaps the biggest reason of all is the additional money I can earn from interest on my mortgage.

I have watched small-town banks become very wealthy this way—so I've long since decided it's good enough for me, too! Let me show you a few Hillcrest numbers, which helps tell the story quite well, I think!

TOTAL SELLING PRICE... $594,000

EXISTING MORTGAGE debt owed by me when
I sold the properties—nine individual mortgages............ $333,055

PROFITS or EQUITY AT SALE..................................... $260,945

Had I been able to sell my property for all cash rather than monthly payments, no doubt the selling price would have been much less—my guess would be $500,000 tops, assuming of course, I could find a buyer willing to pay all cash. Most cash buyers expect to pay at least 20% less than those who need help with financing. Cash is a powerful motivator for driving prices down. Assuming I had sold for $500,000 cash, my profit would have been reduced to $166,945. With State and Federal income taxes payable immediately, I would have considered myself very lucky to end up with $100,000 after taxes. You can get a good idea how my Hillcrest wrap-around contract works by studying the chart on Page 258.

Wrap Around Mortgage Hillcrest Transaction—$594,000

Schedule of Payments Receivable & Underlying Mortgages Payable

YEARS	ANNUAL PAYMENTS INCOME	ANNUAL PAYMENTS · OUT-GOING UNDERLYING MORTGAGES (9)									NET INCOME TO JAY
		GBRLTR.	DEMPY	ERNIE	BENE #1	BENE #2	RUSS	MAHOY	RICE	BENE #3	
1	48,000	4,800	2,400	4,000	3,752	3,752	2,400	3,504	2,000	5,752	15,640
2	72,000	7,200	3,600	6,000	5,628	5,628	3,600	5,256	3,000	8,628	23,460
3	72,000	7,200	3,600	6,000	5,628	5,628	3,600	5,256	3,000	8,628	23,460
4	72,000	7,200	3,600	6,000	5,628	5,628	3,600	5,256	3,000	8,628	23,460
5	72,000	7,200	3,600	6,000	5,628	5,628	3,600	5,256	3,000	8,628	23,460
6	72,000	7,200	3,600	6,000	5,628	5,628	1,200	5,256	1,000	8,628	27,860
7	72,000	7,200	3,600	6,000	5,628	5,628		5,256		8,628	30,060
8	72,000	7,200	3,600	6,000	5,628	5,628		5,256		8,628	30,060
9	72,000	7,200	3,600	6,000	5,628	5,628		5,256		8,628	30,060
10	72,000	7,200	3,600	6,000	5,628	5,628		5,256		8,628	30,060
11	72,000	7,200	3,600	6,000	5,628	5,628		5,256		8,628	30,060
12	72,000	2,000	3,600	6,000	5,628	5,628		5,256		8,628	35,260
13	72,000		3,600	6,000	5,628	5,628		5,256		8,628	37,260
14	72,000		3,600	3,500	5,628	5,628		5,256		8,628	39,760
15	72,000		3,000		5,628	5,628		5,256		8,628	43,860
16	72,000				1,876	1,876		5,256		2,876	60,116
17	72,000							5,256			66,744
18	72,000							5,256			66,744
19	72,000							5,256			66,744
20	72,000							5,256			66,744
21	72,000							5,256			66,744
22	72,000							5,256			66,744
23	72,000							5,256			66,744
24	72,000							5,256			66,744
25	72,000							5,256			66,744
26	72,000							5,256			66,744
27	30,000							1,314			28,686
	$1,878,000	78,800	52,200	79,500	84,420	84,420	18,000	136,218	15,000	129,420	1,200,022

General Notes:

1. **$594,000 selling price, less $333,055 underlying mortgage debt, equals $260,945 profit or equity.**
2. **After all underlying mortgages are paid in full, the net income to Jay is $1,200,022.**
3. **Profits received are $260,945. Interest or carrying charges received are $939,077.**

INTEREST INCOME INCREASES PROFITS

By using the installment sale method and a wrap-around mortgage, I'll receive a total pay back of $1,878,000—that's 313 payments of $6,000 a month. Net earnings to me, after all my underlying mortgages are completely paid off, will be $1,200,022. When you compare my selling profit of $260,945 to the actual number of dollars I'll receive, you can see that my interest earnings alone are over $939,000. *That is nearly four times more than my original profit!* With numbers like these, it's very easy to understand why banks are more than happy to have their loans paid back with those steady monthly payments.

So far, I've told you about my selling profits of $260,945 and almost a million dollars worth of interest income—all earned from fixing up just two small multiple-unit rental properties. Both properties were fixers purchased in rundown condition, with very little up-front cash. Hillcrest was strictly an equity trade purchase with *no cash down payment*, whatsoever. My point is that you don't need a bundle of cash to make a lot of money, if you're willing to substitute your personal efforts for the normal down payment. I discovered a long time ago that *fixing seller problems is the highest paid job in the real estate investment business*.

Flushing Out Big Bonus Profits

Before I leave the subject of wrap-around mortgages and seller financing, I'd like to show you another exciting way to create additional profits using these methods. Big bonus profits or *PROFIT BULBS* (Chapter 12), I call them, and it involves buying back the underlying mortgages at substantial discounts. This opportunity arises because many mortgage holders (beneficiaries) would rather have cash now than payments for the next 10 or 15 years. This only works when the underlying mortgages are held by private parties. Banks and institutional lenders will seldom discount their mortgages. Some even charge extra for early payoffs! It's called pre-payment penalty.

My Hillcrest wrap-around had nine underlying mortgages to start with. Four of them were bank loans that I had previously assumed. Another was for money I borrowed for fix-up work. The other four were all excellent candidates for future discounting—they all had more than 13 years of scheduled payments remaining, and the holders (beneficiaries) were all parties who had carried back mortgages on the property from previous sales. Quite often these carry-back mortgage holders had wanted more cash when they sold and some still want cash today. The only question remaining is how much cash will they settle for?

A Profit Opportunity You Shouldn't Overlook

My experience has been that most private-party beneficiaries will discount their notes or mortgages somewhere between 20 and 50%, depending on their age and circumstances when you make then an offer. Obviously, a lot depends on the financial situation of the party involved.

Buying your own debt back at a big discount is one of the most exciting ways to make money in this business. Chapter 20 will discuss this subject in more depth. It's truly one of my favorite money-making techniques—or *PROFIT BULBS*.

CHAPTER 15

Investing With Others— Small Partnerships

Partnerships are like marriages—there are some good ones that last a lifetime and many that don't last until they're paid for! Like marriages, partnerships stand a much better chance of working and lasting, if the partners are selected for the right reasons.

WHY WOULD ANYONE WANT A PARTNERSHIP

There is only one good reason I know of to take on an investment partner! It's that you don't have enough financial horsepower to do everything by yourself. In other words, you need some help, and most often, it's financial help you need. However, there might be other legitimate reasons for needing help. We'll discuss a couple as we go along. Equity sharing and time-share contracts are two types of partnership investing. Both are designed for investors who don't have the finances to complete the whole purchase.

If you are fortunate enough to choose the right kind of partner and make it work, partnership investing can speed up your financial plans and help you reach your goals much faster than investing by yourself. However, the downside of partnership investing is that you might end up with someone who doesn't work out and it can be a serious setback for even the best of plans. You must be very careful when selecting a partner.

Always ask yourself if you really need a partner or do you just think you need one, and if it is wise to split the profits with someone else? The answers should be very clear before you look for a partner. Every so often I hear about a "lonely investor" who apparently doesn't have enough confidence to buy real estate by himself. Generally, this type of person lacks courage. He wants a partner for moral support. That's not a good reason to form a partnership, in my opinion. It's kind of like the guy who thinks he's better off crashing in an airplane with 100 people aboard, than crashing alone. Take my advice, do not take on a partner just because you want companionship. Partnerships are tough enough when you have a good reason.

PARTNERSHIPS MUST BE BASED ON MUTUAL NEEDS

Consider the want-to-be investor who knows just enough about real estate to be dangerous. He has loads of confidence, but very little cash. Most often, he will attempt to convince someone with money that, by simply joining together, they can both end up rich. But, instead, they both nearly always end up broke. Stay away from people who have big ideas, but no money!

The questions I ask myself when someone approaches me with a partnership proposition are: **What's in the deal for me?**, **What's the risk to me?**, and **What assurances do I have that a partner will do what he says?** One question you should always ask yourself is: What's the *most I can lose* if I do make this deal? Naturally, I'd be very concerned that my partner and I shared equal risk!

Looking for Partners—The Selection Process

Many small investment partnerships are created almost entirely on the basis of friendships between people who work together, attend the same clubs and churches or perhaps enjoy the same social activities. Regardless of what mutual involvement brings folks together initially, partnerships founded solely on the basis of friendships are most likely doomed from the very start. No matter how compatible people might seem to be as friends and social acquaintances, they will almost always change when their personal money becomes involved.

This change often reminds me of the little old man in the driver education films—he's soft-spoken, shy and well-mannered, until he gets behind the wheel, then suddenly all that changes and he speeds down the freeway threatening anyone who dares to come near him! The point is people don't always act like you think they will!

The process involved in selecting a partner is without a doubt the most important consideration for developing a lasting and profitable situation. Individual needs, ability to contribute and skills must all be considered and balanced effectively. The selection process deserves a great deal of thought. It's far too important for quick decisions or snap judgments. *Get rich schemes,* given little thought or planning, are generally failures from the very start.

Partners Don't Need to Be Friends

Contrary to a popular myth, shared by many, investment partners don't need to be good friends to be successful partners. It might help when getting started, but it's simply not a requirement. Obviously, enjoying the same social activities has nothing to do with a profitable real estate partnership.

My number one consideration for establishing a strong and lasting partnership is based on my own self-interest. Do I really care if my investment partner gets rich? I certainly do—because it means that I'll get rich too!

If it sounds like greed provides the need, it's most likely true. Let me caution you here, don't let your mind wander and over-react! Just concentrate on what I'm telling you, and, later on, I think you'll understand why partnerships work best when there is self-interest or greed.

SAVE A FRIENDSHIP—DON'T DO BUSINESS WITH THEM

Years ago my fascination with retailing led me into the world of women's fashion. I traded a couple houses for a dress shop and eventually expanded to several stores. When I first started out, I thought all my friends and people close to me would be the first ones in my store to support me. *How wrong I was!* The only time I ever saw my friends in the store was on "sale days." Occasionally, they would show up when they wanted me to order something special for them, and sell it to them at my cost! The people who made the store successful and kept the cash registers ringing were total strangers. They wanted what I had to sell and were willing to pay for it.

Real estate partnerships often work similar to the situation in my dress shops! You don't need people around who are willing to take, but give nothing in return! Your best bet is to find a partner who can give the partnership something you need—it can be money or the skills and "know-how" to operate the property. You need to find a partner with a strong desire to make money for himself, because, in doing so, he will automatically make money for you too! Perfect strangers are more likely to make better business partners than any of your friends. Strange, but that's just the way it is!

JAY'S RULES FOR FINDING A MONEY PARTNER

My "no-compromise" business rule for finding an investor with money, when I need financial help, is the same rule I use for landlording. I call it my 60/40 rule. What it means is that I'm prepared to give more than I receive—60% to be exact. Here's the way I apply it to partnership investing.

I've always felt that investors with no money should be willing to give up at least 60% of the partnership benefits in order to attract the money. What this means is, if I'm the broke partner, I must be content with 40% of the deal for myself. You should ask yourself this question about the partnership—Who is most likely to make it on their own—*THE PARTNER WITH THE MONEY OR THE ONE WITHOUT?* Without over-analyzing this question, I think you can get my point. The party with the money will always have a much better chance, than the one with no money! Don't you agree?

Money opens many doors! Therefore, let's simply concede that the most valuable person in a small partnership is the one who furnishes cash—a.k.a., *THE MONEY PARTNER.* If you develop your plans accordingly, it will help you structure a partnership or co-investment with the best possible chance for success.

PARTNERSHIPS CAN SOLVE YOUR MONEY PROBLEMS

Developing partnerships to pool individual resources, knowledge and experience can provide an excellent vehicle for acquiring wealth at a much quicker pace than might otherwise be possible. I've discovered that, in most successful partnerships, the partners themselves will often have very little in common except their strong desires to make money. Sometimes, an accountant will team up with a carpenter or handyman. A doctor might team up with a contractor, or with a school teacher with extra hours and mechanical skills. Quite often a real estate agent, who can manage property, will make a very good partner. One of the best ways to find investment partners with the particular qualifications you need is by advertising in the HELP WANTED SECTION of your local newspaper. Let everyone know exactly what you're looking for.

Many folks would like to create a profitable partnership venture, but don't know how to go about it. The first thing you must do is determine what *you* can provide to the partnership—investment capital, your time or a specialized skill you possess. Include this when you advertise for a partner. There are many people looking for what you have to offer, but since they don't know you exist, they can't find you! You must "speak up" and let them know!

I can still remember my early classroom experience when I was a bit too shy to ask the teacher a simple question in front of my class. I was afraid the other kids would think my question was stupid and laugh at me! Then someone else would ask the same question, and the teacher would praise them for asking such an intelligent question. What I'm trying to say here is that there are many decent partners around, but you must speak up and let them know you're looking for them. The worst that can happen is they'll turn you down—and, I promise you, that doesn't hurt! Just keep on trying until the right one finds you.

BENEFITS MUST BE TOTALLY EQUAL FOR ALL

The biggest problem I've observed about small investment partnerships is that they're almost always engineered or thought up by the person who doesn't have any money! This person has all sorts of wonderful ideas, but in order to make them work, he must find someone with money. The typical arrangement is where one partner is asked to put up hard cash and the other is supposed to contribute an equivalent share of personal services.

I don't know how you think, but I'm very skeptical about anyone who proposes a joint plan using my money, while risking only personal services themselves. The question I ask is aimed directly at the heart of the issue, *"If your ideas are so good, then why is it I've got the money and you don't?"* There may be several good answers—which I consider reasonable and acceptable—to that question, but unless the question gets answered to my satisfaction, I will not consider going forward and neither should you!

The Courting Period Requires Honesty

The biggest mistake *NO MONEY PARTNERS* make in trying to entice a person with money is to over-sell and overstate the benefits the money partner is supposed to receive. If I actually received all the profits I've been promised, I would need to rent the Bank of America headquarters building to store my money. Fortunately—or unfortunately, whichever way you view it—I did not invest my money in most of these proposals, so I'll never really know for sure! But, I can tell you this much—a very high percentage of the deals went bust!

Since I have a lot of experience on both sides of partnership investing, that is, the guy who's broke, and the one with money—I feel I'm qualified to pass on a few tips to help you structure a partnership, which will hopefully survive the high fatality rate. Because there are many more want-to-be investors who don't have money, I'll concentrate on helping them find someone willing to put up the money.

Looking for Partners in all the Wrong Places

Several years ago, I wrote a special article for a "business opportunities" magazine, entitled: "LOVING FRIENDS AND PERFECT COUPLES." I offered several tips for finding suitable investment partners, but, one of the most important tips was, *friends are not usually good partners.* Inexperienced folks often think their best friends are good candidates for investment partners. Frequently, you see partnerships develop between buddies on the golf course, members of the same social club and people who attend the same church or work together. If that's all they have in common, it's not nearly enough to guarantee a successful investment relationship.

To the contrary, the most successful money-making partnerships I've dealt with have partners who are not close friends. They are business associates joined together to make money for each other. *That's all they need to be!* Membership in the same service clubs, attending the same church or golfing together has very little to do with the business of making money! You need to understand this fact right from the very start. Your friends will not make you rich!

A PLAN TO END YOUR MONEY PROBLEMS

When I first decided to invest with someone else, I had no question about how I would fit in or what my role would be! Since I didn't have any money, I would be the *worker partner* and I was perfectly willing to give up 60% in return for my 40%. I was more than willing to do whatever it took. I already knew that as the working partner, I'd be responsible for making the property pay off for the partnership. If you're successful at doing this—that is, your partnership makes money—

and if you are the partner who is performing the personal (work) services, you'll never need to worry about finding money to invest again! "Why?," you ask. *It's because there is no shortage of investors willing to invest their money with a winner.* What is in short supply is partners who can turn the money into handsome profits. Let's discuss a plan that works very well for a small investment partnership. You might call this plan a "Partnership Proving Ground."

A Simple A-B Partnership Plan

Partner A has money to invest. *Partner B* has the time and skills to operate a real estate investment. Let's suppose I'm *Partner B*, and I'm looking for someone to put up $25,000 for a down payment on the some apartments. I find a person with money who would like to invest, but he will first need some assurance that I can operate the property and, eventually, turn a profit. I will need to satisfactorily answer the four basic questions that any investor with money (*Partner A*) will want to know about the investment and me. Here are the questions.

1. WHAT'S IN THE DEAL FOR ME?

2. WHAT IS MY RISK IF I INVEST?

3. WHAT ASSURANCES WILL I HAVE THAT *PARTNER B* CAN MANAGE PROPERTY?

4. WHAT WILL *PARTNER B* LOSE (RISK) IF HE OR SHE CAN'T MANAGE THE PROPERTY?

Like I told you earlier, the benefits must be equal, before I apply my 60/40 rule. I always insist my 60/40 rule is what makes a deal work. I want you to underline this next sentence and read it often! You must never forget it! *The better you can become at solving problems for others, the more success you will enjoy for yourself.* This applies to everything you touch in real estate, but it's especially true when you're a broke partner looking for someone with money." Now let's explore the answers to the previous questions.

QUESTION 1. WHAT'S IN THE DEAL FOR ME?

I'll set up some hypothetical numbers to illustrate my partnership proposal.

$100,000 Purchase price of the "Good Deal Apartments"
 25,000 Cash down by *Partner A*
$ 75,000 Mortgage balance with payments of $650 per month

Rents (income) estimated to be $1,100 per month ($13,200 annually).

I've already advised my partner that there will be little or no cash flow for partnership distribution to start with. I propose to "set up" the partnership agreement for a seven-year term. (five to ten years is normal range). My basic plan is to rent out the property for seven years, then sell it and close out the partnership. We'll then split the profits on a 50/50 basis.

$150,000 Estimated property value seven years from now
 (25,000) Return of capital (down payment) to *Partner A*
$125,000 Balance after return of down payment
 70,000 Unpaid mortgage balance
$ 55,000 Profits to split between partners

Cash distribution: *Partner A* and *Partner B* gets $27,500, each.

The partnership will agree to pay $100 per month management fee to *Partner B*. All cash flow from rents will be divided 90% to *Partner A*, 10% to *Partner B*, when available (projected for years three through seven). Average 5% annual rent increases will raise rents by approximately $5,000 during the seven year term. Cash flow distributions to partners will most likely occur starting after three years of operations.

Profits and losses from operations (TAX BENEFITS) will be allocated to partners same as rents—90% to *Partner A* and 10% to *Partner B*. This can be a money-making benefit for

Partner A, depending on his tax bracket and gross income from regular wages or other ordinary income.

In summary, this plan will allow *Partner A* to "net-out" very handsome profits during the seven-year partnership term. First, the $25,000 investment will return a $27,500 profit. Additionally, there will be income tax savings from his ordinary income (tax shelter) and, finally, the accumulation of cash flow from rents, beginning at year three through seven of the partnership term.

Partner B (no money investor) will benefit from $27,500 profit after seven years, plus $100 per month management fee for 84 months ($8,400), while having no money invested to start with!

This is not an unusual partnership arrangement. In fact, it's quite common. As you can see, there are excellent benefits with earnings that far exceed other types of investing. Imagine if you're the broke investor (*Partner B*) and you are successful at putting together a few of these partnership deals. It's truly an excellent opportunity to build wealth without having any money to begin with. ***You can use your skills in lieu of investing money.***

Extra Opportunities for Handyman Skills

There are many other opportunities that become available to you as you gain experience and expand your creative skills. For example, if you happen to be the handyman type, you can do all repairs and maintenance yourself and bill the partnership for your labor. You'll do fine even if your billing rate is half of what licensed tradesmen charge. That saves the partnership money for expenses and, obviously, earns you more money from the deal. If the rental income doesn't cover your expenses in some months, keep a running log and collect it later. Your flexibility helps the partnership as well as building wealth for yourself.

QUESTION 2. WHAT IS MY RISK IF I INVEST?

Obviously, the money partner will be very concerned, especially if *Partner B* doesn't have a proven track record. He needs to feel safe, and you must be able to assure him. There are a couple methods to accomplish this.

First, you can purchase a property as tenants-in-common with undivided and unequal ownership interest, such as 90% for *Partner A* and 10% for *Partner B*. This arrangement (ownership percentage) should be specified in a co-owner agreement. More about that later.

The problem with this arrangement from a money partner's viewpoint is that all the money invested is their's. *Partner B* has no money at risk. If *Partner B* can't, or for some reason won't, perform his responsibilities, then *Partner A* has a couple of serious problems! First, he has no day-to-day manager or anyone to handle the apartment operation. (Remember, *Partner A* is a passive investor and can't handle daily operations.) Second, he could have a tough time trying to get *Partner B*'s name off the property. It is possible to avoid this problem by having a quit claim deed signed by *Partner B* ahead of time to be recorded in the event that something goes wrong. Still, there is a good possibility for a lawsuit, with attorney fees involved.

If this first arrangement is not satisfactory to a money partner, perhaps my second method might offer a better solution.

Alternate Partnership Plan—Option to Purchase

You could allow *Partner A* to be the sole owner. That is, purchase the property in his name only! *Partner B* can protect his interest with an option-to-purchase the property at the end of the partnership term. Let's say the option-to-purchase says that after seven years I have the right to purchase 50% interest in the property for the initial purchase price of $100,000. I can do this by putting up an equal cash down payment of $25,000, same as *Partner A* did to acquire the property. In reality, *Partner B* won't put up a dime! It's done in the form of a credit. The net results will be the same as shown below.

$150,000	Selling price (value)
(25,000)	Return of initial down payment to *Partner A*
$125,000	Left over
25,000	*Partner B*'s Cash down payment (exercise option-to-purchase)
$150,000	Balance
(25,000)	Return of cash down payment to *Partner B*
$125,000	Left over
70,000	Mortgage balance
$ 55,000	Balance to split between partners

As you can see, the bottom line or profit remains the same. Each partner still receives $27,500 as his share. However, with this arrangement, *Partner B* is not on the title until his option is exercised. The right to exercise the option can be tied to specific performance. If performance is not accomplished, then the option cannot be exercised and *Partner A* does not have the problem of trying to clean up the title. He is already the sole owner of the property.

QUESTION 3. WHAT ASSURANCES WILL I HAVE THAT *PARTNER B* CAN MANAGE PROPERTY?

With the option arrangement above, *Partner A* can quickly get rid of *Partner B*, if terms of the **CO-OWNER AGREEMENT** are not met. This plan is generally quite favorable with money investors, because it represents less liability to them if something goes wrong. The worst that could happen is that *Partner A* would be the sole owner of a rental property without anyone to help him manage. That's not so bad, as he can easily hire a professional manager to do the job. If I were *Partner B* in this hypothetical case, I would have no problems with this arrangement. After all, I'm the partner who must prove myself since I have no cash to invest! Again, don't forget my 60/40 rule. *You must be willing to give more, and it will make you very wealthy in the end, believe me!*

QUESTION 4. WHAT WILL *PARTNER B* LOSE (RISK) IF HE OR SHE CAN'T MANAGE THE PROPERTY?

If I'm the money partner, I'm naturally concerned about my own risk. However, I'm also concerned about what my partner stands to gain or lose on the deal. Obviously, it's not cash, since he has none invested initially. However, in our hypothetical transaction, he stands to earn $6,400 annually. Naturally, he will lose that much if he blows the deal. Quite frankly, $6,400 is a very respectable incentive when you're broke. Here's how I figure *Partner B*'s gain or earnings.

$27,500	Profits at the end of seven years (termination or sale)
8,400	Management – $100 per month (84 months)
8,400	Labor billed to partnership – $100-per-month average for seven years
500	10% share of cash flow, years four through seven
$44,800	

$44,800 divided by seven years equals $6,400 per year, for terms of partnership.

A partner without money who can earn $533 per month for seven years has a lot to gain. Even though he has no "up-front" money invested, chances are very good that he or she will consider a profit of $44,800 worth the effort. I always want my partner to make big profits, because in doing so, he'll automatically earn big profits for me too!

A PARTNERSHIP DESIGN IS NEGOTIABLE

The hypothetical partnership we have been discussing is only to show you how it could work! There are dozens of ways to structure a small partnership to meet the needs of the parties involved. *Also remember, there is absolutely no limits to how creative you can be.*

The terms of the partnership—who gets the cash flow, how profits and losses are allocated to the partners, who does what work, and how to divide the profits at the end—are negotiable and strictly up to the individuals who set the partnership up.

What is most important for the success of any partnership, no matter how it's structured, is that everyone gets at least what they're promised! I cannot over-emphasize this point! That's why my 60/40 rule works so well.

When I'm involved with a money partner, I work extremely hard to make sure he gets all his benefits, before I even consider mine. If you follow my advise, you'll never have to worry about down payment money again. Investors with money will start calling you after awhile!

THE PARTNERSHIP PROMISE

Once negotiated, everything the partners agree to must be formalized. That means you need a written document—THE CO-OWNERSHIP AGREEMENT. Do not run your business on verbal promises. You'll end up on the short end of the stick if you do.

The co-ownership agreement doesn't need to be very large, especially for small two-party associations. You must, however, spell out your investment plan—for example: How will you purchase the property? How much money will it cost? Who pays for what and when? You'll also need a statement about the purpose of the partnership *(what you're doing)*.

IMPORTANT TERMS—RULES of thE PARTNERSHIP

Next, you need to write up the *Rules of the Co-Ownership Agreement*. These are the important terms (rules) the partners must live by. I will list the important terms you must make sure to address in the co-ownership agreement.

1. NAMES OF PARTNERS—Place of business, dates.

2. FORM OF BUSINESS—For example: General partnership, corporation or limited association.

3. CAPITAL CONTRIBUTION—Who puts up the money, when and how.

4. SHARING INCOME, EXPENSES, PROFITS. How they will be divided.

5. WHO IS IN CHARGE of daily management of property.

6. WHO CAN SELL, ENCUMBER, TRANSFER or PURCHASE new property.

7. WHAT TO DO ABOUT DISAGREEMENTS.

8. WHO WILL KEEP BOOKS, records and do taxes.

9. DEATH OF A PARTNER—What will you do.

10. INDEMNIFICATION of each partner against debts of other.

11. VIOLATION OF TERMS of agreement. What will you do.

12. RESTRICTIONS OF THE PARTNERS. Don't allow borrowing, for example, or pledge the property.

13. TERMINATION—When, how, for what reason. Is extension okay?

14. WITHDRAWAL PARTNER—How to split assets and in what manner.

These are your primary considerations. When you agree to invest your skills and money, take the time to write down the rules you'll conduct business by, it pays. You'll find an example of the co-ownership agreement I use on Page 289.

FINDING MONEY WHEN YOU DON'T HAVE ANY

One plan that has worked very well for me is another variation of having a partner or co-ownership arrangement.

When you can't swing a deal by yourself, your best bet is to find someone with money! Let's say you are like many other "short-of-cash" investors trying to get your investment plan off the ground. You have that "big dream," plenty of personal drive, a willingness to work hard, but your bank account is empty.

We'll say you've already found a good income property, but the seller wants the normal 20% cash down payment. Let's assume that even though you have a regular 40 hour per week job, you've made up your mind you are willing to work evenings and weekends to build real estate wealth! You don't have much money, but you're willing to work. You'll do whatever it takes to get your investment plan in motion. To say it another way, you're ready to contribute yourself—you're willing to substitute your personal time and energy in lieu of the cash down payment.

NO MONEY—NO PROBLEM—ONCE YOU PROVE YOURSELF

What you need now is someone who does have money. A partner who is ready, willing and able to put up the cash down payment. As I said, you've already found a good solid investment property with great potential, but you need the down payment money fast. *Let me ease your mind by telling you that lots of people are willing to invest their money in these kinds of transactions—they are more plentiful than you might ever imagine.* However, there is a catch—they can be very difficult to find when you're starting out. The problem is that all investors like safe bets. They like to deal with winners. The ones I know are always eager to invest their money, but only if they think you can make them more money.

YOU MUST DEVELOP GOOD TRACK RECORD

Hands-on **working investors**—who develop the ability to find rundown properties, fix them up and make healthy profits for their cash investing partners—soon develop good reputations. *Good reputations are called track records.* The problem for new investors is obvious, they have no track record yet! So when you're new, there is no evidence that you know how to invest someone else's money and earn a profit. That's the biggest obstacle you must overcome. It's not an easy task. However, many others, with less abilities than you, have done it; and so can you.

Author Offers Special Assistance

If you are a new investor just getting started, do it alone if you can. After a couple years investing on your own, you'll be in a much stronger position to set up a small partnership or co-investor arrangement. Make no mistake about it, investing with a partner who can supply extra cash will add to your investment future.

My own investing career has been greatly enriched using the financial strength of well-healed co-investors. If you have any doubts about how partnership leverage can accelerate your earnings and profits, Chapter 17 should help you understand how rewarding a co-investor can be.

Once you decide a partnership or co-investing is right for you, then I can help you with my specialized training materials. Also included with the material are copies of my personal contracts and agreements to *protect you*. Do not participate in any partnership arrangement without a **written contract**. Ask for my training products #2110 or #2115. It pays to start out on the right foot! See the Appendix Section (Resources) for additional information and instructions for ordering.

CHAPTER 16

Sell Half the Property to Increase Your Income

50% SALES TURN NEGATIVE INCOME INTO POSITIVE

*T*here are some very worthwhile advantages to fixing up real estate first, then finding an investment partner to help bail you out of debt. *One advantage is that you can sell your "sweat equity" for a premium price or a very high markup!* The following is an actual case I became involved with. The names have been changed.

Allen, a student of mine, recently came to Redding for counseling and a tour of my fixer houses. He wanted to see for himself exactly what kind of properties I recommend for most investors. He also wanted advice about how to generate more cash flow without selling the properties he'd worked so hard to develop!

During the past several years, Allen has managed to acquire several properties with low down payments. Today, he owns two single-family houses, a five-unit "leper property" (a property no one else wanted to touch), and a seven-unit apartment building. He's done very well with low down payments; however, he suffers from the affliction common to most all *LOW-CASH DOWN* buyers—elephant-size mortgage payments every month!

Low Down Payments Often Spell High Mortgage Payments

Allen has managed to fix up the properties himself as he goes along, but he's in a cash bind now, because his fix-up expenses have basically been paid from his personal bank account, which is running on fumes! Nothing stops a good investment plan faster than running out of money. Allen has reached the point where he fully understands the "flip-side" of *no down*, or *small down*, payment buying. It generally means no cash flow or a very small amount at best.

Allen came to Redding well prepared to discuss his financial dilemma! He brought along six different sets of computer spreadsheets to show me why he was losing money every month. None, however, offered a solution for how he might stop losing money! Allen saved that problem for me to solve.

Best Computer in the World Doesn't Help Broke Investors

My seminar students understand I'm not a computer person when it comes to earning money. I use a rather simple technique that employs what I call the "MLO formula"—that stands for "MONEY LEFT OVER." It works like this—every month I deposit all my rents and other income from carry-back notes in my bank account. During the month I write checks to pay my employees, property expenses and mortgage payments.

Finally, when the last day rolls around, I check the balance in my account. If there's something left for me, that's MLO! See how simple accounting can be? I might add, there's an extra $10,000 in my account today, which I've saved over the years by not buying a computer to figure out whether I'm making money or not!

50% SALES CAN GREATLY IMPROVE CASH FLOW

One of my favorite methods for generating income quickly, which of course is what Allen wanted to hear about, is to purchase a rundown property with good upgrade potential—fix it up, then sell 50% of the ownership to a passive investor. You might be thinking—Why sell only 50%? Why not sell the whole property?

First of all, if I can relieve my cash flow problem by selling only 50%, I'm more than happy to keep the other half for myself. As you shall soon learn, there's extra money to be made by the 50% owner who manages the property in addition to the profits from the sale itself! Let me show you how a typical transaction might look. We'll call this property "El Dumpo Villa."

EL DUMPO VILLA – TWO DUPLEXES AND THREE SINGLE-FAMILY HOUSES
Seven rental units on a large city lot (older, shabby and rundown)

FINANCIAL DATA

Monthly income equals $375 rent per unit, times seven units.... $2,625
Annual income equals $2,625 times 12 months.......................... $31,500
Purchase price (offer) 6.5 times gross rents ($31,500).............. $204,750
Purchase price (accepted)... $205,000

Terms: $15,000 cash down (6 to 10% is average for rundown properties)
$190,000 unpaid balance – seller agrees to carry back the mortgage at 9% interest amortized over 20 years
$1,709.49 equals monthly payments (principal and interest)

TYPICAL PROPERTY **INCOME** AND **EXPENSE** SET-UP AT BEGINNING

Gross income (rents)..	$2,625.00	100%
Vacancy and credit losses (uncollectables)..............	265.00	10%
Gross operating income...	$2,360.00	

Monthly Expenses (owner is the manager)

Taxes..	$180.00	7%
Management...	-0-	10%
Insurance..	105.00	4%
Repairs...	265.00	10%
Maintenance..	130.00	5%
Advertising and accounting.........................	50.00	2%
Utilities..	105.00	4%
Total expenses...	$ 835.00	
Net operating income................................	$1,525.00	
Mortgage payment......................................	$1,709.49	
Cash flow (positive or **negative**)...............	($ 184.51)	

THE TASK IS TO QUICKLY FIX UP THE PROPERTY AND ADD VALUE

Let's assume the fix-up costs (labor and material) will be 10% of the purchase price, or $20,000. Don't forget, however, the buyer will provide the labor for this project. On average, the cost breakdown between labor and material is 70% for labor ($14,000) and 30% for material ($6,000). Therefore, the buyer's out-of-pocket costs for material will be $6,000.

Depending on how fast the buyer can accomplish the fix-up task (I always estimate 12 to 18 months), he'll probably lose a few tenants who don't see any need for improvements. They also see the handwriting on the wall, which tells them that in the very near future, when things are cleaned up, rent increases will likely follow. Tenants living in junky properties, who are paying under-market rents, already understand the reason rents are low. They know very well, rents won't stay low for long once the property gets fixed up!

Rents and Gross Multipliers Go Up Together

Finally, the big day arrives and the property is all fixed up and looks great! Along the way, we had some tenants come and go. The new ones moving in should naturally be paying higher rents. A fixer property of this size, purchased at 6.5 times gross rents, with $20,000 worth of fix-up work completed, should easily command 25% higher rents! That means total rents of $2,625 per month can be increased to $3,285 when the fix-up work is done! A 25% increase is very modest, in my experience. More often, I'm able to achieve 40 to 60%, even higher increases, especially if my fix-up job takes longer than a year—new tenants will see the big improvement and are willing to pay higher rents.

With **new looks** and a **higher income**, the property is now worth much more than the lowly 6.5 gross rent multiplier. For example, in my area, a 6.5 gross multiplier is what you will most likely pay for ugly rundown properties. After they've been fixed up, however, they will generally sell for about 8 times the gross rents.

Buyers Will Pay Extra for No Cash Down Deals

The secret to selling 50% ownership in a fixed-up property for the highest price is **make the buyer an offer he can't refuse**! My suggestion is to offer a NO-DOWN DEAL to a qualified buyer. Naturally, if you need to retrieve your fix-up costs in order to keep buying groceries, a NO-DOWN sale may not work for you! However, folks with good jobs (the kind of buyer you want) are always eager to purchase investment real estate for *no down payment*, especially if you'll manage it, and they are happy to pay you monthly installments. Most of them will not even question your selling price. From a personal income tax standpoint, most buyers can generally offset a good portion of their monthly payments with tax write-offs (savings) from their regular wages or salary.

HOW TO MARKET A FIXED-UP, FIXER PROPERTY

No, I haven't forgotten about the $21,000 cash that's tied up in this deal ($15,000 down payment, plus $6,000 materials). However, the income we can generate from a 50% sale will quickly pay that back. Let's take a look at a marketing strategy for the fixed-up property. You should understand that no-down buyers are willing to pay top retail prices in exchange for an easy buy-in (no cash). In this example, selling for 8.5 times gross rents should work. Here's how the numbers would look:

New annual rents equal $3,285 per month, times 12 months	$ 39,420
8.5 times gross rents equals selling price of (rounded)...	$335,000
Existing mortgage balance	$189,000
Total equity	$146,000

Notice that the selling price of $335,000 represents more than a *63% mark-up*, which is pretty good when you're calculating profits.

Obviously, a 50% buyer will be entitled to half of the income. He will also assume half the existing debt and share expenses on a 50-50 basis. What the buyer will be paying you for with new monthly installments is 50% of the equity! In this case, 50% of $146,000 is $73,000. That's his share. I generally draw up a promissory note, amortized over ten years at 10% interest. That works out to 120 monthly payments of $964.71. *Certainly, 15 years would be acceptable to me if my buyer insists on smaller payments.* The sale gets reported like this for setting up the buyer's 50% purchase:

Total Sale Price	$167,500
Buyer to assume existing mortgage	$ 94,500
Buyer to execute new promissory note in favor of seller	$ 73,000

The transaction can be done with a DEED transferring title or a LAND CONTRACT without a title change, which is often done with *no-down* transactions—either way it's okay, it's strictly optional.

When the 50% sale is completed, here's how the new operating numbers will look. They show who gets what—and who pays for what!

INCOME AND EXPENSE SET-UP—AFTER 50% SALE

	Seller 50%	Investor 50%
Total monthly income	$1,642.50	$1,642.50
Vacancy and credit losses (deadbeat tenants).	164.50	164.50
Gross operating income	$1,478.00	$1,478.00
Monthly expenses		
Taxes	$ 177.00	$ 177.00
Management (your share free)	-0-	165.00
Insurance	66.00	66.00
Repairs	164.00	164.00
Maintenance	82.00	82.00
Advertising and accounting	33.00	33.00
Utilities	66.00	66.00
Total expenses	$ 588.00	$ 753.00
Net operating income	890.00	725.00
Mortgage payments	854.74	854.74
Cash flow (positive or negative)	**$35.26**	($129.74)

CONVERTING NEGATIVE CASH FLOW INTO POSITIVE

As you can see from the after sale setup, the seller's negative cash flow—originally -$184.51—has been eliminated. Instead, the seller now has a $35.26 positive cash flow. That's not much, but we're just a beginning. Each month for the next 10 years the seller will receive $964.71 from the buyer until the $73,000 equity purchase is paid off.

It's starting to get more interesting, wouldn't you say? So far the seller is earning $35.26 positive cash flow from operations, and $964.71 from his note receivable! That increases the seller's monthly income to $999.97. But, we're still not done yet! You'll notice the $165 management fee on the investor's side of the after-sale setup. The law says that owners can't pay themselves a fee for managing properties they own, but they certainly can receive management fees for the portion they don't own! If you are the manager for the 50% you don't own, then you are entitled to the $165 management fee. Now, your monthly earnings are up to $1,164.97. It's getting better—wouldn't you say?

Do you recall the 70-30% labor/material cost breakdown? That's where I told you 30% of all repair and maintenance expenses go for buying materials. Once again, if you perform labor on the 50% share of the property you don't own, you are entitled to earn the labor fees. In this case, the investor's setup shows $164 for repairs and $82 for maintenance—a total of $246—70% of $246 equals $172, which goes to pay you for the labor associated with repairs and maintenance.

Now, let's review exactly what we've accomplished in terms of improving our cash flow. When the El Dumpo Villa property was acquired, the income and expense setup showed a bottom line loss of $184.51 per month. Now, after selling 50% of the property, you'll see the seller's *monthly cash flow* has improved dramatically. Take a look at what you'll now receive each month if you manage the property and do the repairs and maintenance work.

Cash flow from operations	$35.26
Mortgage payment from investor	964.71
Management fee from investor	165.00
Payment (repairs & maintenance) from investor	172.00
Total MLO, after expenses	**$1,336.97**

In case you forgot, MLO means **MONEY LEFT OVER**.

You'll notice the seller's cash flow is a whole lot better after selling 50% of the property. The ancient Roman, Julius Caesar, said: "If you can divide them, you can conquer them." That same strategy works just as well dividing real estate, as it did for the early Romans!

JOINING TOGETHER FOR PROFIT OPPORTUNITIES

Using other people's money to acquire real estate is an excellent way to build your own wealth faster than you could do it alone, however small informal "mom and pop" type partnerships are not without their problems. To begin with, partnerships involve at least two people with different ideas. If you've ever invested in a partnership that loses money, you'll quickly find out just how different those ideas can be.

Still, joining together to make money can and does work, if you can discipline yourself! You must be *tolerant, understanding* and *very patient*. You and your partner must understand that your mutual success is totally dependent upon each other. You must stay focused on the idea that partnership investing takes full cooperation by all the parties involved—anything less will most likely cause failure.

Sound Advice With A Harsh Bite

With *partnership investing*, you definitely need a **partnership agreement**, but it doesn't have to be expensive. I will show you exactly what kind of agreement you need to keep you and your money partners on the right track! You do not need to be formal partners to invest together. There's a simpler way, as I shall explain. First, however, allow me to offer a few suggestions and some personal philosophy, since I've been a participant in many small informal investment groups.

This advice may sound somewhat harsh, however, it was passed along to me years ago by a friend, and it has worked very well for me. The advice was from a marriage counselor who told a young couple contemplating their wedding, *"Before you get married," he said, "First, consider what you plan to do when you get divorced."*

I'm not sure how long the counselor lasted in the counseling business or how many couples sought his advice, but from a purely objective standpoint, his advice makes a lot of sense. In fact, it's exactly the same advice I pass along to anyone who is thinking about buying real estate with another party. *The best way to guard against failure is to be prepared for it and then do everything you can to prevent it.*

Buying Properties as Tenants-in-Common

Co-ownership investing can be an excellent way for a couple of individuals to acquire real estate together. Often two invest-ment-minded people will get together and decide it would be a great idea to pool their resources and buy rental houses. They figure they can sell them, later on, and split the profits. With this sort of informal plan, the best way to acquire property and take title is as TENANTS-IN-COMMON. Don't confuse this with being a renter—called a tenant—or with joint tenancy, which is generally how husbands and wives take title to jointly-owned properties.

Formal partnerships have several disadvantages for smaller "mom and pop" type investors! First, it can be more expensive to set up the bookkeeping. A formal partnership is like a separate taxpayer when it comes to income tax report-ing. Federal taxes are prepared on a special Form 1065, with a Schedule K-1 for each partner. Aside from being more formal and expensive, they can be a pain because 1065s are excellent sources for government tax audits. Also, partnership state-ments and fictitious business name documents must be filed in each county where partnership property is owned. That means you'll always need copies and updates for bank loans or refinancing.

Perhaps the most serious reason to stay away from formal partnerships is because you cannot exchange partnership interests using tax code Section 1031. For this reason, alone, you should give serious thought to keeping your association informal. Most banks or institutional lenders will not make direct loans to small partnerships. Individuals generally have a much easier time borrowing money. Having information on file will most likely require you to purchase a local business license to operate rental properties.

Tenants-in-common investors can benefit from 1031 exchanges. Tax reporting is done by each partner on his individual tax form 1040, for Federal taxes. There are no additional tax forms requiring information about the partners. As tenants-in-common, each partner can own any percentage of the property agreed upon—it can be 50/50, 20/80, 90/10, or any other combination, but whatever it is, it should be specified on the property deed when the ownership is recorded. When a tenant-in-common dies, his share or percentage of ownership in the property is passed to his heirs. That's different than in joint tenancy, where the surviving joint tenant gets the share of the party passing on.

NEVER INVEST WITHOUT A WRITTEN AGREEMENT

The agreement to buy real estate and deciding how to conduct business and pay taxes is the easiest part of partnership investing. Not unlike new marriages, the real test always comes after the initial promises. **THE BIGGEST MISTAKE ANY TWO PARTNERS CAN MAKE IS TO BEGIN BUYING PROPERTIES WITHOUT A WRITTEN AGREEMENT.** You must decide who is responsible for what, who does what, how the money is allocated, and—last, but perhaps most important—how to terminate the investment, in case things don't work out.

A hypothetical transaction between **Fixer Jay** and **Sam Money-Bags** is shown on Page 289. The **co-ownership agreement** outlines the terms agreed to by both investors. The agreement can be recorded at the local county recorder's office. I always prefer to record a memorandum of the transaction instead, without specifying the terms on public documents. Memorandums are my choice, because I don't like telling the general public about by business affairs.

CO-OWNERSHIP AGREEMENT

1234 EASY STREET—GOLDEN CITY, CALIFORNIA

THIS AGREEMENT is made effective as of the first day of April, _____, between Fixer Jay and Sam Money-Bags.

1. *Transaction*: Fixer Jay (Jay) and Sam Money-Bags (Sam) will join together as co-owners for the purpose of owning and operating that certain real estate located at 1234 Easy Street, Golden City, California (the Property), for the mutual benefit and profit of each. Each party agrees to perform fully under this Agreement for the success of both parties herein.

2. *Acquisition of Property*: Sam and Jay have purchased the Property for a purchase price of Seventy-Five Thousand Dollars ($75,000). The cash down payment of Twenty Thousand, Five Hundred Dollars ($20,500) was paid by Sam. Both parties will take title subject to the existing mortgage lien in the amount of Thirty-One Thousand, Five Hundred Dollars and 00/100 ($31,500). The seller of the Property has agreed to finance the balance of the purchase price, approximately Twenty-Three Thousand Dollars and 00/100 ($23,000), with installment payments of Two Hundred Dollars and 00/100 ($200) or more per month, including seven percent (7%) interest until the entire principal is paid in full.

3. *Cash Distributions from Rental*: All excess cash derived from rental of the Property, after payment of all expenses and debt service, shall be divided eighty percent (80%) to Sam and twenty percent (20%) to Jay.

4. *Cash Proceeds from Sale or Refinancing of the Property*: Net cash proceeds derived from sale or refinancing of the Property shall be shared as follows:
First, Sam shall receive back all of his capital invested in the Property by way of initial down payment, fix-up expenditures and operating expenses made pursuant to Paragraphs 2, 11 and 12, hereof. Thereafter, all remaining proceeds derived from sale or refinancing shall be shared eighty percent (80%) to Sam and twenty percent (20%) to Jay.

5. *Management*: All decisions regarding the management of the Property shall be made upon the joint approval of both Sam and Jay provided, however, it is agreed that Jay will have primary responsibility for the day-to-day management operations, such as rent-up, property maintenance, repairs, cleaning and the like, in order to conduct an efficient rental business. Jay shall receive a 6% management fee per month (fee based on income) for managing the property. In addition, Jay shall be reimbursed for his actual out-of-pocket costs and expenses incurred in connection with such management.

6. **Books and Records**: All books and records will be kept at the office of Jay. A statement of operations will be provided to Sam on a monthly basis. This statement will be prepared by Jay as part of his management duties.

7. **Bank Accounts**: Jay shall maintain a commercial checking account at Gold Street Bank, 2930 Silver Lane, Golden City, California or at such other banking institution that shall be approved by Sam, for the purpose of operating the Property.

8. **Indemnification**: Each party shall indemnify and hold harmless the other party and the Property from and against all separate debts, claims, actions and demands of said party.

9. **Termination**: This Agreement shall terminate upon the sale of the Property or by mutual consent of Sam and Jay. Sam shall have the sole right to determine when the Property is to be sold provided, however, that Sam shall first offer Jay the right to purchase the Property for the same amount and upon the same terms and conditions as Sam is willing to sell the Property pursuant to a bonafide offer received from any third party. Jay shall exercise said right of first refusal within ten (10) days after receipt of notification from Sam of his intention to accept said third party offer. Jay shall consummate the transaction within sixty (60) days after exercise of his right of first refusal.

10. **Death of Parties**: Upon the death of Sam, Jay shall have the right to either purchase Sam's interest in the Property in the manner described in Paragraph 9 hereof, based upon a bonafide offer received by Sam's estate or, in absence of such an offer, Jay shall have the right to cause the Property to be sold and the proceeds divided in accordance with Paragraph 4 of this Agreement. In the event liquidation is elected, Jay shall proceed with reasonable diligence to liquidate the Property within twelve (12) months after Sam's death,

11. **Initial Fix-Up Expenditures**: Initial fix-up funds for rehabilitation of the Property will be contributed by Sam. All work will be performed by employees of Jay. Employee time sheets and material invoices shall be part of Jay's record keeping.

12. **Operating Funds**: All expenses for repairs, improvements, taxes, insurance, maintenance and other operating expenses, deemed necessary for the operation of the Property, shall be paid first from rental income derived from the Property and, thereafter, from additional funds to be contributed by Sam.

13. **Business Address**: The official management office for the Property will be Jay's One Stop Rental Company, located at 2020 End of the Trail Drive, Golden City, California 96001. Mailing address is c/o Fixer Jay, P.O. Box 492029, Redding, CA 96049-3039.

14. *No Partnership or Joint Venture*: The relationship between Jay and Sam under this Agreement shall be solely that of co-owners of real estate and under no circumstances shall said relationship constitute a partnership or joint venture.

IN WITNESS WHEREOF, the Parties have executed this Agreement as of the day and year first above written.

_____   _____
        Sam Money-Bags   *Fixer Jay*

STATE OF_____   County of_____

I, _____, Notary Public in and for the State of _____, do hereby certify that on this _____ day of _____ 19____, personally appeared before me _____ to me known to be the individual _____ described in and who executed the within instrument and acknowledged that _____ signed and sealed the same as _____ free and voluntary act and deed for the uses and purposes herein mentioned.

GIVEN UNDER MY HAND AND OFFICIAL SEAL this _____ day of _____ 19____.

Notary Public in and for the State of _____ residing at _____ _____ in said County.

ALWAYS FORMALIZE THE TERMS YOU AGREE TO

This contract is not meant to be exactly the same for every transaction! But, it is typical of the kind you might use for co-investing. The contract is what you write up to formalize and document the terms you agree to when you make a deal. Remember, the important terms we've already discussed when you write up your own contract. Never forget that, as an investor in real estate, if it isn't in writing, *it just ain't so.*

CHAPTER 17

Jay's 90/10 Money Partner Plan for Wealth

My 90/10 money partner plan is a shared ownership arrangement. It means that I and my partner will both invest in a property. The partner will contribute 90% of the cash down payment—making him the money partner—and I will pay the remaining 10%. For this plan to work, the down payment should be at least 20 to 25% of the purchase price. I prefer not to scrimp on the cash down payment. The reason is that when the mortgage debt exceeds 80% of the purchase price, there is great danger of having negative cash flow, while operating the property. The down payment should always be large enough to reduce the principal balance so that the income generated from the property will cover both the mortgage payment and operating costs.

Jay's 90/10 Money Partner Plan for Cash-Poor Investors

When you locate a property that seems promising, thoroughly check out the **income** and **expenses**. You don't want any hidden costs of operation to show up after you close the deal. It can be very awkward explaining to your partner that, because you underestimated operating costs, he must now come up with additional money every month. It's also not very good for the track record you're trying to establish.

HOW THE 90/10 PLAN WORKS

The 90/10 plan works like this—I find a money partner who will put up 90% of the cash down payment after I locate a good solid money-maker property and agree to do the daily management. The money partner is basically a "hands off" investor. I will be the field manager and operate the property. I agree to do the renting, cleaning, fixing, advertising and whatever else is needed to make our joint investment profitable for both of us.

Small Ownership Cost Buys Big Returns

To illustrate how my 90/10 plan works, let's go through some numbers. Assume we locate a $90,000 rental house that we can purchase for $80,000. That shouldn't be too difficult to do when we can pay $20,000 cash down. The money partner in this case puts up $18,000 (90%), of the down payment cash and becomes a 90% owner. I contribute $2,000 (10%) and get 10% ownership in the property. Remember, I'm only talking about the down payment here. I continue to contribute my services throughout our ownership period, which is normally set up for a period of ten years on my agreement. *By the way, don't make these deals without a co-ownership agreement.* The agreement doesn't need to be long and complicated; however it must cover the important terms, which are the same for any co-investment venture. See Chapter 16 for an example of how to use my co-ownership agreement.

For Just 10% Cash, I Receive 50% Profit

Although, I'm only a 10% cash investor, I will earn 50% of the profits at the end, plus there's an excellent chance I'll make a few dollars from cash flow along the way. At the end (the date is specified in our contract), the profits are split like this. First, the 90% investor and I will both get our down payments back (90% investor gets $18,000, and I get $2,000). After that it's split 50-50 on the balance, after subtracting what we still owe on the mortgage. For example, let's say the $90,000 property we buy now for $80,000, sells for double that amount ten years from now—that's $180,000—and the original $60,000 mortgage debt is paid down to $50,000. HERE'S HOW THE NUMBERS WOULD LOOK:

Original purchase price..	$80,000
Cash down payment..	20,000
Mortgage balance to start..	$60,000
Selling price after ten years..	$180,000
Mortgage balance after ten years................................	50,000
Gross profit...	$130,000
Original cash down payment (returned to investors)......	20,000
NET PROFIT to split 50-50......................................	**$110,000**

I realize there will be selling expenses (escrow fees) and, perhaps, real estate commissions to pay. However, as you can see, there is still a sizable profit left over.

Depending on how long the contract agreement is and how much the property appreciates in my particular investment area, I expect to earn a substantial return on my 10% portion of the down payment. Naturally, I'll get my $2,000 back, too.

HIGH RETURNS AND BUYING POWER ARE KEYS TO PLAN

The reason I like the 90/10 plan so well is that I get a maximum for my money. High percentage returns and leverage are the *big wealth-builders* in real estate. This plan allows a skilled investor, with a very small amount of up-front cash, to generate the same "horsepower" as a much larger, traditional down payment. A $2,000 down payment would not be enough cash, in most cases, to bargain with sellers about their price. Most would not be willing to sell for such a small down payment.

This is the kind of deal that works very well for folks who don't have much money to invest, but who are willing to invest their time and professional skills. I always view single ownership (me only) as the best kind. However, lack of cash makes this plan an attractive alternative, until you do have more money. Real estate investors need to be creative. *This plan creates big opportunities for ambitious folks who are temporarily short of money.*

Balancing the Accounts

The 90/10 plan provides for clean accounting of the transaction. The money partner who contributes 90% of the cash down payment also gets 90% of everything else. That includes income, expenses, depreciation and tax credits. He or she is also responsible for 90% of additional contributions, if needed, for upgrading or repairs. The 10% partner (field manager), or worker, is responsible for the remaining 10% of everything, including any shortages during operations of the property. He also receives 10% of the cash flow.

Length of Ownership Should be Determined at the Start

Investment partners should always decide up-front how long they wish to be partners. Always use a co-ownership agreement, as I said earlier. I personally like to state a firm termination date, then provide for extending if we should choose to continue.

How does this deal stack up in terms of equality? What are the values of each contributing partner? Obviously $18,000 and $2,000 are not equal cash contributions. To determine the *non-cash values*, let's make several assumptions. First, we'll say the property we decide to purchase will rent for $800 per month, or $9,600 annually, to start out. We'll also say our co-ownership agreement specifies a term of ten years, and that the working partner will be allowed a 10% management fee.

The ten years of management fees will be calculated at 120 months times 10% of the $800 for a total of $9,600. Also, rents are estimated to increase at least 3% each year during the ten-year term, so that amounts to roughly $400 in additional management fees. After ten full years, my contribution will be the $2,000 cash down payment, plus $10,000 worth of management fees, for a total of $12,000.

Okay, that's much closer to the money partner's contribution, but still less. However, the money partner gets some tax benefits. Assume the $80,000 property will have $60,000 worth of depreciation. For this purpose, let's only consider the 27.5 year depreciation schedule ($60,000 divided by 27.5 years equals approximately $2,200 annual depreciation). The value of a 90% share of depreciation to a 28% tax payer, who qualifies for the deduction under current tax rules, is $554 annually, or $5,540 for ten years.

Now, you see how the contributions of both partners are starting to balance out. The money partner's net contribution after taxes is $18,000 less $5,540 for a total of $12,460. The 10% remaining depreciation is worth about $600 to the working partner if he has adequate income to use it. Also, there's the distribution of cash flow, particularly in the later years of ownership, to be considered. The cash investor will be entitled to 90% of all rental income, over and above expenses. This rent money will reduce his net cash invested even more before the partnership ends. Basically, the contributions end up about equal for both investors.

NEVER FORGET THE GOLDEN RULE OF INVESTING

When you're first getting started, a money partner will always be the most important member of your investment team. I call my 90% money partner "THE GOLDEN PARTNER," and with all golden partners, the Golden Rule applies! The rule is simple—*HE WHO HAS THE GOLD, RULES!* Look at the situation this way—no matter how important you value your personal skills, or how much money you think you might earn for the partnership, it isn't worth a hill of beans unless you can first acquire the property to apply those skills. If you don't have the money to purchase property, nothing else matters much. I hope this point is clear. That's why a money partner is truly a *"Golden Partner."*

Leverage—the Investor's Hamburger-Helper

One of the biggest reasons why small-time investors can end up filthy rich is because partnership investing combines the use of a very powerful law called LEVERAGE. It is written that the Greek mathematician, Archimedes, who discovered leverage, once claimed he could lift the entire world all by himself, if he had a long enough stick and a place to stand.

Real estate investors can do the same thing with the assistance of **other people's money**, often referred to as "OPM."

PROOF OF SKILLS NEEDED FIRST

As you become more successful at investing and in managing properties, it will not go unnoticed by folks with money around you. People will pay you big money for proven skills. Naturally, these skills must be demonstrated before they are marketable to others. A miscalculation made by many start-out investors is believing they have marketable skills to offer, before establishing any kind of track record. Develop a track record first, *AND THE MONEY WILL FIND YOU!*

Rundown Apartment Offers Big Profit Potential

First Main Street was a 92-unit studio apartment built in 1927. After many years of neglect and of being rented to the wrong kind of occupants, the three-story building was badly in need of some TLC—or *TENDER LOVING CARE.*

Structurally, the building was very solid, but many pipes were leaking, the paint was peeling, and the carpets smelled like a thousand cats were locked inside. The seller had great ideas when he purchased the property, but renting it out was his downfall. He had no screening requirements for tenants, other than they had to pay the first month's rent! *To get a key required no applications, no interviews and, obviously, no rules for living in the building.* It was a perfect plan for disaster, and, of course, that's exactly what happened. After four years, he ended up in bankruptcy and was forced to sell the property.

I negotiated with the seller for almost a year, making offers that, if accepted, would have been very beneficial to me; however, since none of them were, I was forced to make my offers a bit more realistic. The seller was broke, but he wasn't stupid. The building had a lot of potential—we both knew it—even though it would take a lot of money to get the apartments fixed up and rented out to decent tenants. I estimated the studio apartments, once fixed up, would rent for at least $250 a month. That's potentially $23,000 per month—more than a quarter of a million dollars annually.

IN SEARCH OF INVESTOR CASH

The big problem for me was finding the immediate cash I needed to fix a building of that size! I had never attempted to fix up so many apartments all at one time and I knew I would need some financial help to get the job done. I decided to schedule an appointment with a local physician who had purchased several properties from me over the past several years. When I told him about the potential income and the very generous depreciation allowance for partially furnished units, he was ready to write me a check! We agreed to set the deal up using my **90/10 INVESTMENT PLAN**. Here's how it worked.

The doctor would put up 90% of the required cash, which would include the *down payment* and *all the fix-up costs*. Naturally, the fix-up money would not be required all at one time since the bills would be paid as we went along. I estimated the job would take about three years from start to finish! It was my plan to improve the cash flow and use the extra rent money to help pay for some of the fix-up work! I also knew that many of the tenants we inherited would most likely not survive too long under our much stricter renting policy.

The building had only 53 tenants on the day escrow closed, and I expected to lose some of those as fix-up progressed and I began systematically weeding out undesirable renters. By fixing up vacant apartments first, we could bring in new tenants at a higher rental rate.

THE WRITTEN CONTRACT A MUST

My 90/10 plan is an ideal two-party investor vehicle for the active investor, with the ability to fix-up and manage the property, and for the passive investor, who can supply most of the cash that's needed. I have, generally, found it best to conduct business as co-investors or tenants-in-common, rather than create a separate tax reporting partnership. Co-investing doesn't mean you should be sloppy about written rules or agreements, merely because it's not a "formal partnership." In fact, let me once again emphasize that you should never invest with anyone without first preparing and executing a written agreement detailing exactly who does what and when! Refer to the example of my CO-OWNERSHIP AGREEMENT on Page 289. You can rearrange the terms of the agreement to suit your particular transaction.

Co-Investors are Tenants-in-Common

My basic agreement for FIRST MAIN STREET was quite simple, which is the way agreements should be. Both investors will take title as *tenants-in-common.* The 90% investor (money partner) will receive 90% of the total benefits, including rents, credits and depreciation, during the period of fixing up the property. The 10% investor (operator)—that's me—will receive 10% of the benefits, plus, in addition, a management fee equal to 5% of the gross monthly rents until the property is sold or traded.

The co-investor agreement should specify a future date when a sale is planned! Naturally, there should be enough flexibility in the written agreement to allow for selling whenever you can take best advantage of a good seller's market. When the property is sold, each investor will be fully reimbursed for his total cash contribution first. The net sale proceeds will then be split on a 50/50 basis.

SELECTING THE RIGHT PARTNER IS CRITICAL

I cannot over-emphasis the importance of a good investor match-up if this plan is to help you create wealth! *In my view, the only time partnership investing makes good sense is when both investors are in need of what the other possesses or can immediately provide!* It will only work if both are made stronger by joining together. Quite often you will see investors of equal means attempting to work as partners; however, they are seldom successful because they are no stronger together than they were as individuals.

First Main Street required my fix-up expertise and the doctor's money in about the same proportion! To say it another way—without a lot of money, all my fix-up skills could not earn me one thin dime! And, from the doctor's viewpoint, he could do absolutely nothing to take advantage of a high-profit opportunity without my fix-up skills and ability to manage the building and the tenants. **This is what I mean when I say that both investors are stronger together.**

Fix the Building, Then Up the Income

The building was acquired for approximately $450,000, which was an excellent price, even though it was a mess! It would require several years of fix-up work and enough time to rid the place of its "flop house" reputation. As I had planned, much of the fix-up work was paid for from the monthly cash flow. *Tenant cycling* and *higher rents* were initiated almost immediately and new tenants moving into the upgraded apartments had no objections to our $60 per month rent increases. Painting and cleaning made a tremendous improvement to the looks. Most tenants had nothing but praise for our fix-up efforts.

HANDYMAN SKILLS ARE WORTH BIG BUCKS

It is not necessary to have a lot of money to make a lot, but you must have a good substitute if you don't have it! Fix-up skills, and the ability to acquire bargain properties and then manage them, are CASH EQUIVALENT SKILLS.

They are as valuable as cash once you prove yourself. As I said earlier, money will find you when you can show investors a successful project or two—projects where you've made money for yourself and your partners.

SEPARATION OF DUTIES IS ESSENTIAL FOR SUCCESS

Although it is clearly spelled out in a 90/10 plan agreement who is responsible for what, it's extremely important that each co-investor be allowed to perform his specified task with the least interference from the other. Obviously, a fairly high level of trust is necessary to get the job done. For example, the doctor didn't try to give me fix-up advice, unless I asked his opinion, nor did I have to beg him for money when the big invoices started rolling in for payment. This kind of understanding is absolutely necessary before any joint project is started.

After three years of fix-up and tenant cycling, the First Main Street Studio Apartments were like new again. They were easy to rent, even though the larger units were renting for $125 more than when we acquired the building. Speaking for the operator/manager side, I was quite proud of what we had accomplished—and, because we were able to attract 35 more renters along the way, our cash contributions for fix-up were much less than we had anticipated when starting out.

Dividing up the Money at the End

When we were ready to sell the First Main Street apartments, my total cash contribution added up to $16,000, and the doctor had invested $145,000 as 90% co-investor. However, an opportunity to sell didn't present itself for another year and a half down the road. Finally we sold the apartments for $300,000, over and above the total amount of money we had invested. When escrow closed, the doctor's check was $295,000, more than double what he had invested. He had also enjoyed substantial tax write-offs in the first three years of ownership. That benefit sheltered lots of doctoring income.

I chose to carry back a promissory note for $150,000, secured by the First Main Street property. The terms provided monthly payments of $1,500 (12% interest-only), with the principle all due in ten years. My total compensation for the use of my fix-up skills and property management was very pleasing to me, as the following numbers show.

52 months of management fees....................................	$ 40,820
120 months at 12% interest payments..........................	180,000
Principle amount from note receivable (end of 10 years).	150,000
Total earnings...	**$370,820**

Leverage is about investing a very small amount of your own money in order to earn a very large amount from someone else. First Main Street is a perfect example of how maximum leverage can help your bank account attain higher status.

In terms of personal hours spent on the job, I've estimated it's somewhere near 2,500 hours for the period of time I owned the apartment. Some days, the work required 10 or 12 hours, but many days it was only a few minutes. According to my calculations, that gave me hourly earnings of almost $150 for this project, you can quickly see that *personal skills* are worth a great deal more than having a regular job down at the sawmill like we discussed back in Chapter 1.

GIVE MORE OF YOURSELF THAN YOU EXPECT IN RETURN

I have had many experiences working with "well-healed" partners. Their buying power and ability to obtain quick credit has allowed me to build personal wealth much faster than I could have ever done it alone.

Again, I will repeat my personal philosophy about working with others because it has much to do with being successful. It also has everything to do with repeat business—that means investors who will keep reinvesting with you, because you make money for them! Always give more than you expect to receive *(Jay's 60/40 rule)*. If you set up a 50-50 partnership arrangement, or a 90/10 investor plan, don't merely be content to contribute your portion. Instead do a little more. When the word gets out, you'll find more cash investors than you can find deals to include them in. **Just a small 10% extra will buy you lots of super deals and much faster wealth, believe me!**

BUYING RIGHT SETS THE STAGE FOR MAKING PROFITS

Every time I start talking about making profits at my seminars, someone always reminds me that all one needs to do is buy properties wholesale and the sell them at retail! You can't help but admire the genius behind such advice, but I often wonder—doesn't this individual realize that's what we're all trying to do!

The big problem is that buying low and selling high is not all that easy to do. In fact, it takes some real sound profit engineering to develop a money-making strategy. A good plan must have several common ingredients, such as PROPER TIMING, EQUITY CREATION, GOOD FINANCING and a REASONABLE METHOD TO EXTRACT THE PROFITS. None of these can be left to chance if you intend to make any serious money investing in real estate.

BEWARE OF NO-DOWN-PAYMENT TRANSACTIONS

Many novice or inexperienced investors have made the mistake of buying *marked-up houses* for *no money down*. They automatically assumed they could earn a profit because no cash was invested! With high mortgage payments and short-term balloon notes, their dreams of becoming rich quickly turned into nightmares, instead. The "free lunch strategy" may work well for selling slick-covered tapes on cable TV, but, in the real world, you won't buy much value for nothing!

The important thing to remember is you can purchase properties **with money** or **with your personal skills—but you must always pay something!** When you are negotiating to buy a property, stop and think about the deal as if you were the seller. Would you sell your real estate for nothing down if you thought someone would pay a normal down payment? I don't think I need to ask what your answer is. In most cases, acquiring properties for no money down means you're paying too much to start with! That's the wrong way to make profits in this business.

PROPER TIMING IS ESSENTIAL

To maximize profits you must buy houses when buying them seems like the wrong thing to do! Buying during a buyer's market (that's when many properties are available, with very few interested buyers) is generally worth at least a

20% discount to those who have the guts to go against the flow! For example, an $100,000 property should sell for $80,000, without much haggling. Conversely, during a seller's market (opposite from a buyer's market), the same $100,000 house will sell for as much as $120,000 (20% more).

As you can see, being synchronized with the real estate up-and-down cycles can be worth $40,000, when you understand that *THERE'S A RIGHT TIME TO BUY, AND A RIGHT TIME TO SELL!* If you follow the crowd, you'll most likely end up doing the opposite of what I'm suggesting here. To prove my point, try and find someone in the crowd, who is rich! I'll bet you can't. When major newspapers and the financial reports begin recommending real estate investing to the public, shrewd owners immediately quit buying and "polish up" their properties to sell for maximum prices to the dummies who will happily pay whatever the market will bear. Remember, the $40,000 price difference we're discussing here has nothing to do with ADDING VALUE or APPRECIATION. You earn the money simply by buying and selling at the proper times.

EQUITY CREATION

When you purchase average properties in average condition, you can expect to pay *average prices* and get *average terms*. Equity creation, or build-up is somewhat difficult when everything is just average. Equity build-up comes from two sources. The first is very insignificant. It's the principal portion of each mortgage payment, which adds to your equity in the property each time you make a mortgage payment.

The second kind is called **ADDING VALUE**. It comes from fixing up a property or straightening out people-problems by skillfully initiating better management. *This kind of equity is forced equity.* The owner makes it happen.

One of the best ways to create equity (my way) is to improve the financial performance (raising rents) of a property! For example, if I'm able to fix up a rundown property and increase rents from $20,000 annually to $30,000, that's FORCED EQUITY CREATION. If the property is worth eight times the gross rents, I've increased the value from $160,000 to $240,000. That's an $80,000 equity increase. It has nothing to do with normal appreciation. It was forced to increase by my fix-up work. If the building appreciates 5%, that will add another $12,000 in value to the $80,000 equity I've already created by force.

CHAPTER 18

100% Financing With Seller Subordination

*F*requently folks tell me, "I've tried to purchase properties the way you suggest, but my bank always says 'No'! What should I do now?" My answer is to keep on trying! If one particular technique doesn't work, just hang on and don't give up—*there's lots of other ways that will work*. Right now, however, let me tell you about a subordination technique that works particularly well, if you have a decent job and a good credit rating.

Subordination means the OWNER/SELLER will allow the bank to make a new loan against the property in front of or senior to the seller carry-back mortgage. This means, if the property should ever be foreclosed, the seller could lose his equity or part of it because senior priority mortgages would be paid off first from the foreclosure sale proceeds. However, without seller subordination, the bank could not fund the loan that's required to make this particular type of sale work.

A NO MONEY TECHNIQUE THAT WORKS

In case you're wondering if this method really works, let me assure you I've made about a dozen of these deals since I started! I've purchased $1.5 million worth of houses using a combination of *subordination* and *owner financing*. I call it my "30-30 PLAN." Let me show you how it works.

First, you must find a seller who truly wants to sell. Not just a "lukewarm" seller who has little or no motivation. I have found this method works best with sellers who have average, medium-grade rental properties. You don't want "trashed-out" junkers and you don't want pride-of-ownership properties. You want a property that can stand a few improvements, but not one that is seriously rundown! The reason is because most lenders won't loan money on junky-looking investment properties.

The typical lenders in my area are thrifts like Beneficial Finance, AVCO Thrift, Chrysler First and Fireside loans. These are the old personal property lenders (chattel) with an add-on license to do real estate loans. Many folks call them "Godfather Loans," because their interest rates are generally higher. However, unless you've done business with these lenders lately, you may not be aware of all the changes that have taken place over the past few years. Today these lenders make real estate equity loans combined with chattel mortgages. Their licensing allows them to write loans securing both real and personal property. The extra personal property security allows them to be more liberal with borrowers than regular banks, so it's generally easier to qualify for their loans.

LOAN TERMS ARE MORE IMPORTANT THAN INTEREST COST

It's true, these thrifts do charge higher interest rates; however, not as high as you might think. More often than not, their equity loans have fixed-interest rates and are generally amortized over a 15 year term. Another very important consideration with these lenders is that they're seldom concerned about junior priority loans on the secured property—that is, loans that are recorded behind (or after) their own. Banks quite often prohibit additional loans on the secured property even though the security is junior to their own loan. Remember this about financing—**It's much more important for investors to borrow money from lenders who will give flexible terms than to borrow strictly on the basis of the lowest interest rate**—within reason, of course! Flexible terms will add value to your properties when it's time to sell, because you can pass along the good terms to your buyer.

An Ideal 30-30 Purchase Plan

My south-side property consisted of four rental houses located on a large city lot. The property was in average condition. The sellers had owned the houses for many years and had done extensive upgrading, like black topping the driveways and building privacy fences. They had also added carports and installed several new roofs. Generally speaking, the property looked in pretty good condition when I bought it. On a scale of one to ten, I'd call it about a seven. *It's important to remember, lenders like properties that look good.*

The seller's motivation was a strong desire to retire. The owners had operated a small travel agency for many years, now wanted to close shop and travel themselves. Like most owners who decide to sell their average-looking properties with a large amount of equity, they wanted a rather substantial cash down payment from the buyer. At least 20 to 25% was the amount they would take, according to their real estate agent.

Lenders Don't Make Loans on Ugly Rentals

Back in the days when I bought the south-side houses, I was accustomed to buying mostly ugly, rundown, "trashed-out" properties for *very low cash down payments!* The sellers were not the least bit proud, and most were highly motivated. They had very few choices when it came to buyers willing to purchase their ugly properties. In short, they couldn't afford to be choosy if they truly wanted to sell. Most were already mentally prepared to accept small down payment offers.

This was not the case with south-side. The property looked good, and, of course, the competition is always much keener when it shows well! Looking back now, I must admit that *the good looks* impressed me, too! I paid a bit too much for the looks, and the result was that south-side turned out to be less profitable than most of my other deals—*but that's another story*. We're talking financing now. We'll look at bigger profits another time.

JAY'S 30-30 SELLER SUBORDINATION PLAN

The south-side property was an ideal candidate for my *NO-MONEY DOWN 30-30 subordination plan*. First the seller must agree to finance (carry back) at least 30% of the purchase price. Second, the existing financing on the property should not exceed much more than 30% of the total purchase price. It must also be assumable.

Here's how the numbers looked when I made the offer to purchase South-Side:

> Asking Price = $115,000 (Reasonable for rental income and condition of the property.)
> $105,000 My offer to purchase (**was accepted**)
> $ 34,650 Existing mortgage (assumable)
> $ 70,350 Seller's equity

The seller agreed to accept my offer of $105,000, and allowed me to place a new second mortgage (loan) on the property for $37,500. The loan funds were disbursed as follows: $32,500 went to the seller, $1,500 went to pay escrow closing costs and $3,500 came back to me at closing. I not only accomplished a NO-CASH DOWN Purchase, I also got money back on the deal, to boot!

Hypothetical $100,000 No-Money-Down Plan

Ideally, an existing mortgage on the property you're after should be approximately $30,000 to $35,000 (30%). The seller must agree to subordinate to a new loan (mortgage) for about the same amount (30%). And, finally, the seller must agree to carry back a third mortgage for the balance (30 to 35% range).

The obvious question you are probably asking yourself is, Why on earth would a seller surrender his equity, by allowing a new loan to be recorded on the property ahead of his interest?

The answer becomes more clear when you follow where the money goes—and understand the benefits to the seller.

WHERE DOES ALL THE MONEY END UP?

The seller gets the money, at least most of it. *That's what makes this deal work!* Some investors try to play games with this type of financing, by attempting to pocket a large share of the new loan proceeds. This tactic is very poor business, because it grossly over-finances the property. It also increases the monthly debt service, which in turn adds far greater risk to all lenders involved. A buyer can be quickly overcome by negative cash flow, caused by high payments, trying to service too much mortgage debt.

I have found this financing arrangement works best when you give all the borrowed money, except expenses, to the seller. For example, using some hypothetical numbers for a $100,000 deal, let's say we can assume a $30,000 existing mortgage. The seller has agreed to carry back a $35,000 third mortgage, and will subordinate to a new bank loan for the balance. Obviously, an appraisal must substantiate the property value, however, most lenders would be willing to loan up to 70% of the appraisal.

In this particular example, a 70% loan means the lender would be willing to loan $70,000 on this property. Since there is already an existing first mortgage for $30,000 against the property, it means any additional borrowing cannot exceed $40,000. ($30,000 plus $40,000 = $70,000.) That's 70% of $100,000.

In this example, I've tried to use very reasonable numbers. In other words, my ratios and loan percentages are well within limits for most lenders. With a decent job and good credit rating, an applicant could reasonably expect to receive loan approval without too much difficulty. Obviously, you'll need an appraisal and, of course, the standard financial information all lenders require from investors.

HOW DOES A SELLER BENEFIT?

The big advantage for a seller with this arrangement is he gets $35,000 cash up front! *That's an extra large down payment for rental houses.*

Typically, sellers are accustomed to receiving only 10 to 20% cash down payments for average, "run-of-the-mill" rental properties. As you can plainly see, this plan will net the seller far more cash at the closing table.

There's also another big benefit for sellers who have owned their rental properties for a long time and, by the way, this 30-30 plan is more or less geared for long-term owners because the numbers work better. This second benefit can really help sellers who have owned their real estate for a long period of time and have large equities or profits build up. Quite often, the extra-large down payment (30 to 35%) will allow the seller to get all, or nearly all, of his original cash investment back. In other words, the large down payment will totally cash him out of the property. He'll have none of his own money left in the deal!

It's a bit easier to finance or carry back a mortgage for the appreciation or growth because you're only financing the profits you've earned. It's always more risky for sellers when their own money is still left in the deal. Most investors are reluctant to carry a mortgage without first getting all of their hard dollars back out of the property. If they can do that, they're generally a lot more agreeable to seller financing for the balance of the sale.

I personally don't object to financing my profits because I like the additional interest income it earns. It's a marvelous recipe for making bonus profits. On the other hand, I need my down payment dollars back when I sell, so I'll have the funds for my next investment.

ADVANTAGES To THE BUYER

The major advantage is that you can acquire real estate with *NO MONEY DOWN*—that is, none of your personal money! What this means is lack of cash doesn't need to stop you from buying income properties. However, as I told you earlier, you must have a decent job and a good credit rating to qualify with commercial lenders. If you do, then my 30-30 plan can work very well to help you acquire investment properties that you might otherwise have to pass up.

To avoid negative cash flow problems with this plan, you must be very careful not to take on mortgage payments in excess of what the property can support! This is where my INCOME PROPERTY ANALYSIS FORM (Chapter 5) can be an extremely helpful tool to assist you. Basically, you'll need to carefully calculate all the expenses necessary to operate the property, then subtract them from the gross income. The remainder is what you'll have left to pay the combined mortgage payments.

The most serious problem with *no money down deals* is that the total purchase price must be financed, unless, of course, other trading is also involved! You must carefully negotiate the mortgage payments so they don't exceed whatever amount of income the property is capable of earning. *Effective use of the 30-30 plan requires special planning and skillful negotiating to avoid negative cash flow!* As I told you earlier, the potential for a "quickie" upgrade, followed by increasing the rents is not likely to be an option with this kind of property.

Let me say it another way! When I purchase ugly rundown houses occupied with deadbeat tenants, the rents are generally far below the existing market rates. In these situations, I'm able to acquire the property below market value because the price is based on the rundown condition. When I acquire rundown properties, my basic plan is to quickly fix it up and rid myself of the deadbeat tenants. After a short period of time, I can generally increase my rents from 40 to 80%. This type of property could never be a candidate for my 30-30 plan, because bankers simply don't have the stomach to loan money on ugly rundown properties.

Lenders Want Clean, Sweet-Smelling Properties

Lenders are like 98% of the population. Whatever they do is generally based on how things look! They will gladly loan money if your project looks good, but they don't want anything to do with the "ugly duckling" properties. Believe me, looks count for everything—loans included! It will serve you well if you understand this basic human characteristic, because it's exactly how lenders think, too!

If your property fits the situation—that is, it's reasonably clean with no visible signs of a problem—chances are quite good that most mortgage lenders will finance (loan) up to 70% or more of the purchase price or the appraised value, whichever is less. In our hypothetical case, we already agreed that the seller is willing to carry back a $35,000 mortgage and will subordinate to a new second mortgage loan to be placed on the property. The buyer agrees to assume the existing first mortgage that exists on the property at the time of purchase. Right about now is where we (the buyer) must ask the seller to cut us a little extra slack. It should be obvious by now that three loans on this property will cost more than the property can afford to pay back under normal circumstances. To illustrate what I mean, let's review the numbers once again in our hypothetical transaction.

Full Purchase Price	Mortgages on Property	Monthly Mortgage Payments
$100,000	$30,000 Buyer to assume (existing)	$242.66
	35,000 Subordinated loan (new borrowing)	420.06
	35,000 Seller carry back mortgage	(Continue reading for explanation of amount)
	$100,000 Total financing	

The scheduled income for this property is $350 per unit, or $1,400 per month in gross rents. For the purpose of planning, let's assume it will cost 40% of the income each month for expenses to operate the property. That equals $560 per month. The debt service (two mortgage payments) will cost $662.72. The total cost for expenses and mortgage payments is $1,222.72 per month. Obviously, there's little money available to pay another monthly payment on the seller's carry-back mortgage. Therefore, I propose asking the seller for some special consideration for the pay back of his note. I would tell him that $175 to $200 is all the property can afford to pay him.

REAL-LIFE CASE HISTORY OF SOUTH-SIDE

Back to my south-side houses—they presented a similar problem for me at the time. Here's how I explained things to the seller—I said, "Look, Mr. Seller, I'm more than willing to place a new loan on the property in my name! No risk to you because the loan won't be in your name. Also, I'll give you all money from the new loan except closing costs and the $3,500 for repairs that you've already agreed to do. It's much easier for you to deduct the repairs from loan funds than spend your own cash out of pocket to do them! I'm sure you realize that my 35% down payment is more cash than most investors are willing to pay! Also, most buyers would certainly want you to finance (carry back) a much larger share of the sale. I'm more than happy to use my good credit to get you the most money, but I do need a favor from you. I'm willing to work at the property and do the repairs and manage the tenants, without taking any money from the property for myself. However, as you can see from the income and expense information **(income property analysis form)**, I'm not going to have enough rent money coming in at first to pay you monthly mortgage payments. I need you to give me a little extra time 'til I can raise the income a bit.

MY PROPOSAL TO AVOID RED INK

I told the seller I'll pay 10% interest on his note in five annual payments. That's $250 per month ($30,000 times 10% equals $3,000 divided by 12 months equals $250). This way I'll have the first 12 months without any mortgage payments. It will give me breathing room and some time to get the rents up a little.

Also, I'll have the opportunity to gain one full year's worth of tax deductions. Assuming an 80% improvement ratio and $12,000 worth of depreciable personal property, an investor in the 31% tax bracket would realize a $1,600 annual savings. That makes up more than half the $250 deficit. A $30 rent increase would take care of the difference. Also, don't forget when I figured out the expenses (40% in this case), approximately $110 per month was allocated for *maintenance*, *repairs* and *management*, that's my job! Obviously, owners can't pay themselves, but saving those expenses amounts to the same thing as getting paid for it. It's still my money, because it's my property.

PLEASE TELL ME—WHERE'S THE BEEF!

Right about now you might be thinking to yourself, "Have I missed something here? I can't possibly see how I'm going to get very rich with this deal! This program is tighter than a banjo string. If I lose one month's rent, or a toilet breaks down I'll end up paying money out of my pocket. There's no safety margin!" Let's be realistic here, I'm not telling you how to get rich. *I'm telling you how to get a free stake in the game.* If you agree with me that owning income-producing real estate is the right way to go, and, further, that rental properties continue to become more valuable, year after year. I would ask, "How can you go wrong?"

You've got a winning hand. First, you're name is on the deed. You are the owner and that's good for you! Secondly, as almost everyone agrees, income real estate appreciates. Rents continually go up, therefore, so does the value of the property producing them. Thirdly, your return on a no-down investment can be phenomenal. Everything you take out of the deal is pure profit! That's because you have nothing in the property to begin with. You could lose the property, but never any money! In terms of risk, I'm sure you would agree, the odds will never get better.

I might just mention that south-side was purchased for $105,000, with none of my own money down! Today it's worth $205,000 and rents are $2,300 per month. Counting appreciation and rents, I earned a profit of $9,400 for every year of my ownership. It also shoots down the theory that it takes money to make money. In the case of south-side, *good credit* and a *decent job* was all I needed!

No Limit to Creativity in Real Estate

Subordination by the seller, like we're discussing here, is not a new idea! Folks who sell empty building lots do it all the time. They sell the lot with an agreement that they will record their mortgage (seller financing) behind a new first mortgage, on the property to serve as collateral for the bank's construction loan. That's how a new building gets financed!

Ray Kroc, founder of McDonald's, was moving along at a snail's pace trying to franchise his hamburger chain until he met up with Harry Sonneborn, a "Real Estate Wizard," who understood the power of subordination. Prior to meeting Harry, Kroc's expansion dream was "bogged down" for a lack of construction funds. Harry did the same thing for Kroc that I'm telling you about here. He asked lot sellers to subordinate to McDonald's lenders in order to get construction loans to build hamburger stands. Obviously, it worked quite well!

Variable Rate Mortgages Offer Another Option

Another transaction, similar to south-side, came to me several years back. The seller insisted that he receive monthly payments. However, he agreed to my variable payment plan. This allowed me to have reduced payments to start with and gave me some extra time to build up my rental income as the mortgage payments went up. The owner carry-back mortgage loan was $42,000, payable at 9% interest-only payments, and all due in 13 years. Interest-only payments at 9% amounted to

$315 per month. However, we structured the note to start with payments of $210 per month (6% interest) and ended up in the 12th year at $420 per month (12.0%). This arrangement worked for both of us. The seller got a 9% average interest rate, beginning at 6% the first year and ending at 12% in the 13th year. I had low payments—$210 to start with, which was about all I could manage when I first took over the property.

I've had other transactions where I've secured the seller's carry-back note to other properties I own. Obviously, you must have multiple assets to make this arrangement work.

As I said in another chapter—you won't need 100 ways to buy real estate. Half a dozen good techniques will most likely do the trick. I also told you that good credit will be one of the most valuable tools in your investor kit. Naturally, all banks and commercial lenders will approve or deny credit based on the kind of records you develop. If you've paid your bills in a timely fashion over the years, no doubt you have a good track record established. If not, you'll need to begin the necessary repairs to get it fixed. Fortunately, that can be accomplished over time.

INVESTOR'S SUCCESS REQUIRES BORROWED MONEY

Many people feel that money is what makes the world go around! I can't claim much knowledge about a spinning world. However, I can comment about money and real estate. Without it, investors like myself would soon be out of business.

Stated another way—Investors who cannot borrow money for financing and fix-up with a little left over for groceries will quickly find their investment plan won't get very far. Borrowing money is the only way most of us can ever achieve our financial goals in a reasonable period of time.

Money borrowing rules are changing almost daily. It doesn't take a rocket scientist to understand that finding money to do ugly fix-up projects can be difficult, especially for investors who are just staring out and plan to do most of the work themselves. Still, it can be done. So keep on reading—you'll see!

MAKING YOURSELF A BETTER BORROWER

There are several self-help measures available to ease borrowing difficulties. The first, obviously, is to protect your credit rating if you have a good one. If you don't, I'll show you a few things you can do to make it better.

You'll also need to keep your personal financial and accounting records up to date and on-line. This should be done at least annually, so you develop good history—sometimes more often is better and will generally be required if you are an active buyer/borrower. I will discuss the financial records (tools) you need for every lender, both private and institutional. *Having good financial tools is just as important for the do-it-yourself investor as having good plumbing tools and a sharp saw!*

BANKERS LIKE HOMEOWNERS WITH A STEADY JOB

When I worked at the telephone company years ago, I had an excellent credit rating. Banks were willing to loan me money even when I didn't need it. The first two questions bankers would always ask me were, "*WHERE DO YOU WORK?*," and "*FOR HOW LONG?*" When I answered, "At the telephone company," and, "20 years," the loan manager relaxed into his easy chair and said, "How much do you want, and when do you need the check?"

I began buying rundown houses long before I quit my telephone job. The first time my banker drove out to see my fixer houses, I thought he was going to throw-up!

Still, he remained impressed with my 20 years of telephone employment. But he also gave me some personal advice about buying anymore junky properties. He even hinted about not making anymore loans. Soon afterwards, Beneficial Finance became my fix-up property lender. They didn't mind junky houses quite so much, as long as I was still employed at the phone company. Quite a number of these early 18 to 20% interest loans got paid back over the years, which made the folks at Beneficial smile from ear to ear.

BANKER ENEMY NUMBER ONE—A NON-EMPLOYED LOAN APPLICANT

If there is anything a loan officer hates worse than an unemployed deadbeat, *it's an unemployed house fixer who wants to borrow money!* The news that I had quit my telephone company job turned every lender against me. If I had been playing monopoly with my banker, he'd have sent me directly to jail without passing Go. One thing you need to understand is that bankers don't like loan applicants who can't produce a copy of a W-2 form. It's their only evidence that someone else thinks you're worth spending money on.

I never realized how important a regular job is to bankers. Before I quit my job at the telephone company, my Beneficial mortgage payments alone were about three times more than my monthly telephone paycheck. Still, Beneficial seemed totally unconcerned until the day I quit the phone company. My 20-year, squeaky clean credit record didn't mean anything as far as additional loans with them were concerned. When the paycheck stops, it's like starting all over again. We must somehow prove to lenders that we still have the ability to pay them back. Bankers can only visualize two kinds of customers—LOW-RISK EMPLOYED FOLKS and UNEMPLOYED DEADBEATS.

HOW TO BUILD YOUR FINANCIAL INTEGRITY

Borrowing money is much easier when you can show lenders good financial records about yourself and your real estate business. It's very important to demonstrate that you know exactly where you stand financially. *Lenders admire organized applicants.* They don't like borrowers who show up asking for money, but can't explain exactly how much they need and, even worse, how they'll pay it back. These problems can be partially overcome with good sound financial records.

JAY'S FIVE BASIC FINANCIAL DOCUMENTS FOR BORROWING

Financial records are the same whether you own one property or fifty. Obviously, if you start while your investments are small in number, you'll need less paper. However, the documents themselves will always remain the same. I think every investor should prepare his own records and keep them updated on an annual basis—more often, if necessary, for loan activity.

I will briefly describe each of the five financial documents and, hopefully, give you enough information so you can develop your own records. Remember, nothing is magic or sacred about these forms. You can simply draw them up yourself on plain white paper or obtain financial statement forms from a bank or the local stationary store.

Another important benefit you get from preparing these forms is it gives you an excellent financial picture of yourself. Many investors don't have the slightest idea about their net worth. Some are afraid to find out! Following is a description of the basic financial tools I suggest you prepare and start using. I promise they'll help you a great deal next time you're ready to borrow money from the bank, or you need evidence to show a seller who is contemplating a carry-back mortgage for you.

Schedule of Real Estate Owned (Form Set-Up)

This form should be typed up on plain white paper with headings as follows:

Property location	Type units	Market value	Mortgage liens	Lender/ bank	Mortgage payment	Taxes ins.	Gross income	Misc. repairs	Net income

Almost every conventional loan application form requires this information in the same order. Therefore, when you keep this form updated, it can simply be made part of any loan application package you plan to submit. Standard 8 1/2 x 14" paper (horizontal) is best to use because it matches the size of most bank application forms.

Schedule of Real Estate and Notes Owned (Form Set-Up)

This form is very similar to the previous one. However, it contains some additional information. I have found it very helpful, especially when private lenders (sellers who agree to carry back financing) ask for financial data to consider financing a sale to me. This form also contains the information you will need to fill out the ASSET/LIABILITY section of any financial statement you prepare. The following headings are used to prepare this document. It should be typed horizontally on plain white, 8 1/2 x 11 inch paper.

Location description	Market value	Your equity	Mortgage balances	To whom payable	Address lender	Scheduled rents/note income

A Personal Financial Statement (Blank Forms Available)

This form can be standard stationary store copy or a form obtained from your local banker. Often, the various lending institutions have their own special forms with their names printed on them. Sometimes their forms have special or unusual questions that they want answered—for example, How do you plan to make your payments in case you die? All financial statements are pretty much the same. You list all your assets on one side, and liabilities on the other. The difference equals your personal net wealth. You might be surprised when you find out what you're really worth!

Profit and Loss Statement (Form Set-Up)

This form should show your total income at the top. Then you will list all operating expenses. After that, list mortgage payments and depreciation. The bottom line will show a profit or loss for the period of time involved.

I always prepare a profit and loss statement, annually. However, more often may be necessary if you are aggressively shopping for loans.

You can type this information on plain white, 8 1/2 x 11 inch paper.

Income should include all sources, such as rents, deposits (if you mix them with rents), coin laundries, notes receivable and management fees (if applicable). Expenses are all your operating costs, which generally include payroll, licenses, insurance, maintenance, repairs, supplies, utilities, advertising, telephone, taxes, legal fees and accounting. Remember, owner draws are not to be mixed in with employee payroll. Draws can be a separate item if you choose.

NET OPERATING INCOME is what's left over after you subtract the operating costs from total gross income. Mortgage payments should then be subtracted from the net operating income to determine positive or negative cash flow (profit or loss). Depreciation can also be shown as an expense item. However, I like to keep it separate from my regular expenses, because, it's really only a paper expense (you don't write a check to pay it). Lenders will often ask what your profit or loss is before depreciation? You'll really impress them when you know the difference and have the correct numbers.

Good impressions, along with good records, will often make the difference between loan approval and loan denial.

BUSINESS FINANCIAL STATEMENT (FORM SET-UP)

A financial statement about your real estate business is almost exactly like a personal financial statement about yourself. If you're just starting out in business, chances are your personal statement is all you need. However, if you have several properties already, or perhaps a real estate partnership interest, I would suggest you prepare a Business Statement to better demonstrate your financial capacity.

In my real estate business, the management division—ONE STOP HOME RENTAL COMPANY—owns trucks, special tools and furniture, which are not included on my personal financial statement. As you expand your real estate activity, you may find it's better to have several separate business entities. Each should have it's own financial statement.

FIXER JAY'S LOAN KIT FOR BORROWERS

After many seminars and thousands of words teaching about my *FINANCIAL TOOLS* (forms)—Students still ask for copies of my **5 BASIC FINANCIAL DOCUMENTS**. Apparently my forms are favored over those from banks and stationary stores—or it might be because I provide (filled out) examples using my own properties to show how all the $$ numbers fit together.

I also provide a sample form showing how I set up my **TRADE ACCOUNTS**, which is an excellent tool for convincing hardware stores, carpet cleaners and appliance dealers to extend monthly credit terms and contractor discounts to me. *NO, YOU DON'T HAVE TO BE A CONTRACTOR TO GET DIS-COUNTS!* But, you must look like a legitimate and thriving business. A professional credit application (also included) and my *TRADE ACCOUNT SET-UP* will make it happen for you, I guarantee.

Naturally, I've recorded a 60 minute audio tape to remind you how all these forms should be presented to a lender, both private folks and bankers.

If you feel you need additional help or you just want to take advantage of the work I've already done—use the order form in the back of the book. You'll get your money's worth! **Order Product No. 2114.**

CHAPTER 19

Free Fix-Up Money From Uncle HUD

I have always been a firm believer that the world is divided into three distinct groups of people—those who wish something would happen, those who make things happen, and those who wonder what happened! Do-it-yourself real estate investors must make things happen if they intend to be successful within a reasonable period of time. Stated another way, you probably won't live long enough to reap the financial rewards of success if you wait for them to come to you.

"WAIT AND SEE" WILL TAKE YOU NOWHERE

Often during real estate down cycles, many inexperienced investors tend to use a "wait-and-see" approach. This slowdown or stalling can severely cripple the best-laid plans. I can assure you that IF YOU ARE NOT MOVING FORWARD, AS AN INVESTOR, THEN YOU ARE SLIPPING BACKWARDS. It isn't difficult to figure out that if you don't acquire properties now, you won't have anything to sell later! And without property, you can't collect monthly rents and you won't benefit from fix-ups *(ADDED VALUE)*, tax shelter and appreciation! Also, don't forget those extremely profitable seller carry-back notes to ensure that your retirement years are comfortable! Obviously, you must own the property, first, before you can benefit from seller carry-back financing.

MORE THAN ONE WAY TO PROFIT

The message I have for investors is, **nothing is going to happen for you unless you make it happen!** It's always been that way, and always will be! Because real estate investing offers such a wide variety of profit-making opportunities beyond simply buying and selling properties, investors should never find themselves bogged down in a position where they can't do something to improve their net worth! For example, I don't concentrate on selling, when it's a buyer's market—meaning buyers have all the advantage. There are too many properties available for too few buyers who want to purchase them! When the situation reverses, I start to think about selling. Obviously, fewer properties, with more buyers, means a higher selling price for me. Meanwhile, I can work on other ways to improve my real estate wealth. One of my favorites is the **HUD GRANT-FUNDED, RENTAL-HOUSING PROGRAM**, sometimes called the matching funds program.

UNCLE SAM PROVIDES MONEY FOR FIXING AFFORDABLE HOUSES

Quite often when normal real estate activity has slowed down, smart investors can still find high-profit investment opportunities. One such opportunity is government-assisted *LOW INCOME RENTAL-HOUSING REHABILITATION*. The program is available nationwide in various formats.

Grant funding is the most active program the government uses to assist landlord owners in fixing up substandard rental properties, making them available to lower income, and subsidized tenants. What makes this program so attractive to property owners is that grant funds are **FREE MONEY**. That's a huge difference from loan funds, which must be paid back. Free money is the government's "dangling carrot" to attract property owner participants. If you learn the ropes and do this right, I will assure you it's well worth the time and effort it takes working your way through the so-called government "red tape."

Landlord-Owners are Requested to Participate

If I told how to get your rental properties fixed up for half the normal price, would you be interested? You should be, it's a very good deal, and it can be accomplished with the use of funds from community development block grants. Funds are administered by local city and county housing departments. In my town it's called the Public Housing Authority (PHA).

In addition, once the rehab work is completed, a Section 8 subsidy can be given to the first family to occupy the unit under HUD low-income guidelines. This can be the existing tenant who is living there during the rehab, if he qualifies for HUD assistance. Subsequent vacancies—after the original tenant of the rehabilitated unit has left—must be filled by open market renting.

AVOIDING VACANCIES—KEEP THE RENTS COMING IN

Before I get too specific, let me say something about "keeping your units rented," I'm talking no vacancies, here. In areas where housing is scarce, keeping your units rented is seldom a problem. However, apartment construction goes in cycles, and during easy-money times—when loans are available—it tends to accelerate, causing vacancy rates to increase as well. Many tenants move to the newer units, if they can. This is where city housing and subsidized rents can really help landlords who have older properties.

I've found HUD tenants stay much longer—they don't move around as much as non-HUD tenants; therefore, I'm a lot better protected against high vacancies. Owners of newer units can't stand too many vacancies, otherwise they won't be able to pay their mortgage payments. HUD programs can work quite well if cash flow is your chief concern!

I don't recommend that all your rental properties be occupied by HUD tenants. However, guaranteed rents can be a real life saver if you're just starting out. About 20% of my current renters are HUD tenants. The city and/or county pays about 75%, on average, of their monthly rent directly to me. I must also point out that it's never late. HUD checks are very dependable, and I can always count on the money on the first of each month, no matter what else might happen.

HUD Goals—Provide Safe, Affordable Housing

The purpose of this program is to financially assist owners of rental properties in expanding the availability of safe, economical housing for the benefit of low and moderate income persons living in the community. Often, the program targets specific areas within a city or community to prevent slum conditions or eliminate blight and further deterioration of a neighborhood. Each city or community uses Federal Block Grant funding, along with State Housing monies and sometimes special assessment district funds to improve the supply and quality of housing within its boundaries.

OWNERS OF RENTAL HOUSES PAY 50¢ ON THE DOLLAR

If your rental house needs $10,000 worth of allowable fix-up work, here's how you can do it for half price, or for only $5,000 cost to you. First, visit your local city or county housing department, where you must fill out an application and several other forms with information about your property and the tenants.

After the housing department receives your application, they will inspect the property and make a thorough fix up and repair list of the work that needs to be done. All code violations —plus worn out or damaged items, such as roofs, paint, floors, appliances, windows, doors, etc.—must be fixed. The house must be "brought up" to an acceptable condition, where everything is safe and works properly.

After you and the housing representative agree on what work needs to be accomplished, a cost estimate is developed and bid specifications are drawn up by the housing department. There is, generally, a cost per unit limit—such as $10,000, for a three-bedroom house. If the estimate is $10,000 for the rehab work, you will be required to pay $5,000. The housing department will grant (that means give) you the other

$5,000 to do the work. Remember, this is not a loan! There is no pay back involved. It's a free grant from the government to you for your participation in the housing program. It kind of makes you feel better about having to spend so much money making repairs!

HOW TO GET STARTED FROM SCRATCH

The first step is to visit your local city or county housing department, often called the Housing Authority, or Housing Assistance Office. Incorporated cities generally have their own housing department under the direction of the public works official. In rural communities, the county performs the same function.

In my area (Shasta County, California)—with a county population of approximately 165,000, and a city population of approximately 70,000—I deal with the city primarily because most of my properties are within the city limits. Outside the city limits it is county jurisdiction. Quite often there are big differences between the two government agencies, even though their funding sources are the same. *One significant difference is worth mentioning here*—within the city limits, property owners are not allowed to perform grant-funded rehab (fix-up) work on their own properties, unless they're licensed contractors and are approved to bid on city housing projects.

In the county jurisdiction, the rules are not nearly as strict. I am allowed to do all rehab work on my houses, as long as I can convince the county housing authority that I am responsible and that I have the necessary "know-how" to complete the work. The county is also much more liberal when it comes to obtaining building permits associated with low-income housing projects. Only extensive work would require a permit for HUD rehab jobs.

The main reason I'm passing this information along, is to make you aware that local housing departments differ a great deal in their methods of administering housing rehab funds. It will be to your advantage to search out all the pertinent information and rules concerning your own particular area, before you formulate a plan of action. In summary, first visit the housing authority, and ask for all information about grant rehab funding for landlords, to find out how it works. Finally, get a map of the area where the city or county wants to apply their funding. I've found it's usually in the older, rundown sections of town. That's where rehab is needed the most.

Dealing with the Local Housing Authority

Often times, city housing won't give non-owners the time of day. The housing department considers it a waste of time to discuss rules and regulations with anyone who cannot participate. Local real estate agents are responsible for this, as many agents have a tendency to badger the city housing staff in an attempt to secure loan information or rehab commitments for their listed properties. Obviously, available funds would make properties easier to market. In my town, the very first question city housing will ask is, "Do you own the property?" That's the first requirement!

Getting a Feel for the Program

Your initial visit to city housing can be a bit awkward the first time. At best, you'll come away with mixed feelings, and you won't know whether to thank them or resent them. Milling around public agencies can be a particularly frustrating experience for those who think of wasted time as wasted dollars.

There's a standard clause in real estate contracts that says, "time is of the essence." That clause has little meaning at city housing. You'll see what I mean after a few visits there. Housing folks are a very patient bunch. They view the world as always having lots more tomorrows! The only exception I've ever experienced was with a project where I was asked to convert my two-bedroom houses into three-bedrooms. The housing department needed six three-bedroom houses to maintain their spending level—and thus preserve their annual budget. You can benefit a great deal by knowing this kind of information. I suggest you make friends or develop a good contact within your local housing department. It pays big profits in the long run, as you shall learn later on in this chapter!

NO MONEY DOWN DEALS ARE VERY POSSIBLE

Some cities and/or counties have more money than others to spend on housing. Some also have "pet projects" and will offer more incentives, such as loans, for these particular projects. This too is valuable information. It's another reason why you need a contact in the housing department.

I've found, the financial climate is always changing, especially when it comes to federal block grants. Some days you can't find any money and on the very next visit they seem to throw money at everyone. However, unless you stay in close contact, you'll never know exactly when you have an advantage.

On the day I first noticed the "un-yellowed" squares, I had no idea that I'd accidentally stumbled onto a gold mine. I found out by asking questions, which most housing folks were happy to answer. They generally give you lots of information if you just keep asking, and that's all I did when I saw the "un-yellowed" squares on the bulletin board (more about squares later).

I had no idea they would loan me City and State of California housing funds, until they told me it was available. They explained that city housing had an emergency fund for special uses and that they would seek approval to use it for converting my two-bedroom houses to three-bedrooms. On that particular project it cost me only 20¢ out of each fix-up dollar to completely upgrade six houses. The balance was funded with "low-cost" housing loans.

Since that time, I've done a variety of jobs (with matching grant funds), where the city or county housing departments have paid for everything (total costs). First, they gave me 50% free grant money, then they loaned me the balance for my 50% owner contribution. The bottom line is that my units got completely upgraded and I was not required to come up with any out-of-pocket money. Was that a great deal or not? I certainly thought so!

HOUSING AUTHORITY NEEDS LANDLORDS TO PARTICIPATE

Even free money is not without a price! Many property owners will have nothing whatsoever to do with city/county housing programs, because of the seemingly endless "red tape" and regulations that's associated with the bureaucracy. It's true, you must put up with some "nit-picking"; however, I've found that it is more than offset by the benefits I receive.

After many years and a good number of projects, I'm financially better off from participating with the local housing authority. My long-range "profit goals" have been greatly enhanced. I've done moderate rehabilitation conversions, grant funding projects and I continually participate in the rental certificate and voucher programs. I highly recommend that all landlord/ owners learn about the various programs in their own communities and use them.

Another important point I must mention here is that with all local HUD programs—including the annual rental assistance contracts—*the city or county will always do their best to see that you achieve cash flow.* They want you to make a profit. It's in their best interest to have you profitable! They thoroughly understand the value of your staying in business. Remember, they need you too. Knowing that CITY HUD will help you keep your rental units full is very satisfying, particularly if you're just getting started in this business and lack of cash flow is one of your biggest problems!

Selecting the Right Property

If you already own rental property in an area where city housing wants to commit rehab monies, that's good! You're ready to submit an application and get started. If you don't own a suitable rehab candidate, yet, then it makes good sense to visit your local housing office and discuss where they are doing grant funding jobs with landlords. Better yet, get a map of the area, if they have one. Then start looking for deals.

Each city and/or county will set maximum dollar limits for their matching fund projects. Limits are similar in most towns. The maximum amount of a grant is based on the unit size. Generally it is tied to the number of bedrooms, and it applies to both apartments and detached houses.

Dollar Limits for Maximum Grants in My Area

Studio units............................. $5,000 grant maximum

1 bedroom unit........................ $6,500 grant maximum

2 bedroom unit........................ $7,500 grant maximum

3 bedroom or more................. $8,500 grant maximum

Matching grants can only be used for serious work—not cosmetic stuff! Eligible repairs or rehab means basic housing improvements. Swimming pools or servant quarters will not qualify under this program. Eligible items include roofing, plumbing, electrical upgrading, painting, floor coverings, fencing and most anything having to do with energy conservation—such as double-pane windows, new doors, weather-stripping and better insulation.

WATCH OUT FOR THE HIDDEN COSTS

In addition to basic housing fix-ups, cities and/or counties often include their pet projects with grant funding approvals. Some local agencies will insist on the construction of sidewalks and gutters, storm drain extensions, sewer and water hookups to municipal utilities and the resurfacing of parking lots and driveways. These items can be very expensive!

If you assume these things won't cost you, I assure you, they will. While it's true these items improve your property in the long term, making it more valuable, it must always boil down to present economics. For example, how much more rent can you collect from a low-income tenant who has 65 feet of new storm drain and a sidewalk added along the side of his two-bedroom apartment? You're right if your answer is "none." What about selling the property sometime in the future? You may have a more valuable property because of the additional sidewalk and drain pipe, but I doubt it! It's really just a wild guess at best, I think.

The purpose of this discussion is to make you stop and think. Sure, grant money is free, but the matching funds are not! That's your money! If sidewalks and drain pipes add $12,000 to the cost of the project, then $6,000 of that amount must be paid by you. So, you need to figure out how much more rent you'll receive, how long it will take to recover your cash outlay, and make sure it's worth doing.

MULTI-UNITS EARN YOU MORE PROFIT FOR EACH DOLLAR SPENT

Applying the maximum grant fund limits described above, let's suppose you own a single-family, three-bedroom house you would like to rehab. The maximum amount of work you could do under the matching grant program (50-50 split) would have to be accomplished for $17,000—$8,500 free grant money, plus your $8,500 cash contribution—that's the maximum amount allowed. From experience, I can tell you that $17,000 gets used up very quickly. Let's say the housing authority estimates the total job (based on whatever rehab they determine must be done) at a total cost of $25,000. Now, the math works against you. The city still contributes $8,500, however, your share jumps to $16,500. You see, it's no longer a 50-50 deal because it exceeds the maximum grant allowed.

For this reason multiple units or several houses clustered together on a single property (lot) works much better. It gives you more free grant dollars to work with. With multiple units, chances are that every unit will not require the maximum grant allowance. That means, if some exceed the unit limit and some cost less, there's a good possibility that overall you'll stay within the grant limits for the project.

Several year ago, I completed a matching funds project on an 11-unit property called Viola Cottages. Each unit had one bedroom. At that time, the maximum grant allowed for one-bedroom units was $5,000. Because there were 11 cottages involved, the maximum amount of free grant funding available to me was $55,000 (11 times $5,000 equal $55,000). As it turned out, the project actually cost $90,000. I received $45,000 of free grant money, and, of course, was required to contribute an equal amount of matching funds. Two of the cottages cost more than $10,000 each to rehab. However, the cost of the other units was much less than the $5,000 limit.

MORE HELP FOR BROKE OWNERS WITH KNOWLEDGE

Assume that you own or plan to acquire property in the middle of your city's "hottest target area." This is where local housing politics can often benefit you financially if you make it your business to know what the city wants. The reason is that, quite often, the city government wants rehabilitation work done even more than the property owner does. This is particularly true in cases where promises are made to the Federal government by local agencies (city and county) to clean up specific areas as a condition for receiving large federal block grants.

In these situations, it's not the least bit uncommon for cities to go beyond simply approving your 50% grant fund application. In addition, they will often loan you your share of the matching funds, assuming, of course, you don't have the money to contribute yourself. Most of us don't. Wouldn't you agree?

Why would the city give you free grant money, then loan you an equal amount without requiring you to put up any money at all? The answer is, they want to see the job get done; therefore, they are willing to provide additional assistance. What owner could ever say no to a rehab project that costs him nothing. That was exactly the situation with my Viola cottage project. The local housing department wanted to see the project done even more than I did, so they loaned me the money to make sure it happened.

CITY LOANS WORK IN TANDEM WITH GRANT FUNDS

You are most likely to receive *city HUD loans* by going the "extra mile"—that means owning property and being willing to participate with your local HUD housing in special target areas where the city is highly motivated to clean up older rundown neighborhoods. Also, single bedroom units in locations where a shortage of affordable senior housing exists will cause the housing folks to push a little harder. Often, that translates to making easy loans. We've already discussed proper timing, which is always an important issue when negotiating with HUD representatives, or anyone else for that matter.

City loans are the very best kind of loans, because they are generally "tailor-made" deals. Most are designed to fit a particular need or situation. For example, city housing will always insist that you make a profit—generally about 8% or so, but it varies. Let's say the city gives you 50% grant funding and then agrees to loan you the other 50% (which is your 50% contribution).

The loan will be designed with a special interest rate and terms so you'll be able to make loan payments and still make a profit at the same time. They do this by lowering the interest rate to fit the deal, based on projected income. They also set the rents you can charge, so they know exactly what income and expenses should be. Most loans are amortized over 10 to 15 years and sometimes they even forgive payments if you will rent to special hardship tenants who are on the HUD waiting lists.

City Housing Is Like a "Life Boat"

It's important for me to point out here—that when cities and/or counties design easy-pay loans for you and guarantee that you can make your monthly profits for doing rehab projects with them, *there's a flip side*—they will require periodic inspections of the property. Plus they will expect you to manage units well and keep them in good condition.

Remember, the sole purpose of give-away grants and low-interest loans is to provide incentive to owners and investors to create safe affordable housing for lower-income citizens who might otherwise be left out in the cold.

FREE FIX-UP MONEY FROM UNCLE HUD   *337*

"Too Big to Fail" Philosophy Applies to Duplexes

The Federal government often "bails out" large banks and businesses to prevent them from going broke when they are characterized as "Too Big To Fail." It's only fair that this same courtesy is extended to landlords who own smaller properties, occupied by low-income HUD tenants. I have known investors who have had their housing loans forgiven or the payments reduced, when the reasons were justified.

City housing is not your typical "Hard-Ball" lender. They are flexible like owners who finance their properties with carry-back loans. Cities never want to own the rental properties, therefore, it's comforting to know—when you're working with city HUD, and the economy falters—you won't be foreclosed on or you won't lose your investment property. It's more likely that city housing will adjust your mortgage debt to fit whatever amount of income you have. Safety nets like this are very valuable, especially for landlords who are concerned about cash flow.

The Extra "Red Tape" is Grossly Overstated

All government operations use more paper than they should! That's especially true when HUD is involved. However, most of the paper goes in files, never to be seen again after you sign it. Your application for a city HUD loan or grant funds won't be much different from any bank application. Financial information is about the same for all borrowing, even though the forms may vary slightly.

For every housing loan I've obtained, my standard personal financial records are all I've ever needed to satisfy city or county agencies. Financial information is common to all borrowing. You need five basic forms or documents to provide an adequate picture of your credit and financial capacity to housing lenders. Refer to Chapter 18—in the section entitled **"JAY'S FIVE BASIC FINANCIAL DOCUMENTS FOR BORROWING"**—for a detailed explanation of the five basic financial tools I use. Just prepare the forms yourself and you're ready to go.

THE EASIEST LOAN IN TOWN

It goes without saying, you should have your financial tools updated on a continuing basis. That way, when an opportunity presents itself, you are ready to respond. HUD-assisted loans made by housing agencies are quite easy to obtain. For example, in my area, the city will loan up to 90% of the property value. Banks who write loans (mortgages) in second or third position for rental properties normally require at least a 30% safety margin of owner equity. Also, your personal credit is seldom any major issue for housing loans. *Don't forget, these agencies want to give you a loan if it fits their own goals.* They generally assume your credit is good if you already own rental property. To summarize, housing agencies will require you to submit all the financial forms we've mentioned. However, they are not likely to scrutinize them with a fine-tooth comb, because they want you to qualify!

Remember the key to getting the most help is—You must own or acquire properties in "so-called" target areas. You will get this information by asking your housing department. My approach to the grant funding program is the same one I use in all my real estate activities. First, determine the needs of others, then "tailor-make" your project or plan to fit that special need. By doing this, you'll make more profits than all your competition, *I'll guarantee it!*

Steps to Take With Properties You Own

The application process—"red tape," as some will call it—is really not that difficult to complete. The housing department will give you their standard application forms, and you can simply fill in the blanks! You will need to provide each tenant's name, rent amount, size of family, estimate of tenant's income (information from the tenant's rental application you have on file). You will also need to provide data about the property itself, such as location, size, number of bedrooms, upstairs, types of heating and cooling, and information about utilities servicing the property. I always provide a sketch of the property (prepared by me at the time of purchase). You can fill in the required information with different colored pencils. Sketches are a very useful tool. Refer to the example of Jay's typical property sketch in the Appendix section.

Housing agencies will want to know financial information, such as your personal estimate (guess) of value, number of mortgages, balance owing and the monthly payments (debt service). You must provide proof of insurance and a copy of your latest property tax bill. If you owe back taxes, you can sometimes get them paid up to date with grant funds—always ask first. After you complete the paperwork, submit it to the agency. They, in turn, will review your request and visit the property to determine what needs to be done. Next, they call you to set up a feasibility meeting to discuss their findings and to inform you whether or not the property fits under their program guidelines.

At the feasibility meeting all questions will get answered. For example, *you might be wondering if you can do the rehab work yourself.* Now is the time to ask! The agency will inform you if your current tenants qualify for rent subsidies (more about this later). In short, the feasibility meeting is the time for all parties to ask anything and everything they'd like to know about the proposal before moving forward.

If everything gets satisfactorily resolved and you receive approval, you will need to provide a termite report (not over two years old). If you don't have one, you will need to get one, at your own expense. However, the cost of any recommended work will become part of the rehabilitation project.

Once the termite report is completed and submitted to the housing agency, they will inspect the property thoroughly and proceed to write up a complete itemized repair list (bid list), along with cost estimates and specifications for doing the work. After costs are established, most housing agencies are required to submit their plans to a loan committee or housing review board whose members will make a decision on the agency's recommendations. Assuming all goes well, and the project gets approved, the next step is the tough one. Once approved, the owner must put up his share of the total cost—as I said earlier, normally 50%.

UNDERSTANDING THE MOTIVATION AT CITY HALL

Just as price and terms will be different—depending on whether you're buying or selling—so it is with the housing folks at City Hall. Certainly they have rules and regulations to follow, but in the final analysis, housing departments are both judged and funded on "how much" housing they provide. For example, let's say my city has been allocated funds (block grants) to rehabilitate 50 houses during the next fiscal year, but, due to a variety of reasons, they only complete 25 houses.

What happens is they lose the money! Normally, the funds are transferred to another city that has a faster, more efficient housing department. Also, it's quite likely that next year my city won't get funding for 50 houses. Instead, they'll only get enough for 25. I don't think I need to tell you that giving back unspent tax-payer money is the worst thing that can happen to any government-funded agency, like city housing.

"YELLOWING" THE SQUARES

At this point, I'll pass along some personal philosophy so you can better understand exactly how important "good timing" is. I've accomplished several very profitable rehab projects (fix-up jobs) on my houses, obtaining easy-to-pay, low-interest city loans, because I took the time to learn exactly what buttons to push and when to push them. *Timing has everything to do with getting your jobs approved and making the highest dollar returns at your local housing department.*

Before my city housing department used computers they kept track of all rehab jobs on a large cardboard poster mounted on the wall. Houses were represented by squares drawn on the poster. Each square represented a house that was funded for rehab. Some squares were for two-bedroom houses, others were for three-bedrooms, and so-forth. As the budget year progressed, and as the various houses were rehabed and completed, the housing folks would "yellow in" the squares with a felt-tip pen. Obviously, those squares not yellowed meant the houses were not yet scheduled for rehab.

IT PAYS TO LEARN WHAT MAKES CITY HOUSING TICK

Several years ago, late in the housing budget year, on a routine visit to City Hall, I discovered an entire row of un-yellowed squares—each representing a single three-bedroom house. More importantly, the houses were funded, but fix-up work had not yet been contracted or even scheduled. For some reason landlord-owners with three-bedroom, fix-up houses were not responding to the program in the numbers the city had expected. Less than 90 days remained for the housing department to contract and rehabilitate these nine three-bedroom houses or face the loss of grant funds to another faster-spending agency.

The city housing did not wish to lose funding, no matter what, and I clearly sensed a panicky-type urgency, so I proposed converting my two-bedroom houses to three-bedrooms by making the garages into third bedrooms and adding a carport. Normally, the city won't do this, because it's much too expensive per unit. At the time, the cost limits were $7,500 per house, and I would be required to furnish $3,750 (my 50% share) plus everything over the $7,500 limit for each house. The bid estimates were $12,000 per house and I advised the city I could not come up with $8,250 for each house.

The Thought of Money Left Over is Totally Unthinkable

Quickly the city became very creative. They said, "Okay Jay. How much money can you provide?" I said, "$2,500 per house is my limit." They didn't like it, but they didn't say "no."

What finally happened was that they made me a city loan at 5% interest for 15 years of amortized payments, combined with a State of California housing loan at 3% for 15 years with deferred monthly payments (that means none). They also provided 15 years of housing subsidies with rents 20% higher than normal for each of my converted three-bedroom houses. I now have cash flow and guaranteed rents for 15 years, even when the houses are temporarily vacant. It's a very good deal for me because they always provide positive cash flow.

YOU HELP YOURSELF MOST WHEN YOU'RE HELPING OTHERS

The moral to this story is a simple one. Find out what city housing needs, then deliver it to them in a timely fashion. Think not what city housing can do for you, but rather what can you do to assist city housing. You'll find, as I have, it pays big dividends.

I recommend that all landlords pay a visit to city and/or county housing departments. Find out exactly how the these programs work in your own area. Each agency uses various combinations of Federal Block Grant Funding along with local loans. I have found that each city has its own specific goals for upgrading community housing. They also have particular sections of town (generally older) that they target for major fix-up funding grants. Often this information can be extremely helpful when you're considering where to purchase your next property.

You'll find it's good business to learn all you can about local HUD programs in your area. It will give you insider knowledge about where you might consider investing, plus you'll be able to find out how much rehab money you might expect to receive if you acquire fix-up properties within the city's special "rehab" target areas. Over the years, my work with city housing has been very profitable for me—and it can work the same way for you!

CHAPTER 20

Buying Back Mortgage Debt for Bonus Profits

We've already discussed wrap-around financing and the opportunities to buy back underlying mortgages at substantial discounts. However, buying mortgage debt is not just limited to wrap-around mortgages when you sell! It can work anytime when you own properties and are making mortgage payments to private parties. Let me show you how this is done.

LOOK FOR PROPERTY WITH PRIVATE MORTGAGES

The first step involves buying the right kind of property! The right kind of property is one where the potential exists for adding value like I was able to do at Hillcrest and with other properties. It must have one or more private mortgages or promissory notes secured by the property.

To illustrate, let's say we locate an older rundown apartment building and discover it's had three different owners during the past few years! Further research reveals that with each of these sales, new owner carry-back mortgages were created to facilitate the transactions. These mortgages, commonly referred to as *owner financing* or *purchase money notes* are long-term and they are assumable to a new buyer.

The seller is asking $300,000 for the building, and that price seems reasonable based upon the income. The seller will take $30,000 cash down payment (10%) and allow the buyer to assume the three existing mortgages with remaining balances of $85,000, $50,000 and $35,000. To complete the transaction, the seller is willing to carry back another mortgage for the balance of the sale price. The new purchase money mortgage will be for $100,000, secured by the apartment for a term of 20 years.

These are excellent terms for a buyer. Obviously, the seller must have a fairly high motivation level to make this deal. In most cases the properties I find with these kinds of terms have been "milked"—that is, allowed to run down because the owner is not willing to spend any rent money for routine upkeep. High expenses and management problems are top motivators for selling, I've found.

I realize I haven't presented a complete picture of this apartment building—things like deferred maintenance, bad tenants and the location. However, I've told you the rents (income) are reasonable for the asking price! That's really all you need to know to move forward. Don't negotiate too much over the asking price, if it's reasonable. Here's where you need to spend your time and efforts. First, ask for copies of all three existing notes (mortgages).

One of the things you'll want to find out about each of the notes or mortgages is the face amount. That's the amount when it was originated. For example, let's say the note with the current $50,000 balance started out at $57,500, when the sale was made that created it. That means $7,500 of the principal amount has been paid so far. Obviously, $50,000 is still owing. Having copies of each note will also give you the names of the buyers and sellers, the exact amount of the payments, interest rates, remaining time until pay-off and any special provisions you should know about. Private party notes can have a wide variety of provisions. Always read them very carefully.

I once had a note that stated, *"The Beneficiary may increase the interest rate from 9% to 11% beginning January 1986, if the duplexes are not properly maintained."* I seriously doubt if it's enforceable, but it's still one of the terms stated in the note. The only way I would ever know it's there is to see the note for myself. Notes drawn up between private parties often contain very unusual terms. My advice is to get copies and read them several times! The term you don't want to see is a *"Due on Sale"* clause, meaning you would have to pay the note or mortgage off if you sold the property.

Knowing the Players Can be Very Profitable

When I first began buying real estate, I had no idea about how much money could be made indirectly from my transactions! **Strategies that go far beyond the simple wisdom of acquiring properties below market values can double and triple your profits, even after you already own the property.** If this sounds too good to be true, believe me, it's not! Let me explain what I mean by "making money indirectly."

Let's continue with the apartment example! It's the right kind of property, because it's property that's been around for many years. It's had enough longevity to have accumulated multiple sales and, thus, has multiple notes (mortgages) secured by it. Newer buildings are not as likely to have accumulated these notes from multiple sales. Let me clarify, I'm not talking about notes or mortgages where cash money was borrowed against the property! Those are often called hard-money loans and you can forget about discounting them.

Purchase money notes are the kind we're looking for! These are notes where no money was actually given to anyone. They are actually a form of credit extended to the buyer to facilitate the sale. These are the kinds of notes or mortgages that can earn you big money! And, this is where having knowledge about the players (beneficiaries) can earn bonus profits for ring-savey investors. The more you can learn about the folks who receive the payments, the better your chances for plotting a strategy that will hit pay dirt!

Let's say, for example, Mr. Jones is receiving monthly mortgage payments from the sale of his apartment building. Jones sold the apartments several years back, because the tenants were about to drive him completely bonkers. At a weak moment, Jones rented to a rock band who played loud music all night, slept all day and only paid rents when Jones stayed up late enough to catch them! Mr. Jones is now 67 years old and retired. He sold the apartments and carried back a note for $90,000, which now has an unpaid balance of $85,000. Jones is scheduled to receive payments of $780 per month for 25 more years.

Most Sellers Would Have Rather Had Cash

Jones would have preferred a cash sale for his apartment building. He wasn't keen on financing the deal himself. However, because he got behind on the maintenance and suffered rent collection problems with the local rock band—there was only one buyer that showed any interest in purchasing his property. It was pretty much a case of take the offer or forget the sale. Jones took a small cash down payment and carried back the balance of his equity on a note for $90,000. It's the only way he could get anywhere near the price he was asking!

Life expectancy for Jones is 73 years. That's only six years from now and his note will still have 19 more years worth of payments. If you're trying to purchase the property, the time to talk to Jones about discounting is after you already own the apartment and have assumed the note. If you attempt to discount the note during negotiations, Jones is quite likely to insist on more cash, because he feels he has more bargaining power! Waiting until after you own the property, before asking about discounts, will generally get you better results.

I always like to meet with most mortgage holders at their homes. You can learn a great deal about people by seeing how they live. For example, you'll be able to answer questions like: What kind of cars do they own? Are they old junkers? What's the condition of their furniture? Is it expensive? Do they have college-age children that cost extra money? This information helps me determine what their needs might be. If I can learn what they might need or want, then figure out a way to give it to them—I will nearly always come away with a sizable discount. This personal information can be very valuable, believe me!

JAY'S RED MUSTANG STRATEGY

I always like to tell the story about my red mustang transaction! It illustrates the importance of understanding human nature and how it can be used to make big money in this business. Susan and her boyfriend unexpectedly inherited my mortgage when Susan's mother passed away. The balance was $48,000 and my payments were $441 per month. The note had 11 years to go before the balance would finally be paid. I learned of the mother's death when I received an address change for mailing in the payments. Shortly after that I contacted Susan with an offer to buy the note.

I explained that I could refinance my property, but would only be able to "net-out" about $25,000 after expenses. That's what I offered! Susan's boyfriend was a typical "deadbeat," who always needed cash, but, surprisingly, he didn't like my offer. Reducing my debt by $23,000 was a little more than he could stand to see me get, otherwise I think Susan would have gone for the $25,000 cash. However, they stuck together and refused my offer.

IF AT FIRST YOU DON'T SUCCEED—TRY HARDER

Quite often you won't get your first offer accepted! Don't give up, just try a little harder! The day I drove by Susan's apartment and saw her boyfriend's junky Subaru sitting on wooden blocks with the engine missing, I knew exactly how to re-do my offer. Human nature being what it is, I knew that young people living in a tiny one-bedroom apartment with no transportation will soon self-destruct! When I saw them again, they were a lot more motivated and willing to talk about selling the note. They were on the couch watching TV when I drove up in a shiny new red mustang convertible.

My credit union had already agreed to loan me $20,000 so I could pay cash for a the mustang. It was a beauty! My loan payments would be $381 per month. The dealer also gave me a trade-in allowance for my 21-year-old pickup. Had I purchased the car on a regular installment contract from the dealer, the total cost would have been around $25,000, including the $2,000 credit for my pickup.

I left the motor running when I went to the door and knocked. When it opened, they both saw the car at the same time. It was love at first sight. They both immediately fell in love with my shiny new red mustang convertible.

I let Susan drive around the block a couple of times and squeal the tires! When she got back, she wouldn't get out! It was as if she were glued to the seat. Anyone can own a new mustang convertible, I told her. All you need is $25,000 and they'll deliver it to your door.

As we talked, I was watching the boyfriend. Several times he glanced over at his jacked-up Subaru, then back to Susan, who showed no signs of getting out of my car. I could tell by now, the boyfriend was hooked! Can we both take a ride, he asked. I agreed, but only around the block—one time!

Reeling in the Big Ones

Marketing folks call it "the hook." Others use the term "set-up." But whatever the name, it doesn't matter much. I will tell you this much—IF you can figure out how to structure your proposals in a manner that will fit the wants or desires of a particular client, you can "hook" them with a remarkable rate of success!

Susan and her boyfriend needed a new convertible like they needed a hole in the head. Once they saw the car, they wanted it very bad! Seeing it in their own driveway, holding the keys and then finally driving around the block was a bit more temptation than either of them could stand.

In less than 24 hours the deal was done! They had traded the note to me for much less cash than my original offer. You recall my offer was $25,000 cash. The deal they finally accepted was this new convertible and $3,000 cash. To me this deal was much more attractive than my original cash offer because there were no appraisals, no points, and no new deeds on my property. I paid $20,000 cash to the car dealer with money borrowed from my credit union, so the only out-of-pocket cash I needed was $3,000 to close the deal.

The bottom line is this! My $48,000 debt is gone, along with the $441 monthly payments scheduled for the next 11 years. Instead, I now have $381 monthly payments for only six years. If you forget the payments for a moment and just consider the trade itself—it cost just $25,000 ($20,000 borrowed funds, $2,000 trade-in and $3,000 cash) to buy back $48,000 worth of long-term mortgage debt.

THERE ARE MANY WAYS TO SKIN THE SAME CAT

I'm sure you can see for yourself that buying the right kind of properties—like the kind we're discussing here, with private party notes attached—can offer some exciting high-profit opportunities, once you get the hang of using this technique!

I recall being a part of one particularly long and drawn-out negotiating battle where four or five different counter offers were exchanged back and forth between myself and the sellers. We argued for weeks over a $4,000 difference in the selling price! My offer was $146,000. The seller wouldn't budge below $150,000!

Let's suppose this property had three $25,000 private notes attached, all of which could be assumed! Say we finally purchase the property and eventually are able to buy back the debt! With something on the order of my mustang deal (about 50¢ on the dollar). That means three $25,000 notes (mortgages) discounted by 50% would be $37,000 worth of discounts.

When you match this kind of potential profit against the small $4,000 difference in the seller's asking price, you can begin to see where your negotiating skills will earn the biggest profits.

Again, let me emphasize—THE PROPERTY YOU BUY MUST BE THE RIGHT KIND OF PROPERTY. You obviously won't be able to do what I'm suggesting here if you purchase property with the normal 20% cash down and finance the balance with a bank loan. There's simply no room for this kind of creativity in deals like that!

WHAT YOU SHOULD KNOW ABOUT BUYING DEBT

We've discussed buying the right kind of properties—the ones with good assumable debt! I also told you this particular strategy can be best utilized after you become the new owner. Quite often I'm asked, "Why not buy the existing notes (mortgages) at the same time you purchase the property? For example, during the escrow period, couldn't we simply negotiate with each mortgage holder, asking them if they would be willing to sell their notes for a cash discount? It seems reasonable that if the note holders would agree to sell while the property is in escrow, we could wrap everything up in one glorious transaction! We could buy the property, obtain our discounts and close the escrow all at one time! What a time-saver that would be!"

Don't Forget About Greed—It's Human Nature

Remember what I told you earlier about human nature? Real estate knowledge is not what makes this idea work. It's much simpler than that. It's the human nature factor inside all of us!

I well remember trying to do everything in one glorious transaction, as stated in the question above, and it turned out a horrible mess! I told the seller I had just enough money to pay 10% cash down for his property! I then contacted both private note holders who had mortgages on the property totaling $78,000, and got their approval to sell me both notes for $53,000 cash. My plan was to refinance the property to get the pay-off money! Everything was moving along fine until the seller got wind of my $53,000 cash offer. He demanded a bigger down payment—plus he told me, "Since you plan on refinancing my property anyway, I'd rather not carry back any mortgage myself."

I've never made that mistake again. The lesson I learned was to nail down *first* things *first*, then move on to the next step! Remember there are many sellers out there who are perfectly content to accept your 10% cash down offers. They're willing to carry back all of the financing as well—*that is, unless they get the idea you're holding back on them*. If that happens, greed takes over and you may lose the entire deal!

FACTORS THAT MOTIVATE DISCOUNTS

I keep a file on every private note holder I'm making payments to. I can't tell you exactly what this information is worth, but a modest guess would be about $50,000 a year. In other words, that's the amount of profit I expect to earn annually from buying back my own debt. My files contain information on every beneficiary (the folks who get my payments) including number and ages of family members and when children will be ready to start college, so I'll know when to start sending them letters pitching my $35,000 cash tuition plan!

One of the best opportunities occurs when note-holders pass away. My mustang deal is perhaps the best example. Through my beneficiary files, I had been tracking Susan since she was only 13. You saw the results! Most people who inherit valuables, like mortgage notes, tend to lust for cash instead. Keep a good eye on the obituary column. I always cut out the notices for my beneficiaries, then I make it my business to find out who will inherit my payments.

Joint note-holders who divorce are naturals for substantial discounts. Keep good track of them too! Divorced folks seldom want payments. Most often they need cash—and quickly! It's also good business to send note-holders a Christmas card around November 30th. Lots of good discount deals pop-up when people are in the mood to buy gifts or take a winter cruise. Of course, you must spell it out—**PAINT THEM A PICTURE** when you send out your cards. Let them know you have cash available, usually about half of what you owe!

I hope *you've* got the picture, by now! Next time you're negotiating to purchase a property, take a good hard look at the various ways you might profit from the deal. You never want to miss the opportunity to deliver a Mustang. The experience can be quite exhilarating, believe me!

Making Money by Accident Hooked Me

When I first began buying real estate, I had no idea that buying back my own debt was something I could do, let alone how to make it happen. The very first time I made money doing it was completely unexpected! The mortgage holder who I was making payments to, asked if I'd be willing to pay off my mortgage, immediately, if he reduced the amount I owed him. I said, "Yes." He knocked off $16,500, and I was never so pleasantly surprised in all my life. What an easy way to earn $16,500, I thought!

Quite by accident, I had unknowingly stumbled into my first discount buying experience, and I quickly realized I was onto something big! I remember thinking to myself, "If I can make $16,500 on just one medium-size mortgage, how much could I make if I had a hundred mortgages to work with?" That's when I first began to realize that the more private notes or mortgages I could assume (take over) when I acquire the property, the greater the odds of buying some of them back for a discount, *after I become the new owner.*

SETTING THE STAGE FOR DISCOUNT PROFITS

If you want to make big bucks buying your own debt, the first thing you must do is locate and acquire the kind of mortgage debt that creates an opportunity for discount buying. As I told you earlier, the more common institutional financing—the kind of mortgages that banks and savings and loan lenders make on properties—are not what we are looking for here. Newer properties (non fixer-type) are almost always financed by institutional lenders, rather than by sellers who carry back a mortgage. Therefore, if you insist on acquiring newer real estate, chances are this lucrative opportunity will not be an option for you.

Older properties are where you'll find the private notes or mortgages. Since many banks will have nothing to do with financing older properties, especially non-conforming, multi-unit rentals, it means sellers of these types of property are forced to carry back the financing themselves, in order to sell them! Quite often these properties will sell every few years, with each seller, in turn, creating a new mortgage or promissory note for a portion of his equity. I have personally bought properties with as many as seven different personal notes secured by the property. Three or four seller carry-back notes are not the least bit uncommon with older rental properties like we're discussing here.

DON'T THROW THE GOLD AWAY WITH THE SAND

New investors have told me they won't buy properties with more than two mortgages on them! If there happens to be more than two, they insist the mortgages be consolidated (paid off), leaving only a single loan or mortgage when escrow closes. They claim having just one mortgage payment simplifies everything, as there's only one payment and one lender to worry about. It makes a nice, clean manageable transaction, they say!

I must tell you that having clean and manageable transactions should never mean tossing your money or profit potential out the window! If it does, you're in for big trouble as an investor! Never lose sight of the target—otherwise you'll be passing right over some hefty profits that could easily be yours for the asking. It's much like having big gold nuggets in your pan, then throwing them out with the sand because you fail to recognize they're really the gold you're looking for!

REASONS WHY MORTGAGE HOLDERS SELL FOR DISCOUNTS

Why in heaven's name would anyone sell their mortgage for less than the unpaid balance? Why would the former property owner be willing to accept a lot less money now than he previously would take when he sold the property? Obviously, there is no end to the list of answers, but here are some of the most common reasons I've found.

1. **Note (mortgage) holders want cash instead of monthly payments, for many different reasons. Perhaps they wish to**

   a. Purchase a new car, boat or recreational vehicle,

   b. Take a vacation,

   c. Buy something that requires a lot of cash,

   d. Start up a new business venture, or

   e. Send the kids to college or help them buy a new home.

2. **Partnership breakups, and divorce or separation.**

   a. Partners need to split the money, go their own ways.

   b. Divorce often creates a need for immediate cash.

   c. Investors who sell rundown properties often carry back the mortgage to facilitate the sale. They would have rather had cash to begin with!

3. **Lifestyle changes—health or death of beneficiary.**

   a. Older mortgage holders may want cash for "one last fling."

   b. Mortgage holder has sudden need of cash for hospital or retirement home.

   c. Younger relatives, who inherit mortgages after death of beneficiary, often want cash now, instead of payments over a long period of time. Many folks are impatient just like Susan and her boyfriend.

FINDING THE RIGHT MORTGAGES IS WELL WORTH THE SEARCH

If I could pick an ideal property to purchase, it would be one that has at least four private-party purchase money mortgages that I could assume (take over). Purchase money mortgages are the kind that result from sales where the seller carries back the financing. They are different from mortgages where cash is loaned against the property. Like I told you earlier, they are an extension of credit. No cash is dispersed.

Here's a typical example—an owner sells his $100,000 property on terms. Both parties agree to a $10,000 cash down payment and a seller carry-back mortgage or promissory note for $90,000 at 10% interest, with monthly payments of $800 (including interest and principal), until the balance is paid.

The $90,000 mortgage or promissory note is a *PURCHASE MONEY MORTGAGE* and it's the kind of mortgage you should be looking for. First of all, it's a long-term note—it will take about 28 years to pay it off, which means you will have many opportunities along the way to contact the holder (property seller), or his heirs, about selling the note back to you for less than the balance owed, more commonly referred to as a discount.

Naturally, the mortgage should not contain a due-on-sale clause in the terms! However, the big percentage of seller carry-back notes do not have due-on-sale clauses. They don't have pre-pay penalties or late fee charges, either. This makes them weak in terms of their value as commercial paper. Mortgage brokers and professional note buyers want these kinds of terms when they purchase notes for cash.

A $90,000 note with 25 years of payments remaining, no due-on-sale clause, and created just three years ago with only a 10% down payment is almost *non-marketable*. The cash amount anyone would likely pay for such a note *(soggy paper)* would not make the note-holder very happy. In reality, the person most likely to pay the highest price for this weak note is the property owner who is making the payments on it (mortgagor).

VALUE, LIKE BEAUTY—IS IN THE EYE OF THE NOTE HOLDER

There are many books written about buying notes at discount, but the business of buying and selling notes is not what we're really discussing here. We're talking about buying back debt on properties we purchase without much regard for the standard yield computations. I know I'm going to hear from a few note buyers, but let me just say this anyway—I am more than happy to buy back my own mortgage debt and pay a little more money than most professional discount mortgage buyers would pay—and you should be too! Trust me on this one and you'll end up a happy camper.

Note Buying Strategy is Designed for Payees

My note-buying strategy is based more on events and circumstances in the lives of my payees, as opposed to being based on the yield charts! For example, when I locate a property I'm interested in buying, I always make it my business to see copies of the promissory notes to determine if they are good candidates for my future buy-back strategy. Notes are not public information like recorded deeds, so you must ask property owners (sellers) to produce them.

After I determine that terms are acceptable, I try to dig out information about each of the beneficiaries, using my "Detective Columbo" techniques. Remember, beneficiaries are the folks you mail your payments to! I want to know if they're rich, poor, young, old, how many kids and what ages the kids are, especially when they're about the right age for college. You must force yourself to be a little "snoopy" to do this right, but it will pay off over time, believe me!

JAY'S "CHRISTMAS LETTER" GENERATES PROFITS YEAR-ROUND

Once I discover that my note holder is an average family man, with *a ten-year-old car* and *three teenage kids*, I'll send him my "**SPECIAL CHRISTMAS LETTER**," as follows.

November 15th, any year

Dear Average Note-Holder:

My, how time flies! Only six more weeks until Christmas. As you know, I've been sending monthly payments to you since I bought your property on Ugly Street nearly two years ago. Today, I just received word that I'll be getting a substantial cash distribution from my late Aunt Lucy's estate. It's not quite as much as the balance I owe you ($56,570 as of 11-95), but I'm writing to everyone I make mortgage payments to, in order to find out who might need the extra cash. According to the terms of our promissory note, I still have about 15 years of payments left until it's finally paid off.

In the past, several people I send mortgage payments to have asked me if I could pay them cash, instead of monthly payments. Naturally they are willing to reduce the amount I owe them in exchange for immediate cash! Because my income is mostly from rents, I seldom have enough cash to take advantage of their generous offers, but I do appreciate them offering, just the same. Naturally, I'm happy to pay cash when I have it.

The way things look, I should have my money a couple weeks before Christmas, but certainly not later than January 1st. If you have a need for extra Christmas cash, or perhaps for Jimbo's college tuition this coming spring, now would be the time to let me know, before I spend the money on something foolish! Also, if you could let me know how much of a discount you'll give for cash, I'd appreciate it very much. The best way for you to determine the true value is to call several professional note buyers in the yellow pages or newspaper ads. They will give you a quick estimate of the value—free of charge. I'll assure you, I'm always willing to pay more than professional mortgage buyers because I don't have a fancy office or the overhead to pay—and, of course, there are no fees dealing with me!

Sincerely,

FIXER JAY

INVESTORS NEED A HEALTHY FINANCIAL DIET

Once you get a taste of buying back your own notes or mortgages at whopping discounts—you might just decide to expand your note-buying business and leave fixin' toilets to someone else.

Before you sell your toilet auger, however, let me remind you that income earned from note-buying is not the same as income from rentals. For one thing, you will get no depreciation or capital gain treatment. In short, you won't get any tax shelter—and tax shelter has been a big part of my fortune-building strategy. My suggestion is to do both. *Operating real estate and buying notes make excellent bedfellows.* Check out the tax treatment with your local advisor.

ADD A PROFESSIONAL TOUCH

My good friend, investor teacher, Jimmy Napier, is a journeyman discount mortgage buyer and seller. In fact, Jimmy wrote the book on note-buying. It's called "**INVEST IN DEBT**," and I strongly urge you to get your own copy. Jimmy explains, in very clear language (a bit southern), how to figure exactly what to pay for notes and mortgages—also, how to protect yourself. For information, write: **JIM NAPIER, INC.**, P.O. Box 858, Chipley, FL 32428, or call 800-544-4488. Tell 'em Fixer Jay sent ya!

CHAPTER 21

Managing Tenants is Key to Investment Profits

Why in heaven's name would anyone of sound mind and body, who is capable of doing regular work, choose to become a landlord? That's a tough question for many folks, because it seems like many people have had bad experiences with rental units and tenants. Their tales of woe travel quickly and are often exaggerated to add more color.

WE DO IT FOR THE MONEY

The short answer is quick and easy—we do it for the money. Certainly there are other reasons, however, I know many "do-it-yourself" landlord-owners who consistently earn $100 per hour or more doing landlord work. *Landlording is the key ingredient to making big money for "Do-It-Yourself" investors.* Owners of small rental properties, who don't learn landlording, often throw away a fortune because they allow the tenants to run their buildings.

You Can Avoid a Crash

There is absolutely no need for this to happen. In many instances, I have discovered incompetency is to blame more than anything else, because it takes more than owning a rental property to be an effective landlord. In fact, I would guess *amateur landlords* cause almost as many problems for themselves as their renters do. Income property owners, who don't know anything about supervising tenants or landlording responsibilities, are in the same kind of situation as the man who buys an airplane, but hasn't learned to fly yet. In both situations a crash is almost inevitable and very predictable.

THE DREAM—WORKING FOR YOURSELF

Besides making money, being a successful landlord offers some major personal benefits. For example, you can provide an excellent quality of life for yourself and your family. You can live wherever you choose and set your own work schedule. You can spend more time on hobbies and doing the personal things that never get done when you work for someone else. Being your own boss is the envy of every wage earner, because almost everyone dreams of making lots more money working for themselves. The good news is, investor landlording is a profession that will allow almost anyone to achieve financial independence and the highest quality of life, if they really wish to participate. My good friend and investor, Richard Epley, says:

> *"IF YOU ARE COMFORTABLE WITH THE ROLE OF OWNER-LANDLORD, AND DON'T ALLOW YOURSELF TO BE INTIMIDATED BY THE RESPONSIBILITY OF SETTING UP THE RULES BY WHICH TENANTS MAY LIVE IN YOUR PROPERTIES, YOU WILL FIND, AS MANY OTHERS HAVE, THAT OWNING AND OPERATING SMALL RENTAL PROPERTIES PROVIDES A VEHICLE FOR SELF-EMPLOYMENT AND SELF-EXPRESSION THAT IS DIFFICULT TO MATCH IN OUR SOCIETY."*

Dealing With People Can Be Nasty

I certainly agree with Rich that it's a great way to go once you get the hang of it. However, like anything else in life, you must pay your dues up front. **Landlording is the price of admission for investors who operate income properties.** Landlording and managing rental property is a people-intensive job. Everyone knows that dealing with people, especially one-on-one, can be nasty work. It's only natural in situations where one person always pays money and the other always receives it, that disputes can surface quickly. Landlords, with any experience at all, understand it doesn't take very long on rent day for normal conversation to erupt into a shouting match. With low-income tenants it is usually more apt to happen because they run out of money faster. Still, there are methods you can use to avoid most problems.

YOU MUST UNDERSTAND TENANTS

The first thing you need to do is understand tenants, what makes them tick, what they really want, what they should expect, what their goals are, and, of course, what they are entitled to when they pay rent for living in your properties. Let me explain some lessons I've learned over the years. One of the most important things I've discovered as a landlord is that there is no such thing as a $400 per month apartment that rents for $400 per month. I can show this very easily by inviting 25 tenants into one of my apartments and asking each, individually, what they think it's worth? Most of them will answer $350, the others will guess $375. No one is likely to support the current $400 rent I charge.

The sooner new landlords understand this basic adversary relationship between the payee and the payor, the better they fare. It's easy to understand why it doesn't usually work out for tenants to be close friends with landlords. After all, tenants feel landlords charge too much, with very few exceptions.

TENANTS CAN MAKE OR BREAK A PROPERTY

It's my opinion that small-time real estate operators fail much too often because they never learn or don't understand land-lording. Almost all investors I deal with are owners of houses and small apartment buildings, so I've had plenty of opportunity to observe the problems. Even if your basic investment strategy is to buy, fix and sell—tenants will normally be involved in the process. Tenants can make or break a property, because owners who pay little attention to landlording skills are basically allowing their tenants or hired property managers to have control over their investment dollars. Remember, *control of our money is one of the MOST IMPORTANT GOALS IN THIS BUSINESS*.

PERSONAL INVOLVEMENT IS THE KEY

One major advantage small operators have over limited partnerships and syndicators is personal involvement. Owners who are personally involved are much more sensitive to what is going on around them. They can act quickly when the need arises, with no front office approvals, no voting, no politics. It's strictly a *single-party control*. Later on you will understand why complete and absolute control is such an important part of making big money in real estate. In the meantime, you'll just have to take my word for it.

Hands-on operators, the do-it-yourselfers, can put the big guys to shame in terms of building net worth. Call it "tender-loving care" if you wish, or whatever else, nevertheless it works. There is indeed a great deal of truth to that old Chinese proverb which says—*"The best fertilizer in the world comes from the owner's shadow."*

SUCCESS MEANS WEARING MANY HATS

It's important to understand that as an investor you will need to perform many different functions, in order to be successful. Landlording, although extremely important, is only one of these functions. You must also learn how to determine the right price to pay for properties and the reasons why. You must learn how to structure the financing. If you choose to do fix-up, you'll also need to develop those particular skills. Learning market trends, negotiating techniques, judging locations, understanding true costs of operating rentals, hiring workmen and developing good usable records and a bookkeeping system are also necessary parts of being a successful owner-investor.

MONEY DECISIONS MUST BE YOUR DECISIONS

In my opinion, when investing, you cannot leave any step out, otherwise it causes a weak link in your investment chain. Under the slightest strain, that link always breaks first. One weak link can wreck your whole plan very quickly. When small owner-operator investors decide to hire property management companies to handle the dirty work—like interviewing and renting to tenants, collection of rents, evictions and anything else they do for their monthly fee—they give up far too much control, in my opinion. *One of the most important benefits you need for making money as an investor is total control.* Never allow your ability to make money or earn profits be governed by others.

In other words, the owner is the one who sets the rents, works on the buildings as much as he thinks is necessary, and finally decides when it's time to sell. In short, he must be the one responsible for success or failure of the property. No other investing is quite like owner-operator investing, where all the control is in the hands of one person. If you can learn to do these things well—great success will be yours!

Landlording is Top Skill

When working with syndications, the stock market, REITS and limited partnerships, you simply hand over your investment money to someone else to manage. If it works out well, you earn a profit. If not, you lose your money. In the meantime, you have absolutely no control over any part of your funds, like you do with *personally-owned real estate*. When beginners ask me, "What kind of skills are necessary to be successful as an owner-investor who plans to invest in small apartments and houses, like you do?," my answer is this: "It takes quite a number of different skills, however, at the top of my list is always landlording."

FIX-UP SKILLS WORTH BIG MONEY

Landlording is not my top listed skill simply because you must do it first. It's first because it ultimately involves more than the word might suggest. For example, would you guess that my landlording-tenant skills play an important role in my being able to buy properties cheaper with much better financing terms than most other buyers? *It's true!* You see I have the ability to fix-up "problem properties" that competing buyers are reluctant to purchase. I have learned that dirty houses filled with deadbeat tenants are much easier to clean-up and upgrade than most of my competition realizes. More importantly, they produce profits much faster than the nicer looking properties.

A People Business

Operating rental properties and dealing with tenants are not separate jobs. The fact is, they're inseparable parts of landlording. I owe much of my real estate success to my landlording abilities. Some folks won't agree with me on this issue, and many argue that professional property managers get paid a handsome percentage of the gross income to relieve owners of this thankless task. "So what's the big deal?," they ask.

PROPRIETORSHIP—A MUST

Proprietorship is the big deal, because no one has the same level of motivation like the owner. Remember, it's owners who borrow money against their homes, who invest their entire life savings in an effort to make a better life for themselves and their families. Owners have a much greater interest in their own success than anyone else. That's what proprietorship is about, and it's something that fees from the rents can't buy. Picture this scenaro if you will—*it's Sunday afternoon*—a tenant (who we'll call Mrs. Jones) calls in a repair to fix the handle on her toilet. It broke off and the lever fell down inside the tank. The flapper chain came loose, so Mrs. Jones can't flush her toilet. Naturally, all the grandkids are there on Sundays. She needs help now! I'll give you ten to one odds, if you have a hired property manager, that toilet won't flush a single lick before Monday morning. *But if it does*, I'll bet it was the owner who drove out to fix it!

My point is this, *if you collect rent for Sundays, you must also arrange for flushing toilets on Sundays.* It's just sound business practice and common sense—*proprietorship is what's most important*, especially for new investors who are just starting out. Professional managers might be all right for later on, but seldom in the beginning.

IT PAYS YOU TO THINK LIKE YOUR TENANTS

If you will teach yourself to think like your tenants, you can stay a step ahead of them most of the time. Believe me, staying a step ahead is very important. *It's always better to act now, than to react later.* A friend of mine, who's been a policeman for 30 years, told me "the best detectives in his department are the guys who could have easily been the best crooks." "The reason," he claims, "Is because these officers just naturally think like crooks. It's like they're able to antici- pate a crook's next move before he ever makes it."

This same uncanny intuition can be used, to a tremendous advantage, by landlords who can develop it with their tenants. Just like the cops, landlords can benefit by predicting tenant moves ahead of time. With practice, anyone can do it. Three years after I started landlording, I realized I had become quite skillful at it. I might add, you are never finished developing this skill—you simply keep getting better at it, as you go along.

ALWAYS ACT—DON'T REACT

Good landlords always act, they don't react. Don't fight and argue with tenants, anymore than your boss would fight and argue with you about work requirements on your job. Obviously, discussions are all right, and fairness should always prevail. Landlords are the boss of houses, and they make the rules for tenants who live there. Never compromise the issue of who's in charge. *When tenants living in your rental property have the upper hand, you've got big problems.*

How do I know this? At least half of the properties I own today were purchased from highly-motivated sellers who wanted, more than anything else, to get away from their ten- ants. I made that possible for them—at my price, of course!

Most beginners don't fully understand the true value of learning to be good landlords. As you may have guessed by now, it's a serious business. Not only does it involve the daily management of your tenants, it also has a great deal to do with your real estate profits—buying and selling!

GIVING FAIR VALUE FOR A FAIR RENT

Several years after I became a "full-fledged" landlord, I found myself trying too hard to make my tenants extra happy! I didn't realize it at the time, but that will never happen. It seemed like the more I tried to please them, the more my tenants wanted! For example, some of my inexpensive apartment units have only linoleum on the floors, and several of my economy houses have carpets in the living rooms and hallways, only. The tenants were always asking for extra carpets after they rented the property. My houses are always clean and painted with my standard off-white color inside; however, tenants wanted me to furnish them paint of their choice, to repaint because they didn't like my color.

Many of the requests I received would have cost me a ton of money if I agreed. Often, they would ask me for rear yard fences, where none existed when they moved in. What generally happens, in this case, is they acquire a dog or decide to baby-sit small children in the backyard for extra income. Don't misunderstand what I'm saying here, I'm very much in favor of rear yard fences, and front ones, too, for that matter. But there is something missing in all of these requests! It's called "consideration"—the contract term for the extra money. The question is, who should pay for these special upgrades and add-on requests? Houses without carpets are like cheeseburgers without fries. **However, you only get the fries when you pay extra for them.**

Telephone "Repair" Instructions

One of my special techniques to help me cut down on extra repair visits is to ask tenants to thoroughly explain their problems over the telephone, when they call in. This has two positive effects. First, I can get a good idea of what's wrong. I am better able to judge the seriousness of the situation and can, quite often, instruct the tenants on how to fix the problem themselves. Good explanations over the telephone also help landlords decide what type of service is needed. My "how-to-help-yourself" instruction over the phone is especially helpful, when problems involve things like re-setting a fuse or breaker, what to do about noisy neighborhood kids, or how to flip the toilet handle so the water stops running.

Summer months bring on a rash of calls about coolers that don't seem to work. Often, it's merely a case of the control knob being switched to the wrong position. Most evaporative coolers can be set to "vent only" (only the fan operates) and tenants mistakenly think something is wrong. The fact is, many don't bother to read the knob setting, instead they immediately call the landlord. *I have found about 30% of my tenant's repair calls are really unnecessary to begin with.* Getting good information from the caller, along with providing self-help instructions by telephone, can save landlords many trips to the tenants house—and obviously a great deal of money, to boot.

Playing 20 Questions

The second reason for asking my tenants lots of questions is to discourage unnecessary calls. *Tenants don't like to answer lots of questions.* Don't ask me why, I'll probably never know. I do know, however, they won't call nearly so often, when they know I will quiz them. Perhaps some realize how silly the call is to begin with. Next time they try to help themselves first, before they start calling.

I always ask my tenants to describe "what is leaking," and if they can actually see where the water is coming out. By listening to their explanations and descriptions over the telephone, I can generally determine if a *supply line* or *drain pipe* is involved. Obviously, water supply lines, under pressure, are a far more serious concern to me than dripping drain pipes.

DON'T UNDERESTIMATE DANGEROUS SITUATIONS

Electrical problems in older houses and apartments can be very serious. You must always use special consideration with the older-type fuses, service mains and obsolete fixtures. Because they are old, they may be worn out and it's usually best to replace them. Electrical problems can burn houses down. Obviously, you need qualified electricians to perform any major electrical repairs.

I always investigate electrical problems very quickly to determine what course of action is appropriate—and you should too!

LANDLORDING CAN BE LEARNED VERY EASILY

One of the many advantages of owning and operating rental houses is that individuals without any special skills or advanced training can become very successful by substituting personal efforts (labor) instead of spending tons of money to get things done. It often takes new owners twice as long to do a job, but don't forget, it's also training—in addition to cheap labor.

You don't have to agree with me about any of this. You don't even have to like it, but please keep an open mind. Someday, if we meet, you may even thank me for the help. Also, don't forget, *I'm still in training myself,* I have many teachers every single day of every month. They live in my rental houses and apartments. As you've probably already guessed, my teachers are my tenants. They have taught me many tricks of the landlording trade—and the experience has been priceless, believe me!

I continue to receive letters from subscribers and distraught landlords who can't control the people living in their properties! I must tell you now, if you cannot handle your tenants—that is, make them pay rents and follow a few basic rules—you can expect a very miserable existence owning real estate. Worse yet, you probably won't be around long enough to enjoy those big profits I promised you in earlier chapters.

PROPER TIMING REQUIRES THAT YOU RENT OUT YOUR PROPERTIES

Investing in real estate the way I do it—that is, buying fixer houses and so forth—will not be a profitable venture unless you can properly time your acquisitions and sales. Proper timing is worth big bucks! For example, today in Southern California, three-bedroom houses that once sold for $375,000 might be a tough sale at $325,000! Some are down by $100,000 in areas where big defense contractors have laid off thousands of employees. The point is, you shouldn't plan on selling when prices are down. Obviously, if you don't sell now, you will need to rent your properties until the market changes for the better! You're probably wondering when that will happen? My answer is, "I don't know, but history has proven time and again, it will get better someday! Don't forget, die-hard investors must have a little faith!"

Real Estate Prices Go Up and Down

The real estate market is cyclical! It goes up and down, like a yo-yo, and it always has! A 20% swing in the cycle can be worth as much as $40,000. That's because, in a seller's market, $100,000 properties can often be sold for $120,000. This happens when many buyers are chasing too few properties. Conversely, the same properties can be acquired for $80,000, when times are tough and buyers are scarce. It goes without saying, owners who tailor their investment plan (buying and selling) to match the real estate cycles can greatly enhance their bank accounts!

I have long supported the proposition that "mom and pop" investors must learn good landlording skills in order to earn serious real estate profits! When the market is down, you want paying tenants renting your houses. Paying tenants will afford you the luxury of continuous income, while you ride out the low cycles. With income, you can patiently wait to properly time your sales! Many investors continue to ignore this wisdom, hoping to earn big profits, regardless of what real estate cycle they are in! This strategy, if you can call it one, is akin to driving your car through stop signs everyday, hoping that no other driver will be coming the other way.

GOOD MANAGEMENT IS VERY IMPORTANT TO OWNERS

I often refer to landlording and managing as the 10% job that earns the other 90% of every rent dollar I collect! I charge 10% to manage properties (although I only do property management for myself, today). I might just add, however, I'd be more than happy to pay a so-called professional manager 10% of my gross rents to operate my properties the same way I do. The problem is that no manager will ever provide the same tender loving care and quick response to problems like I do! Whoever pays the bills is always the most concerned! That's why it's so important for do-it-yourself investors to learn how to manage and do it for themselves, at least initially, anyway!

Managing people is a difficult job! Managing people, who are also your tenants and have an obligation to pay you their hard-earned rent dollars, every month, can be particularly challenging to your sanity! Ask almost anyone what the worst thing about owning rental houses is, and they will tell you, without even taking time to think, it's the tenants. Even folks who don't own real estate will give you the same answer. It seems quite clear that property owners who wish to survive very long in the business and who expect to make any serious money while they're doing it, are well-advised to learn the management, part of the job early on!

QUICK ACTION KEEPS LANDLORDS ONE JUMP AHEAD

Landlords who learn to act before small problems become big ones will be able to control most tenants. This strategy works very well for collecting rents and, also, for enforcing tenant rules. Speaking of rules, many landlords have far too many rules! It's best to keep your list of rules short and, thus, more easily enforceable.

One of the most important questions all landlords should stop and ask themselves is—*Would I rather be popular or profitable?* You don't have to be a sinister or dishonest person to be a wealthy landlord! What you must be is a good business person. "Good business" means that accounts receivable (rents) are collected in a timely manner and that your assets (houses) are maintained properly by tenants who lease them. *See how simple this is!*

Landlords will not be disliked or hated anymore than the supermarket cashiers when they demand the rent payment due. Everyone knows that cashiers will not allow groceries to leave the store until they collect the money for them. Customers expect that and don't feel hateful toward cashiers for following the rules. Landlords who insist on timely rent payments are no different than cashiers when they collect the rents.

INDECISION CAN COST LANDLORDS BIG BUCKS

Most collection problems arise because landlords are indecisive. Don't confuse this with being charitable or big-hearted. Make no mistake here, indecisive landlords want their rents, but the problem is they don't know exactly what they should do to enforce collection, so they accept excuses, instead. If any landlord wishes to test what I'm saying about being big-hearted and charitable, try this exercise.

First, make sure you collect the rent, in cash, when it's due. Next, fill out a proper receipt for payment and give it to your tenant. Finally, when you've counted the money and given back a rent receipt, reach in your pocket for the money and give it all back to your tenant. Now, that's a true act of genuine charity! Don't you agree?

Think about the supermarket cashier. If you don't pay, you don't get the groceries. State laws provide adequate remedies for landlords who get ripped off! Eviction laws are the rules of the game for tenants and landlords. Owners who don't know these laws, intimately, from start to finish are bound to become the victims when their tenants don't pay. Property managers (owners) must be in a position to act quickly when their income is threatened, otherwise they're throwing money away, needlessly.

DON'T BOTHER ASKING THE DUMBEST QUESTION

To me, the dumbest question for a landlord to ask his tenant is, "WHERE'S THE RENT MONEY?" The reason it's dumb is because both parties already know the answer. The landlord knows the tenant hasn't sent it, and the tenant knows that's why the landlord doesn't have it! *Any question where both parties already know the answer, before it's asked, is a dumb question.* It's only value can be to harass; and harassing tenants is not a good management technique. It's far more beneficial for landlords to simply issue a Pay or Quit Notice, when they don't receive the rents when they're due. After all, if a tenant expects to be late with his rent payment, it's up to him to obtain permission or work out a deal with the landlord. Property owners who chase after tenants for the rents are generally wasting good time. They need a better strategy and, quite possibly, better tenants too!

NON-CONTACT MANAGEMENT WORKS QUITE WELL

Years ago I discovered I could manage tenants more effectively without making "house-calls." It's done with written memos and a telephone answering machine! The answering machine allows me to take repair calls and conduct regular landlord business, and it allows me to monitor all incoming calls. As a general rule, most tenants don't like talking to a recorded message, and tend to hang up a lot, unless, of course, they actually have something important to tell me about! Tenants don't like this arrangement as much as I do, but it works very nicely as long as I handle repair calls promptly and provide good service. Eventually, they get used to the recording.

Folks who attend my seminars and fixer camps are taught how to write LANDLORD MEMOS that can greatly assist with rent collections and help enforce tenant rules. *There are two major benefits that come from using these memos.* First, you can avoid listening to all those tenant "sob-stories," which you'll always hear whenever you make personal house calls! Also, there's absolutely no need to get personally involved in emotional discussions about late rents, extra cars, or too many visitors. If there's any problems with my rent or rules, I simply write a memo to my tenant advising him what needs to be done to fix whatever's wrong!

The second benefit for landlords is having a written record of everything you tell your tenants. Written memos eliminate all the confusion about who said what to whom and when! If you end up in court with your tenants, copies of memos, advising them about what to expect if they don't pay the rent, will go a long way toward winning an eviction judgment.

USE CUSTOMIZED MEMO FORMS WITH YOUR LOGO PRINTED ON THE TOP. THEY ARE 8¹/2" x 11" IN SIZE AND CARBONLESS. THEY CAN BE ORDERED IN DUPLICATE OR TRIPLICATE FORMAT. I FILE MY OFFICE COPIES IN A SPECIAL 9 x 12" BINDER FOR EASY REFERENCE. IF YOU ARE INTERESTED IN ORDERING THE SAME MEMOS I USE, WRITE TO: THE BUSINESS BOOK CO., ONE EAST EIGHTH AVENUE, OSHKOSH, WI 54906. AND ASK FOR THEIR FREE FORMS CATALOG. (I use form 1157.)

HOW TO MANAGE YOUR TENANTS BY MAIL

If you think my memo system has merit and you'd like to try it out yourself, but you need a "jump start" to get you rolling, I can help you! *I've assembled 165 of my best memos, covering almost every landlord-tenant situation you'll be confronted with.* I've also recorded two hours of audio tapes to instruct you on how and where to use these memos, for best results. All memos are labeled, indexed and categorized by subject matter, so you can instantly find the right sample for every situation.

I'll guarantee this collection of actual, already-tested memos, will save you countless hours of trying to figure out what to do, plus they will spare you the anguish of typical face-to-face tenant encounters. To get this collection of memos, order Product #2118, "HOW TO SUCCESSFULLY MANAGE TENANTS WITH A POSTAGE STAMP." See Fixer Jay's Product Information in the Appendix Section for an order, or call toll free **1-800-722-2550.**

CHAPTER 22

Landlording Skills Can Make You Very Wealthy

*O*wning rental properties that produce income is without question a very profitable and worthwhile achievement! Most folks who buy properties and rent them out end up a whole lot richer than those who do not. I'm a good example of this, but I must tell you, there is definitely a catch to owning rental properties.

The catch is you must quickly learn to operate your rentals and handle the tenants who live inside them, or you'll think a lot more about *giving up* than counting your money! Getting rich is a very strong motivation that will carry you through some rough times. But, tenants who don't play by your rules can have a devastating effect on your bank account, as well as your quality of life as an investor.

I don't mean to scare you, particularly if you're just starting out. But, I do wish to call your attention to a very important function of property ownership that usually receives little attention, because it's not a very exciting part of investing. In fact, it's the part of ownership that turns many would-be investors off.

KNOW-HOW IS THE BIG DIFFERENCE

Have you ever noticed how some people perform complicated tasks, while making them look so easy? Then you try the same thing and fall flat on your face? More often than not, the only difference is skill and experience. With average intelligence, you can do the same thing successful landlords do, if you'll dedicate the time to learn how.

Landlording and managing your properties is very important. It's not the most exciting part of ownership, yet it's a path that must be traveled on your way to the gold mine. As I told you earlier, I often refer to property management, including people management, as the 10% part of ownership that's necessary in order to receive the benefits from the other 90%. The 10% figure I use comes from the amount I used to charge (10% of the gross rents) when I managed properties for others.

HORROR STORIES ARE CAUSED MOSTLY BY IGNORANCE

Almost everyone has heard about "the tenant from hell" who trashes the house and leaves owing six months worth of rents and fix-up costs in the thousands. Movies have been made about the subject to further dramatize it. There is no question that this kind of thing does happen. However, it doesn't have to happen to you, nor should it if you pay attention to what I tell you.

Often, you will hear discussions to the effect that "deadbeat tenants know all the rules, which enables them to stay in your property month after month without paying any rent!" Sometimes, the storyteller says this is so, because the tenant has minor children and receives welfare or other public assistance.

The problem with this picture is that the storyteller is giving far too much credit to the tenant for being *street-savvy* and *knowledgeable*. But, chances are the tenant's success is much more likely attributable to an ignorant landlord! Seldom will a tenant stay long in my property if he or she doesn't pay rents. After you've done a few evictions yourself and learned what papers to file, and how to do it, in a timely manner, you'll have enough confidence so that horror stories won't happen at your properties. Again, I'll repeat, ***know-how is the difference between owners who operate their rental properties successfully and those who are simply hoping for the best.***

OPERATING PROPERTIES LIKE A BUSINESS

The first rule of good landlording is to treat it exactly like you would any other well-run business. Managing your property is a business and the tenants are your customers. It's your job as a landlord to provide the best product (houses) and the best service you possibly can. If you do this better than your competition you'll, obviously, enjoy much greater success. Customers (tenants) will naturally beat a pathway to your doorstep, when they feel that you offer the most value for their housing dollar.

To make sure you are ready for the kind of business you want and expect, you must try hard to offer your customer the best value in the rental business. That means a clean house or apartment, priced ideally at $10 or $20 below the competition! This may not seem like much, but for renters who compare houses and watch their expenses closely, it's enough to swing many deals your way. It will also keep your houses full of tenants, while the other guys carry the vacancies.

How can you tell if you're $10 or $20 below the competition? The best way I know of is to study your local classified ads—"Houses or Apartments For Rent." Then drive out and see what the competition is offering at comparable rents. Adjust yours up or down according to what you observe.

When I first started to study my rental market, I quickly discovered that houses freshly painted inside with attractive window coverings, were not all that common in the moderate rent range. I immediately made it my operating rule to keep my houses freshly painted and the windows dressed up with inexpensive curtains, drapes or mini-blinds. That rule has always made money for me and it's a perfect example of a good business decision.

Another decision I made years ago was to build white picket fences in the front yards of my houses, where possible. I discovered that renters are turned on by these homey-looking fences and are willing to pay more rent to have them. By providing picket fences, I not only attract more customers than my competition, but I get higher rents for the same size house.

Extending Credit to Deadbeat Renters is a Losing Proposition

One of the most controversial issues in the landlording business has to do with collecting the rents and knowing where to draw the line on late payers. First, it should be perfectly clear that landlords do not contract with tenants for the purpose of extending credit or providing free shelter. In my own county there are 19 agencies who already do that, and I see no reason to become number 20. On the other hand, I'm a landlord and a business person, and I recognize the need to be somewhat flexible, particularly with long-term tenants who are valuable customers to me!

My policy regarding late payments or split payments is to be fairly strict about enforcing the terms of my rental contract during the first six months of every tenancy. After six months—a period of time that allows me to observe a tenant's paying habit—I will generally become a bit more flexible, assuming their pay record is okay. During the first six months, I routinely issue Three-Day Pay or Quit Notices, no later than five days after the rent is past due—sometimes sooner, depending on how the weekend falls. If the rent remains unpaid, I usually evict the tenant. If I receive at least half of the rent, I generally accept it and re-issue the Three-Day Notice, showing the balance due. *Don't forget, rents are the wheels on your investment vehicle! If you don't collect them, you'll get left behind sitting on your axle.*

Acknowledging Who Pays all the Bills

If you buy income properties correctly—that is you don't overpay for them—you'll soon discover, as I have, your tenants will be paying for everything. They pay all the expenses for operations and all the mortgage payments to finance them! To my way of thinking, you'll need to look long and hard before you find a better deal. Think about it, your renters are paying for everything and, when the mortgages are finally paid off by them, it's you who ends up owning the real estate.

Fewer Rules are Best—But be Sure to Enforce Them

Two of the rules that I insist must be followed are—rent must be paid on time and a late fees must be paid when rents are more than five days past due. If you don't collect late fees, tenants will have no incentive to pay on time. Another rule I follow is that no extra persons are allowed to move into my houses without prior authorization. I don't usually deny permission as long as an additional tenant don't exceed my occupancy limits, he or she fills out a rental application and is qualified to be added to the rental contract.

Parking unauthorized cars on my property (sometimes towed there) is not permitted! I do not allow working on cars (other than minor maintenance), for more than one day. Junky cars with their hoods up—or sitting around on wood blocks—portray an unsightly image of the property. It also sends the wrong signal as to the type of tenants most owners are looking for! If you allow trashy-looking cars to accumulate on your properties, they will soon look like junk yards. Enforcing this rule is a continuing effort, but it pays off at the cash register.

Emotions Should Not Control Landlord Decisions

If you are intimidated by tenant threats or if you are hesitant to begin a non-pay eviction, because your renter has had problems in the family or some other crisis, you'll have great difficulty operating your rental business. Ask yourself how many supermarkets would tell their customers to take groceries home without paying for them, or how many banks do you think would allow you to skip the mortgage payment this month because your tenant hasn't paid his rent yet?

No business could operate very long by trying to accommodate the never-ending, hard luck stories, and neither can you. Fair business dealings are the key to a landlord's success. No jury would ever render a judgment for tenants who make up ridiculous excuses for not paying their rents. Therefore, your decision to quickly evict non-paying renters is based on simple fairness, and should never be based on personal emotions.

One rule that has served me well for many years, in terms of fairness—which is generally thought to be a 50-50 proposition—is to try my very best to give my tenants a full 60% worth of fairness in exchange for them giving me just 40% in return—40% is all you'll ever need to make you very successful and very rich.

At my seminars, I teach a concept that minimizes personal contacts between landlords and tenants. I see no reason for landlords to hang around their tenants, except for maintenance visits and repair calls. Obviously, I don't chase my tenants down to get the rents. This practice lends itself to emotional confrontations, which is exactly what landlords need to avoid.

QUALIFICATIONS OF THE IDEAL TENANT

Contrary to the landlord-tenant "horror stories" we've all heard about, most tenants who agree to rent your property will also pay their rent. This fact eliminates about 95% of your collection worries. Trust me on this, I have enough tenants to prove it. The big problems are caused by the remaining 5% who can practically destroy your life! It's the old story about the rotten apple in a barrel, unfortunately, *"DESTROY YOUR LIFE"* is not too strong a term for what can easily happen when innocent, but ignorant landlords, do battle with "ringwise" deadbeat tenants.

Most Tenants Will Pay the Rent

Human nature, being what it is, works the same way for all of us. Dating back to the cavemen era, shelter has always been very important. There have been countless times during my landlording career where my tenants have literally turned my houses into caves. That's why I understand something about cavemen and their women. Joking aside, literally every human being wants to have a private shelter, their own castle, so to speak. Consequently, housing—no matter how big or small, expensive or cheap—is a top priority for most people, including tenants. What this means is that 95% of all tenants, whether they work or not, will pay their rent. *Housing is a high priority for humans.*

MANY LANDLORDS HELP TENANTS GO HAYWIRE

If you have ever made the rounds to model homes, when developers present their annual home tours, you'll understand what I'm about to tell you. All the houses are "decked out" with attractive furnishings and immaculately landscaped yards. If you're like me and you start the tour looking at houses selling for $75,000, but end up viewing those that cost $375,000, it warps your value system. Your taste is suddenly "light years" ahead of your financial capabilities. Most folks who take these tours like me, are rational-thinking people. Their minds can adjust to their pocketbook, without too much difficulty, but consider what would happen without this adjustment. The results would be similar to a child at the dinner table whose eyes are much bigger than his stomach. Without some kind of help or restrictions, he'll always take more food than he can possibly eat.

Landlords need to use good old common sense when they rent their properties. *Never show $800 houses to renters who are not qualified to pay more than $400.* Obviously, they'll want the house they can't afford. You must always qualify a tenant's ability to pay rent, first. That will dictate the kind of property you show them. It works just like the model homes—if you show a $800 unit to a $400 tenant, I'll guarantee you'll either lose him as a customer or he'll attempt to stretch beyond his financial capacity by agreeing to rent the $800 house. If he succeeds, as many do, you'll be the one who has a problem. Just remember, you're also the one who is mostly to blame!

Many lower income tenants need help with their finances. Some have great difficulty figuring out a balanced budget for themselves. Younger couples are usually overly optimistic about how much rent they can afford to pay. It's up to you, as the landlord, to determine how much gross income is needed to pay the rent you charge. I've found that most welfare tenants would be willing to sign my rental contract for a suite at the Hyatt Regency; however, they can't pay the rent. When that happens, I've got just as big a problem as they do!

Never forget—rent money works something like handcuffs between landlords and tenants. It's a common link! *If your tenant has a problem paying rent, you've got a problem too!*

Renting to Lower Income Tenants

People ask me why I rent to low income tenants? They want to know if it wouldn't be better dealing with tenants with whom I have more in common. At this point I generally try to explain that *IT'S A BUSINESS DECISION!* Years ago, I decided that if I wanted to survive in the income property investment business, I would need steady and continuous income. I did not have any surplus money available in my bank account for rainy days, a couple of rainy days would drown me—and I knew it!

THE FIRST RULE OF BUSINESS IS TO DEFINE YOUR CUSTOMER

When you begin to think of rental houses as a business venture, rather than simply as investing in real estate, it will open your mind to a whole new world of opportunity. The rules of business will help you immensely as an investor-owner of rental properties. Here's an example of what I mean.

In my town, I discovered there are *two primary groups* of people who need rental houses. The first group consists of young folks, 20 to 35 years old. The second group is seniors, ranging from 60 to 78 years of age. My town provides the younger group with mostly service-type jobs—*RESTAURANTS, STORES AND GAS STATIONS*. There is also a large number of single mothers, with children, who receive public assistance. Many senior renters are living on social security, and about half of them receive additional income from private pensions.

Approximately 75% of both groups can afford to pay between $400 and $550 rent per month—about ten percent can pay more, the others must pay less!

Once you develop similar statistics for your buying area—which I recommend you do—your strategy as an owner of rental properties will take a definite direction. Stated another way, in my case, if I intend to rent my houses to the largest number of potential renters in my town, I must develop an investment strategy that allows me to own houses, I can profitably rent out for somewhere between $400 and $550 per month. Obviously, if you buy, fix and sell—you should think along the same terms. Many investors try to figure this out after they buy the property! That's doing business backwards, in my opinion!

REASONS BEHIND MY RENTING STRATEGY

The main reasons for my decision to enter the lower income rental market were based on the economics and it offered the least possible risk of going broke.

The economic reality for me was that I had very little money to buy real estate. Buying older rundown houses and small apartments fit my budget. I found that buying "ugly duckling" properties and turning them into "beautiful swans" allowed me to obtain much higher rent "mark-ups" than my competition. For example, I might pay $20,000 per unit, then spend $4,000 on fix-up, after which I would rent the unit out for $400 per month, or $4,800 per year. That's a 20% annual return on my house ($4,800 annual rent divided by $24,000 value is a 20% return). Contrast this return with a "pride of ownership" property that costs $50,000 and rents for $450. Never lose sight of the reason for being in business—TO MAKE A PROFIT.

Least Possible Risk

Most everyone knows that starting any new business venture is filled with many uncertainties. Perhaps the most important question is: *Will there be enough customers to pay the bills?* Renting houses is not a great deal different than selling widgets or running a restaurant. Both are very much dependent on paying customers.

In order to increase my odds for success, I decided that my rental houses should be priced so that the largest number of available renters could afford to rent them. *Stated another way*, I wanted my houses to be within the price range of 75% of all tenants in my town. To do this I would need to own houses that I could profitably rent out for $400 to $550 per month.

I think you can see the safety part of my strategy. By targeting 75% of the renters, where low income rentals were already scarce, I felt quite confident that vacancies would never be a serious problem for me. Vacant houses, with no means to pay the mortgage payments, could have easily shattered my investment plan. I had no extra cash in the bank for a rainy day. That's why the least amount of risk was very important to me!

Avoiding the Pitfalls of Management

Landlords often find themselves in "hot water" with tenants because they try to inject too much logic and common sense into tenant management. Logic and common sense have their place, but seldom count for much where legal issues are concerned. For example, *it is nearly impossible to effectively force your personal living standards and ideals on your tenants*—a very common mistake for many new landlords. I would advise any rental owner to seriously think about what I'm saying here, because it has a lot to do with sanity, YOUR SANITY. What good would it do you to make a million dollars from your rental properties, if your tenants drive you to the "nut house?"

KEYS TO GOOD MANAGEMENT ARE ACTION AND ENFORCEMENT

Enforcement of the "rules," whether it be the civil code (laws) or your own house rules, is the best way to develop a smooth running management operation. Preventative techniques are as important to managing tenants as they are for the doctor who manages your personal health. One of my main criticisms of *"PROFESSIONAL PROPERTY MANAGERS"* is that they very rarely act, *mostly they react!* They're always ready and willing to fix a malfunctioning toilet—but only after it overflowed and ruined the living room carpet.

YOU MUST ALWAYS GET THE MONEY FIRST

Rent monies are " the life blood" to apartment owners. Yet, I know many property managers and owners alike, who participate with their tenants each month in a silly little "rent collection ritual." The tenant starts the game by saying *THE CHECK'S IN THE MAIL.* Then the landlord begins calling every day or driving out to the property to inform the tenant that he hasn't received the check yet. Sometimes this goes on for weeks.

Playing this game will only eliminate whatever respect one party might have for the other. It generally leads to more bickering about other matters, as well. Don't allow yourself to be part of this game. You'll fare much better if you simply use the "rules" already on the books. I'm referring to your state's landlord-tenant civil laws; and of course, your own rental contract terms, agreed to by your tenant.

Rent collection is a landlord's most serious business. Quick enforcement of the rules, when needed, is the best method I know of to keep your tenants paying as they promised.

No Need to Keep Asking the Dumb Question

As I told you earlier—it's worth repeating again, there's no need to keep asking the landlord's dumbest question—WHERE'S THE RENT? In case I didn't impress you with my reason before, let me explain again. It's dumb because both parties—the landlord and the tenant—already know the answer—*THE LANDLORD DOESN'T HAVE THE RENT, BECAUSE THE TENANT DIDN'T MAIL IT YET!*

No question ever needs asking, when both parties already know the answer. Asking is merely harassment, which does no good for either side. Also, it makes a liar out of your tenant and there's no sound business reason for doing that. The first rule of rent collection is, NEVER CHASE THE MONEY. Make your tenant responsible for paying rent on his own or move him out. It's much easier on your nerves and in the long run, your collection rate will be much higher.

GOOD TENANT RECORDS ARE ESSENTIAL

Some property owners tell me that it's not necessary to have written rental contracts and agreements, if you choose the right tenant to begin with! I certainly can't argue with that! My problem—same as most other landlords—is that it's impossible to do that 100% of the time! Obviously, folks who tell me this have never been exposed to a courtroom eviction drama. If they had been, they would understand that, **if it's not in writing, you don't have a case!**

Several major differences I've found—when renting to low income tenants, as opposed to those with more money in their pockets—is that when a crisis develops, they have no cash reserves to keep them afloat. Many do not even have decent paying jobs. Others depend on housing subsidies, social security and welfare assistance. Another problem, quite common with the younger renters, is their inability to handle money! Many have difficulty living within their means.

People often ask me if it's better to rent houses to tenants with low paying jobs than to others in low income brackets, such as HUD assisted tenants (Section 8) or AFDC—Aid to Families with Dependent Children (Federal Welfare). Frankly, I see very little difference!

Service jobs in my town pay very low wages, and don't include any benefits. Take-home pay, *NET CASH*, is about the same for a full-time service job worker as it is for a single woman with two dependent children receiving federal assistance. Financially, the woman may be slightly better off because she also gets food stamps and free health care. It's better for me if she doesn't have a car to buy gas for. Try to be objective here—remember we're talking about the tenant's ABILITY TO PAY. Personally, I don't like the federal welfare system. However, that's not what's at issue here.

THE APPLICATION FORM—WHAT YOU NEED TO KNOW

It's very important that you learn, everything you can about an applicant before you hand over the keys. This information should be required whether you rent $300 apartments or a suite on top of the "Trump Tower." The first thing you need to find out is if the tenant can afford to pay the rent you're asking; secondly, what can you do to collect the rent in the event your tenant fails to pay during his stay with you?

Let me point out that by design—my entire approach to landlording is to minimize personal contact between myself and my tenants. My emphasis is on providing safe clean properties and giving the very best service I can for a fair rental price. In return, I expect my tenants to pay the rent and abide by the 13 house rules on my rental contract agreement.

The more personal information you can obtain on your rental application, the better prepared you'll be if something goes wrong. You need names and addresses of parents, friends and co-workers. You also need references, including their last two landlords. One question on my application asks, *"Who will co-sign for you if your own credit information is not sufficient to qualify you?"* The answer may lead you to a financially sound backup party (co-signer), who don't mind signing your rental contract with the tenant. Often parents or in-laws will do this if you ask them to. Always have them fill out your application, same as tenants, when they agree to be a co-signer.

There's absolutely nothing in my contract that should make it necessary for much further discussion beyond the initial rent-up. Obviously, when a tenant rents my house, he or she has already inspected it and decided it's the place they wish to live. All I need to know is—*Are they qualified?*—that's it! It's not really necessary to like your tenants. **A landlord's obligation is to provide the best housing for the tenant's rent dollar.**

Can Your Tenant Afford the Rent?

As a general rule, young tenants can pay about 35% of their net take-home money for rent. That means after taxes or payroll deductions, if they work. Older tenants and seniors can often pay 50% of their income and have no problems. The reason for this difference is because they have learned to budget their resources and because they're wiser. Obviously there are exceptions.

As a rule of thumb, the rent-to-income ratio must be lower for tenants with young children, automobiles and pets. These things cost extra money! Don't rent to applicants who do not have adequate income to pay the rent you charge (based on whatever rent-to-income ratio you use). If you violate this rule, you'll end up chasing rents and listening to "tear-jerker" stories every month about why the rents are late or how the mailman lost the check. Clearly the saddest part of the story is, you did it to yourself!

Section 8 Tenants—Rent Assisted Program

Federally-funded Section 8 programs, administered through local city and county HAP *(HOUSING ASSISTANCE PROGRAMS)* offices, provide extra protection and sometimes higher rents to landlords. There is extra paperwork involved. However, the agency will prepare most of the documents for you. You, as a landlord-owner, will still select the tenants. Often they are sent to you by the agency from their waiting lists. Regardless of the extra paperwork required by HUD rules, you should still have every tenant complete your rental application and, if approved, sign your rental agreement or contract.

The major difference with HUD, Section 8 tenants, is you need good cause to evict. The good causes are listed on the HUD lease documents. Obviously, non-payment of rent and property destruction are good causes. The best feature about the Section 8 program is guaranteed rent provision. If your tenant goes "bonkers," you can receive up to two months extra rent. Also, should damage to your property occur, the agency will inspect the property, determine whose at fault and reimburse you for all authorized repairs, minus the wear and tear, of course.

RENTAL CONTRACTS DON'T NEED TO BE COMPLICATED

Five pages of regulations is not worth anything unless your tenant is willing to abide by the terms. Never forget this, *a good landlord-tenant relationship is not better because you have a "killer contract," it's better because you learn to select tenants who will pay you and abide by the rules.* Therefore, learning how to qualify and screen your tenants is the area you should strive to improve. The contract will only be needed if that effort fails.

I have found from my courtroom experiences that if there is any confusion or doubt concerning the paperwork, nine times out of 10, the tenant is automatically given the benefit of that doubt. Therefore, my rule concerning a rental contract is: *"IF IT'S SIMPLE, IT'S GOOD."* Obviously, certain language must be present to protect the landlord, such as who pays attorney fees for a violation of the contract terms. *Mine says the tenant pays*—as I imagine you would have guessed!

Other important terms should include the tenant's liability regarding his personal property, when late fees are due and the amount due, and who is responsible for repairs when things break while the tenant is in possession. My agreement is a legal-size (8 1/2 x 14") sheet, printed on both sides. I have amended it several times over the years. It's very basic; however, it covers about everything.

If you would like a copy of my application, send a stamped, self-addressed envelope with your request to: KJAY, Box 491779, Redding, CA 96049-1779 (marked APPLICATION REQUEST).

Don't Take Shortcuts With Formalities

The tenant interview, as well as filling out the forms (application and contract), does not change because a tenant is renting economy priced rental property or because he does not have a regular job. The same application procedure should be followed, no matter what the applicant's circumstances are. You must have complete tenant files in order to operate a rental business successfully. Don't misunderstand what I'm saying about "tenants not having a regular job." This doesn't mean you should rent houses to unqualified people. Obviously, there must be sufficient income from some source or you don't rent to them, period! Let's be clear on that point!

LARGE DEPOSITS PROVIDE ADDED PROTECTION

I do not collect first and last month's rent from my tenants. However, I do collect larger deposits than many other landlords. It's easy to justify a large deposit when you explain that your house is still cheaper to move into because, unlike other landlords, you're not charging them for last month's rent, in advance.

Typically, a landlord in my town will ask $450 rent for the first month, $450 for the last month, plus a $300 security deposit. So, the move-in will cost the tenant $1,200 total. I think it's much better to charge $450 rent with a $650 security deposit. My total move-in cost is $100 less than others, and the $650 deposit is fully refundable if the tenants leave my place clean when they move. Obviously, I'm much better secured with the larger deposit.

This strategy avoids a "game" tenants often play with landlords. On the first of the month, *they are short of money.* So, they give a 30-day notice to move—"New job," they say. They request that the landlord apply their pre-paid, last month's rent, for the current rent due. The landlord must comply. Then, toward the end of the month, they proudly announce, "I won't be leaving after all. My new job fell through." Obviously, most landlords don't want vacancies so they feel somewhat relieved. The problems is that now the last month's rent is used up, and it never gets put back in the landlord's account. This means the landlord is holding a very small deposit ($300 in this example). This is not enough money to adequately protect the landlord, nor is it large enough to make a tenant think twice about skipping (not paying rent) and staying as long as he can before the Marshal arrives to toss him out.

One final note—when something goes wrong, act immediately! Serve the proper notice to begin eviction. Action makes a believer out of your tenant quicker than any words you will ever say to him.

LANDLORDING SHOULD BE ABOUT PROFITS

Landlording is not supposed to be fun, so don't think of it as a hobby! You must prepare yourself mentally to deal with all kinds of tenants. They are your customers. As such, they deserve the kind of treatment you would normally extend to any good business customer.

Avoid personal conflicts over such matters as lifestyle, inside housekeeping and moral issues. That's not your business! All matters relating to their tenancy should be kept strictly business.

Obviously, it would be a poor business decision to keep an unruly or destructive tenant, because it would cost you money! Making money is top priority in the landlord business, the same as any other business.

TENANT URGENCY—NOT MY URGENCY

It is your business to see that rents are collected, and that rules and regulations stated in your RENTAL CONTRACT AGREE-MENT are abided by. You must never allow a tenant to intimidate you! It will almost always lead to "poor decisions" on your part. *In my view, the most common mistake landlords make is to allow the tenant's urgency to become their urgency.* Let me explain!

After many years of doing this job, I'm hard pressed to think of any situation so compelling that there wouldn't be enough time to think it through. For example, if there's a fire at my property, the community fire department will handle everything. As for me, I'm protected by insurance on the building. Should someone die in my house, I've yet to meet the landlord who'd be of much help. The county coroner is the person you need.

In short, leaky pipes, broken windows and cockroaches are no more urgent to a landlord than dying patients are to a doctor, or raging fires are to a fireman. They are simply a part of doing business!

Pipes will always leak, people sometimes die and buildings occasionally burn to the ground. That's just normal stuff when you own a lot of rental houses! Your job as a landlord is to be as responsive to your tenants as you possibly can. But, never, never allow yourself to be stampeded into making rushed decisions, simply because your tenants think you should.

Landlords Must Know the Law

Every landlord should know and understand the landlord-tenant laws in his own area. Once you know the laws, your fear of renters or of being intimidated by them will vanish. An overwhelming number of owners incorrectly assume these laws favor deadbeat tenants. I can assure you this is not the case, although it appears that way, sometimes! Laws are mostly about equity. It's well to remember, there are unscrupulous landlords, just as their are bad tenants.

OWNERS SHOULD DO EVICTIONS

Landlords should do their own evictions, at least the first few times. This provides "combat training" and some valuable experience toward learning how the system works. Besides experience, there are two other important reasons. As always, the first has to do with money, landlord's money (that's us). The second involves time, but since time is also money, you can see an obvious connection.

The simple truth is that knowledgeable landlords can save themselves a sizable sum of money in attorney fees—and are in a much better position to move quickly and rid themselves of "non-desirable" customers. Quite often attorneys consider evictions as "fill-in" or bottom-of-the-barrel" type work. Unless you're a regular client, your eviction problem will not have a very high priority with most lawyers. They've got more important things to deal with. Tossing out deadbeat tenants is not the kind of work that will get an attorney's name painted on the front office door—besides, most are far too slow, in my opinion!

Capitalize on the Owner's Edge

Naturally, owners have the built-in advantage of knowing all the details about their tenants. Another plus is that owners can give immediate attention to a single problem, whereas attorneys, more often than not, are working on many different cases at the same time. This *immediate-attention* issue is most important, because you don't want hostile tenants living on your property any longer than absolutely necessary.

Here's what happens—the longer an eviction gets delayed, the more angry and hostile most tenants become, and the more your property is at risk for a TRASH-OUT. **TRASH-OUTS** happen when a tenant goes "bonkers," kicks in the sheet rock, plugs the toilets, cuts wires inside walls and paints the interior with black aerosol spray cans. You can greatly reduce the risk of a trash-out by conducting quick, efficient evictions. That means no name calling, and no threats, confrontations or harassment. When you know the laws and understand what you can and must do, you are in the driver's seat. Also, in case you don't already know, the one person most interested in saving you time and money will always be you! You got it?

Learning Paperwork is Not a Difficult Task

Filling out the proper forms, as well as going to court, when required *(90% of my evictions default before ever facing the judge)*, is as easy as changing locks, painting walls or fixing a faucet drip. **The key is you must learn the procedure.** Although court clerks are generally forbidden to give you legal advice, they are quite helpful and will tell you what forms to use. After you've been through the routine a few times, you'll feel like Perry Mason. Who knows, you might even develop a brilliant law career to fall back on in case you flop as a landlord!

Repairs and Customer Service

If you rent houses, you should think of yourself as a business person. Business people provide a product or service for money. You are providing houses for rent. It's basically the same. The point I wish to emphasize here is, if you provide top-grade service to your customers, you'll stay in business a long time. You'll also come out far ahead of your competition, which improves your bottom line (cash flow).

Many part-time landlords get themselves into serious difficulties by stalling or putting off repair call-outs. Here's a typical "putting off" response—*"Yes, Mr. Renter! I know it's only Wednesday, but if you'll just be patient and hold off using the toilet until Saturday! That's my husband's day off, you know."* Ask yourself, if you were the renter, would you appreciate that response. I rather doubt it and neither will most of your tenants.

If you collect rents, you are obligated to keep the property in good working order. Pattern your repair visits after Sears or the telephone company. Both have been in the service business 100 years or so. It's interesting to note that neither company sees the necessity of doing routine repairs during non-business hours. *The key to good repair service is to determine the urgency, schedule the job realistically, and then get it done.*

What will set you apart from other landlords, is how well you handle repairs. Tenants will always complain about high rents. You can't change that; however, when they get top flight service for their money, they will accept it better. As a result, they'll remain your customer much longer, which of course, is what all owners lust after.

Be Snoopy, When You Can—But Tactful

An extra bonus benefit for "hands-on" owners is that call-out repair visits provide the opportunity to "snoop" around inside your property without the normal 24-hour notice formalities. The more things you know about your tenants—like are they decent housekeepers, how many beds (the count should match the authorized tenants), also number of pets living on the property, versus those listed on the rental agreement—the better control you will have, as owner-manager.

OBEY THE LAWS OF HABITABILITY

It doesn't matter whether you rent $100-a-month river-bottom shacks or $10,000-a-month hilltop mansions, the rules are the same! Pro-tenant states, like California, have nasty penalties for what they call "slum landlording." Slumlords are owners who milk their properties, pocket the rents and never contribute a nickel to their upkeep. This is a very shortsighted "game plan" and a straight path to disaster.

CALIFORNIA CIVIL CODE, Section 1941.1, specifies the minimum habitable living standards for California. Other states have similar statutes; however, even if they don't exist, common sense would dictate adhering to these minimum living standards.

Here are the laws of habitability for rental properties in my town. You should use them too.

1. No leaky roofs, doors or windows. Exterior of house must be weather-tight. *Cardboard on windows won't get it!*

2. Plumbing and gas facilities must comply with building codes. They must be kept in good working order.

3. The building must have a properly approved water system, with hot and cold running water. Drains must be connected to approved sewer or septic system. *Hoses running out the back door from a wash machine are not approved, when anyone's looking.*

4. Heating is required. It must be approved in accordance with local codes. Gas appliances must be properly vented and maintained in good working order.

5. Electrical wiring (lights) and other equipment must conform to building codes and must be kept in good repair. Be careful with electrical fixtures in bathrooms near water.

Shocked tenants might forget the rent payment day, which could shock the landlords.

6. Apartments and houses must be clean, when you rent them. No junk piles or garbage lying around. Also, rats, mice, roaches and other pests must be eliminated or, *at least relocated*. Always rent clean properties. It's good business.

7. Owners must provide garbage cans or city receptacles (bins). They must be routinely emptied, cleaned and be in place on the day you rent the unit.

8. Floors, stairs and handrails must be kept in good repair. Also, make sure lighting is provided in stairways. *If your tenant can't see the handrail at night, why bother having one?* Watch for weak spots in flooring.

Don't forget these eight items—they are not optional! They are the law! If you rent with violations, and if your renter turns you in, you'll have the kind of problems you don't need, believe me. Landlords have been known to find themselves in serious trouble for non-compliance. Even worse, they may be sentenced to live in their own units with—guess who?

Paperwork Runs the World—Also Landlords

Always use an **application form** and a **rental contract or agreement**. Ask prospective tenants to fill out your application first. If you approve their application, then prepare your rental contract. My contract specifies the terms (rules) that tenants must agree to, if they wish to live in my property.

I have never discovered any need for a long-winded rental contract with page after page of monotonous terms. For one thing, no one will ever read it—and, secondly, I might be asked to explain it someday.

LONG-WINDED CONTRACTS DON'T MEAN THEY'RE BETTER

Rental contracts should specify that tenants are to keep the property in good repair, live peacefully along side their neighbors, pay rent in a timely manner and allow the landlord to inspect with proper notice. It should also specify who pays attorney fees to enforce violation of rules; when a late fee is required and the amount; and who is responsible for tenant's personal property losses. Make it a point to have all adult tenants sign the agreement, as well as, co-signers, if they guarantee the rents.

In the event, you discover you've rented to a "deadbeat"—a simple straight-forward contract will serve you best as an attachment (exhibit) to your lawsuit (unlawful detainer). Surprising as it may sound, judges want to find out very quickly who has done what to whom, and why. A well-written contract helps them understand. Keep it simple for them, and you'll walk out of court a winner.

You Must Screen Your Tenants Well

You don't need to rent your properties to brain surgeons and oil tycoons. Most folks have enough money to pay their rent. The fact is, about 98% of all tenants pay their rent, even if they don't have a job. We're truly living in the land of milk and honey! Wouldn't you agree?

The key is to fit the tenant to the property. Never show welfare tenants a $200,000 house. They'll rent it quickly, but they can't pay next month's rent, unless, of course, you're only charging $400 a month. If that's the case, I suggest you start looking at another career.

Most Tenants Rank Housing as Their Top Priority

Retail credit bureaus are of limited value (almost none) for checking out lower-end renters. Often, they have no record and, when the do, it's hardly ever good. However, that doesn't necessarily make them bad customers. A point to remember here is that tenants will normally pay what they have to pay.

Most of them rank shelter much higher than doctor bills, credit cards, furniture payments and movie rentals. *I rate customers on their record of paying rent only.* As for those other bills, it's simply none of my business!

Don't believe all the horror stories about welfare and Section 8 tenants being terrible renters. The truth is—some are, but most are not! Generally, these tales originate from next door neighbors, folks who don't own rentals—and misinformed seminar speakers who have never actually met face to face with a real live tenant.

DISCRIMINATION DOESN'T PROTECT DEADBEATS

If you decide to use a three-to-one ratio (income-to-rent), that means a renter with $900 per month income, can pay you $300 per month rent. Obviously, you must be able to verify the income. I always insist on seeing all occupants who will live in my houses—that means the adults and the children. If I don't like what I see, I don't rent to them. Disliking people is not discrimination! ***Discrimination is unfair treatment of renters based on race, color, religion, sex, ancestry, national origin, age, family status and physical or mental handicap.*** You'll notice "deadbeats" aren't protected by fair housing laws. If you don't want them as customers—YOU DON'T HAVE TO RENT TO THEM.

ALWAYS GET THE MONEY

This is my favorite management job. Quite frankly, I suggest you make it yours too! Nothing is more important to landlords than collecting rents. You won't stay in business very long if you fail this part! Forget what you think about tenants and their spending habits.

Just remember, if they were as smart as you about money, you'd probably be renting from them.

Collections First—Love and Kisses Later

Quite often landlords ask me, "Should I accept partial payments for rent?" My answer is generally, "Yes, if you feel that's all your tenant has. Of course, you should always give him a Notice to PAY or QUIT for the balance as soon as you get the money in your hands."

Let's face it, many low-end renters are lousy money managers. Most can't ever save a dime. Do you think you should ever tell this person not to come back until he has all the rent money? Can you imagine how lonely you'll be, waiting for that to happen!

HELPING TENANTS PAY YOU—MAKES SENSE

I rent to a number of welfare tenants who receive assistance checks twice monthly. It's nearly impossible for many of them to pay me the total rent from one check, unless they have other income or housing subsidies. It's my policy to try and work with tenants by splitting their rent payments. I ask for 60% on the 1st, and 40% on the 15th. I also require a larger security deposit in exchange for this favor, just in case something goes wrong. With larger deposits on hand, I reduce the risk of rent losses. Obviously, this arrangement is a compromise, but it seems to work quite well for me.

Never Change Your Paperwork To Do A Favor

An important point here is never modify or change your written rental agreement or contract to show anything other than month-to-month tenancy. **You should rent by the month, period.** Your decision to accept more than one payment is strictly a verbal agreement (nothing in writing) between you and your tenant. *You should explain it to your tenant*—"This is strictly a special favor to help you pay. In the event you don't do as you promise, this favor is immediately withdrawn, and it's back to the terms of the written agreement."

Let me remind you what I said before—rents are the wheels on your investment wagon! Collect them anyway you can or you could find yourself sitting on your axle with a broken bank account.

The Value of Tenant Cycling

What I tell you here applies mostly to fix-up investors like myself. It has to do with TENANT CYCLING. Let's say you purchase six apartments that need fixing up. Rents are $350 each when you buy them. However, you have determined that two-bedroom units should rent for $550 each, once fixed up.

I'll assume you're short of fix-up money, the same as most investors! You figure it should take at least 12 months doing the work yourself. It's important to keep your cash flow as high as possible during the fix-up phase, to pay the expenses and mortgage payment. That's the situation. *Here's how to keep the rents coming in!*

MOST INVESTORS CAN'T STAND NO INCOME

First, understand you can't jump from $350 renters up to $550 renters in one giant leap, not unless you empty the building and do a "rehab blitz." That means doing all the work in all the units, at one time. That costs a ton of money, plus loss of all rental income—$2,100 per month, in this case, for several months. You'll go broke doing fix-up projects that way!

Many low-end tenants can't stay put. They move from place to place like gypsies. Don't ask why! I don't know the answer, and it doesn't really matter. The point is, you need them. "Gypsies" will be your rent paying customers for the next 12 months, or longer. They will pay your bills as you transition from $350 rents up to $550 per month. It will take at least 12 to 18 months to make the change-over when doing most of the fix-up work yourself. Meanwhile, keep the units filled up with "gypsies"—and keep those rents coming in.

Shorter-Term Tenants Work for Rehabers

You can't fix up one unit in a six-unit building and immediately begin charging $550 rent. The other five units will hold you back. You can, however, move up to the $425 to $450 range on the first change of tenants. The reality is, your normal $550 tenant won't live next door to $350 renters, no matter how nice the apartment is. On the other hand, "gypsies" will. They see your apartment as a bargain. They only plan to stay a short time anyway, and don't care who they live next to, as long as the price is right, for now.

Your Broker Will Likely Never Understand This Strategy

During the 12-to-18-month fix-up period, it's customary for some of the existing tenants to move out. Obviously, they can see that rents will soon be going up. Better-class tenants won't show up for awhile because most of them cannot visualize how attractive the completed project will look. Your interim solution to cash flow can be the "gypsies." They come and they go, but they pay their rent and that's what you need most of all during the fix-up stage.

I've found it takes most do-it-yourself investors a little while to perfect this strategy. My good friend and real estate agent, Fred, still doesn't understand how I do it—nevertheless, he still agrees that it works very well for me!

FAIRNESS ALWAYS COUNTS THE MOST FOR EVERYTHING

Let me say this about ***fairness****—*it's what you owe your customers. Many new investors mistakenly think tenants will make good friends. The odds are very much against such a notion. In fact, if you ever expect to have any quality time for yourself in this business, I would advise you, make friends elsewhere and do it right from the very beginning so you can avoid future problems.

I've worked very hard to develop my ***low profile landlording techniques***. As my business has grown in size, my employees deal directly with my tenants. I'm mostly involved in the background. Naturally, I make most important decisions. However, in the background, I can stay neutral and objective about my business—and it works very well for me!

JAY'S 60/40 STRATEGY PAYS BIG DIVIDENDS

Here's some good advice I wish to pass along to landlords. *If you make it your personal business goal to give 60% of yourself; and in return, expect only 40% from your customers (tenants), you'll end up a very wealthy and successful!* Going the extra mile will pay big dividends for landlords, and I will promise you this—a 40% return is all you'll ever need to become a very wealthy rental tycoon. Try it and you'll see!

CHAPTER 23

Cash Flow Keeps You
Green and Growing

*M*ost everything in life has trade-offs! Early in my invest-ment career, I made a searching and fearless examination of my financial situation. It didn't take me long, especially when totaling up my assets, to determine that what I needed most from my investment plan was cash flow. Whatever else I thought I might need would have to line up behind cash flow. *It's very simple, with cash flow you survive, without it you don't!*

CASH FLOW MUST ALWAYS BE FIRST PRIORITY

Over the years, investing for cash flow as first priority has paid big dividends for me. *Cash flow* is what I always advise new investors to think about first. It's the basic rule of invest-ing the way I see it. When you have cash flow (money) coming in, you grow financially. I call this "Green and Growing." When you are green and growing, all things are possible, investment-wise. Without money coming in, nothing grows except discon-tentment and the constant worry about going broke.

DECIDE WHAT YOU ARE—INVESTOR OR SPECULATOR

Going for *cash flow* as a first priority is one of the most important differences between investing—what most of us *think we're doing*; and speculation—what most of us *should not be doing.* Certainly, I'll be the first to agree, investing is not as exciting as speculating! But, it's much healthier financially. Folks who intend to stay in business (investing) for the long haul have got no good reason to speculate, unless their bank account is stout enough to sustain a sudden jolt—*that means, take a loss.*

I have several friends in the San Francisco Bay area who purchase expensive houses, when they think prices are appreciating rapidly. When their timing is right, they make a killing, but when it's not, they usually end up going broke or losing the property. Sometimes, they find themselves stuck in a holding pattern, where mortgage payments and expenses will often add up to much more than the amount of rents they collect.

Not only that, but all their capital is tied up, which means they are effectively shut down. A good friend of mine has been a millionaire six different times. But, each time, he's lost it all. That's what I dislike about speculating. **When I earn the money—I want to keep it.**

SPECULATION CAN DISTORT YOUR LONG-TERM GOALS

One of the most important reasons why investors need to have goals and a good workable plan, is so they can periodically check to see if they're on course, and to make necessary on-going adjustments. If all of your investment money, no matter how much or little it may be, gets used up and you don't have enough income (cash flow) to pay the monthly bills, chances are, you're speculating more than you are investing. With a good investment plan, you should be able to constantly monitor your dollar returns (rents or sale proceeds), so you can project how much cash flow you will have. You should also know when those returns are expected to materialize. That's just sound business practice. Speculators don't usually know when returns are coming. Some are not even sure they will. To me that's simply not good business, and it's much too risky for most do-it-yourself real estate investors, including me.

SORTING OUT DREAMS FROM REALITY

Investors who think like speculators often tell me, "It takes time and hard work to make money." Yet, many of these same folks spend countless dollars for get-rich-quick books and easy-street seminars. Obviously, they need to think a little harder about planning and goals. "Speculator Thinking" often tends to make new investors feel like they can start right out—skipping all the hard work—and be able to multiply their small investments into huge dollar profits and be rolling down the freeway in a new Mercedes within a short period of time.

That's exactly how my Bay Area friend thinks. But, he's had some serious setbacks lately, and things haven't been going so well for him. Like they say in the car business—My friend is upside down in several high ticket properties." He's now driving an old '69 VW bus; however, it was not too long ago he was driving a big silver Mercedes. My friend is a good example of a speculator—out doing his thing!

SEEKING THE THRILL OF VICTORY

Investors who get confused and speculate in real estate are much like gamblers. They often place their bets simply because they feel the dice are hot (the market is right). They reason, "Others have made a killing so why can't I?" That seems to be their only justification. It's extremely important to remember, **gambling should never be substituted for a well thought out profit-making plan.**

Much like a gambler who bets his entire stack of chips on a hunch, speculators often jump head-long into deals that look like easy profits. "What will happen to me if the market changes quickly, and I lose all my money?" Investors should ask themselves this question before they jump! It's very exciting to make big profits in real estate. The best and surest way to do it, however, is to follow a good solid plan and experience a few investment victories (cash flow deals). Success will build your confidence. Several small victories are easier to achieve than one big one. As the late movie actor, John Housman,

used to say on his popular TV commercial for Smith Barney—
"We make our profits the old fashion way—we earn them." Real
estate investors should plan on doing business exactly the
same way.

Most Speculators Seldom Get Richer

Speculators tend to look for instant gratification and quick
turnover profits. The idea sounds great. However, the odds of
doing it are not so great. How many people do you know who
got rich in real estate overnight? Perhaps you know someone
richer than you are who works very hard at investing.
Chances are, you do. I know quite a few. If you watch them,
you'll notice they don't seem to do anything spectacular. Still
they continue to acquire more properties and they keep getting wealthier! How do I know that, you ask? I can see by the
big houses they live in and the frequent trips they take! Poor
folks live in small houses and stay home a lot. Let's discuss
one of the problems I see that tends to go along with specula-
tive-type thinking. It's called "borrowing money from banks."

BORROWING FROM THE BANK IS GOOD FOR BANK

Many investors spend too much time trying to figure out ways
to borrow money from banks and other "hard money" lenders.
The reason, they claim, is to purchase their next bargain
property—then hold it for awhile, and sell it for a big mark-up!
Naturally, their intention is to pay off the lender and pocket a
hefty profit for themselves. In theory, this sounds like a
marvelous plan. However, consider several problems that are
working directly against the goal of making profits for yourself.

To start with—bank borrowing costs lots of money, especially for inexperienced investors, without top-notch credit. It's
even more expensive when you're buying property for non-
owner occupancy (rental units). Normally, there will be points
to pay, higher interest costs, and variable rate payment
adjustments tied to indexes, that are very sensitive to rising
costs. Many bank loans are written with short-term call dates
and very low loan-to-value restrictions. This means that larger

cash down payments will be required. More often than not, there is personal liability included in bank loans—plus substantial pre-payment penalties if you pay the loan off early or wish to refinance it.

Also, there are always document fees, appraisal costs, escrow charges and, sometimes, expensive termite work required. Sometimes deferred maintenance repair funds are held back and deducted right off the top, which reduces net dollars the borrower will receive. I may have missed several other hidden expenses. However, let me ask you this question—***Does borrowing money this way seem like a good idea?*** It probably does if you are a lender. Frankly it sounds like a marvelous plan if you're goal is to make your banker rich.

Borrowing money from commercial banks and money lenders is *more likely* to make you poorer, rather than richer. Obviously, there's a much better way to borrow money for buying the kind of properties I recommend in this book—***seller financing should always be your first choice***.

NEW BANK LOANS SHOULD BE INVESTOR'S LAST CHOICE

If you're someone who underlines important parts—then grab your pen and mark up these next couple paragraphs—**they're important!**

Always try very hard to borrow the bulk of your new mortgage money (the real estate debt) from the owner who sells you investment property. Remember, bulk means "most"—it doesn't mean every situation. Sometimes, commercial borrowing is justified if the deal has positive cash flow. However, that's the exception, rather than the rule. It's okay to assume the existing commercial loans (mortgages) on a property when you buy it—but, your goal should be seller-provided financing for the balance owing, after you pay the agreed-upon cash down payment.

For example, a common situation might be where you purchase a property for $100,000. You agree to pay $10,000 cash down payment and assume an existing $40,000 mortgage—*IF IT'S ASSUMABLE*. Next, you negotiate with the seller to have him carry back a note for the balance, which equals $50,000, in this case. With this type of real estate loan being provided by the seller, you will generally end up with far superior financing than if you had borrowed new money from a bank or hard money lender. Most sellers are much more generous with terms—especially when they are selling rundown properties.

With seller financing, there won't be any points to pay and it's quite likely you'll be able to negotiate an interest rate less than what most banks would charge. You can also avoid a variable interest rate mortgage, unless you and the seller design your own to fit the transaction. The beauty of this kind of borrowing is that there are no set rules to follow. When there are no rules, the Golden Rule applies—*THE PARTY WITH THE GOLD RULES.* In most cases that can be *you*, if you're buying the kind of properties I suggest—*the kind* that looks ugly at first glance, but gets more beautiful with each new dollar it earns.

REAL ESTATE INVESTORS MUST THINK LIKE BUSINESS FOLKS

Can you guess the similarities between real estate investing, cattle ranching and running a railroad? At first thought—you might not think there are any. However, let me assure you, there are. They are all specialized businesses that must be operated using good sound business principles if they are going to be successful. A negative similarity is that if they fail to earn a profit, they'll all go broke about the same way.

A common fault that many speculators have is they tend to see themselves as "dashing real estate tycoons." Operating a business—using time-tested methods and a well-defined plan—sounds like a "real drag" to most tycoons, so they simply don't bother learning. "Let the accountants figure out my taxes and worry about employees and payroll," they say. When the tenant sues a tycoon over some management violation, which, of course, he didn't bother learning about, *it's no big deal!* That's what attorneys are for. Right? And, as for setting up the books necessary for good operating records, tycoons never have enough time to worry about things like that! "Someone can do the records later when we need them," they claim. Speculators and tycoons often think like that. However, it's a very poor plan for real estate investors, who would like to be around *long-term* and *enjoy their wealth*.

BUILDING REAL ESTATE WEALTH DOESN'T HAPPEN BY ACCIDENT

Every so often, speculators will make a killer deal! They need a dump truck to haul their money home. But, I've noticed they never stay rich very long. When they do deals that don't work out, they're always back to zero again. When you study wealth building, you'll quickly discover—as I have—this is not the best way to achieve it! You need a plan that takes you from where you are right now (today), to where you'd like to be later on. It must be simple and easy to follow, because complicated schemes seldom work.

The *"no-money-down"* craze of the early 1980s was based on simplicity, which is exactly why it was so popular. No-down seminars had sold-out audiences everywhere. It was a simple idea all right, but it had a serious flaw. The debt was too expensive for the income! The plan was not achievable for most people. Even the instructors couldn't make it work well enough to keep themselves out of the bankruptcy court.

JAY'S FORMULA FOR MAKING MONEY IN REAL ESTATE

My wealth formula is designed for *do-it-yourself investors*. If your goal is similar to mine—that is, to build long-term real estate wealth for yourself and your family—my strategies can help you. I never speculate unless I have extra money to gamble. Even then, I always insist on favorable odds to tilt the scales a bit my way. For the record, however, let me simply say I'm truly an investor at heart. Once I earn profits in this business, I get very possessive, and have no intention whatsoever of giving them back. That's the way most investors think and it's also the key to staying in business for the long term. Let me tell you the four basic ingredients to my formula—then we'll analyze each one.

INVESTING MY WAY—4 BASIC INGREDIENTS

My formula for successful investing is a simple one, it's easy to do. ***It has only four ingredients to make it work.*** If you use them all, you'll find as I have, they're all you'll need.

1. **A WORKABLE PLAN** – You must have achievable goals and allow yourself a reasonable period of time to achieve them.

2. **ACTION** – The work necessary to execute the plan—Periodic adjustments are okay, but quitting is not an option.

3. **INVEST** – Must be continuous, and investments must be profit-making real estate. Speculation is never allowed in the early stages.

4. **COMPOUNDING** – Money must be continuously kept in the earning mode (that means fully invested). Your knowledge will compound automatically as you follow steps 2 and 3.

Now let's discuss each one individually to make sure you understand what I'm suggesting here!

A Workable Plan

A workable plan does not mean you need a complicated computer study with stacks of printouts, to support some pre-conceived notion that all real estate is bound to make money no matter what—*because that's not true.* All real estate does *not* make money! The simple fact that many banks and saving and loan associations have thick files crammed full of "take back" properties (REOs), should offer sufficient proof of that.

A basic truism you can bet your boots on—*nobody ever gives property back if it's making money for them.* People who give properties back do so because they are *losing money.* Chances are, they jumped into a deal quickly, without doing the proper research. Many new investors start out before having sufficient knowledge to operate the property after they own it. Still others buy properties without the foggiest idea of how to make a profit. In short, the lack of advance planning is the number one reason investors go broke.

Plan Must Be Simple With Achievable Goals

My basic business plan has always been to acquire rental houses that will earn a reasonable operating income. To achieve my monthly income (cash flow), I buy houses that will return 18 to 24% annually. That's a realistic and workable plan.

For example, let's say I purchase four older, two-bedroom houses (or small apartments) on a large city lot for a $110,000 selling price. To get that low purchase price in my town, I'll need to buy rundown junker type houses to fix-up. Assume my fix-up costs are $10,000. That means my total cost, or basis, in the property will be $120,000.

Don't confuse this example with tax basis information. This is for the income analysis only. Once the four rental units are fixed up, the basis for each house will be $30,000 ($120,000 divided by four equals $30,000). After fix-up, my target rents will be a minimum of 18% annual return, which equals $450 per month ($30,000 times 18% equals $5,400, divided by 12 months equals $450). If market rents are $500 per month, my return will be 20%; and with $600 rents I'll earn 24% annually. Once you learn this business well enough to consistently earn 20% plus rent returns, you'll begin to visualize more opportunities—perhaps diversify your portfolio. *This strategy is a major advantage for investors who follow my advice.* The idea is to go for *cash flow first*—then acquire the better grade houses after you've established a steady monthly income. Folks who do this in reverse are always fighting the cash flow battle—and many lose the ranch!

From Fix-Up Properties Too Squeaky Clean Houses

Several years ago, I started acquiring larger single-family houses with my extra rental income. My objective was to buy them for 10% cash down and get the sellers to carry back mortgages for their equity (owner financing). I prefer situations where I can assume existing loans, because I don't like to refinance the property. My plan with these larger houses is to break even on the rents and expenses starting on the day I acquire them. These types of properties appreciate much faster than my small houses and make excellent long-term investments. They can also be sold to users (homeowners) as well as to investors. Users buy with their hearts rather than their heads and will often pay 10 to 15% more.

It's much easier, with seller carry-back mortgages, to customize the payments in order to *balance out* the income with the expenses. ***Owners are much more flexible than banks.*** My longer range plan is to completely pay off one house every year. ***Naturally***, I always allow myself an extra year or so for miscalculations.

Another transitioning plan I have is to slowly sell off some of my management-intense rentals, using INSTALLMENT SALES and 1031 EXCHANGES, when the selling terms are right. This allows me more freedom to pursue other goals. However, any transaction I make will always be designed to minimize my personal income taxes. Wrap-around financing is my favorite method of selling, unless a tax deferred exchange can be worked out! I have written a great deal about wrap-around financing (wraps) in other chapters. My Hillcrest cottages installment sale shows you how to ***earn a million bucks*** using this method. *(Review Page 258 for details.)*

Action—The Magic That Makes it all Happen

It goes without saying, you'll never make any serious money in the real estate business unless you do something—you must make this happen. I have briefly described my plan of action, to give you an idea about how I earn my money. Without question, the most important ingredient in any success formula is ACTION. Nothing moves forward without it, and even the most brilliant idea will fail without "action" to make it happen.

Once I discovered I could make more money than other investors by *fixing and renting rundown houses*, I just kept doing it. I became a SPECIALIST at finding the right properties, fixing them up economically, then renting them out quickly. Folks often say, "Jay, fixing your type of houses seems like a lot of work! How do you fix them so fast?" Two things will automatically happen as you gain experience. First, you get to be very good when you specialize at doing something. You become more and more efficient, as you eliminate needless work. Secondly, with cash flow coming in, you're quickly in a position to hire help. You can furnish the knowledge, while your help does the work. This allows you to expand at a much faster pace.

Investors who don't pay enough attention to cash flow, when they first start out, will severely restrict their potential for growth! Remember, a single horse can only pull a small wagon. Many investors I know could easily stand a bit more horse power to keep their investment wagon from "bogging down" in the sand. *"Horse power"* translated means *cash flow*.

INVEST—BUT KNOW EXACTLY HOW YOU WILL PROFIT

Obviously, in order to make money in real estate, you must invest. However, that's not all there is to it. You must only invest in those properties that will earn income and produce profits. Buying property is the easy part—earning income and profits requires a good thorough analysis before you sign the deal.

So far, I've told you how I started out by investing in the inexpensive rental houses. I did this—not because I'm cheap—but because I discovered that inexpensive rental houses will earn a lot more money than higher-priced houses. What that means is more cash flow! Many people ask me about the long-term profits for inexpensive rental houses. Will they be as good as the more expensive single-family houses? Probably not, from an appreciation standpoint; but, appreciation was not my top priority when I first started. ***Appreciation is a luxury for folks who have extra money in their wallets to buy groceries while they wait for it to happen.***

WRITE YOUR PROFIT PLAN—THEN GET A SECOND OPINION

My highest priority when starting out was to have a little cash left over at the end of every month. Appreciation often takes too long, so I made it—my second priority. That strategy has served me very well over many years of investing. It's tried and tested, and—best of all—it works!

Determining how you will earn profits on a property you plan to acquire, can be done best with a ordinary pencil and yellow legal pad. That means *manually*, or by hand. Writing numbers out longhand helps reveal flaws quickly, I've found.

Doctors will often send their patients out for a second opinion, so why not get one for your real estate deals. I've found that wives, husbands and friends can always spot the things that are wrong with most plans. Write your profit numbers down and show them. Explain how much income you'll be taking in every month from rents. Then, using conservative appreciation numbers, estimate the profits you'll make when you sell 10 or 15 years from now. See if anyone else agrees with you. You don't have to make this exercise complicated. It's only a check-point, but it makes you prove your case—IT'S VERY OBJECTIVE. Outsiders—people other than yourself—are seldom as optimistic as you might be. This is especially true when you must explain to your spouse why you're investing the vacation money on rundown ugly houses instead of a family trip to Disneyland.

COMPOUNDING—THE SECRET TO BUILDING A FORTUNE

Compounding is a silent but powerful wealth-building ingredient that works continuously around the clock to make you richer as you invest, almost without you noticing! A money growing cycle begins when each one of your investment dollars starts earning money from itself. Those earnings, in turn, join forces with more invested dollars to keep repeating the wealth-building process. Rich folks can tell you that real wealth comes from REPEATING THE COMPOUNDING CYCLE, over and over.

Compounding money at extra high rates of return is perhaps the greatest single advantage of investing in real estate. Small investors who learn how to make compounding work for themselves can quickly move up to "wealthy tycoon" status.

I have personally completed transactions that have exceeded 100% compounding. However, it's difficult to do this for very long. Still, experienced investors can keep their dollars earning consistently in the 25 to 50% range. If you were to invest $20,000 today, and compound at 50% annually, after ten years you'd be a millionaire! If you average just 30% annually, it will only take you about five years longer to get there. Compounding is very powerful medicine for your financial health and well-being!

Lack of Cash Doesn't Have to Stop You

Perhaps the most used—or overworked excuse given for not investing in real estate today is, *"I'm waiting until I get enough money together to start!"* Money is not what you need most— what you need is *CREATIVE THINKING.* There are any number of completely legitimate and respectable ways to acquire real estate without waiting for your financial ship to come in, and besides, who's to say if it ever will.

One of my favorite *minimum-cash-down* techniques over the years has been my LEMONADE OFFERS. The name comes from the ingredients in lemonade (sugar and lemons). Obviously, the more sugar (cash) you use, the sweeter the lemonade, but you shouldn't be discouraged if you're short on sugar. Some of the best deals I've made were done with nearly all lemons. Lemons, of course, can be anything other than cash!

MATCHING YOUR OFFER WITH THE PROPERTY

To make this plan work, you must remember to keep things in a proper relationship! What I mean is—don't try this technique to purchase a premium "Grade-A" apartment building. The seller will only laugh at you—owners of premium apartment buildings don't need to accept "lemons." They are very much aware they can get all sugar for their properties. Don't waste your time with them! Lemonade offers are not power offers! As the name implies, they have the strong flavor of lemons. Most sellers are looking for sugar; and, of course, that's what broke buyers have the least of. *Weak offers seem stronger and are more effective when matched up against weak sellers.* Learning why a seller is weak becomes an important key to acquiring good deals using this method.

WEAK SELLERS MAKE WEAK OFFERS WORK

Weak sellers—meaning their property is less marketable for many different reasons—must accept weak offers. Just how weak depends on what's wrong with the property and how "turned off" potential buyers become when they see it! For example, sometimes a property will look so disgusting and the tenants are so ugly that not one single buyer will make an offer to purchase.

When this happens, the seller is in a very weak position. He becomes highly motivated to sell. This is the ideal situation, where lemonade offers work best! It's quite likely you'll be able to negotiate a purchase with lots of lemons and very little sugar—*sometimes none*—if you can accurately determine if the seller is in a real bind to sell and wants out bad. If that's the case, then anything you offer will stand an excellent chance of being accepted.

Ugly Houses Means Less Competition

Fix-up properties are perfect candidates for lemonade offers, because of how they look and the way they are managed. Ugly looks will drive away about 95% of all the potential buyers. Even the ones who have the courage to make a bid, will generally insult the owner with a low-ball offer, thereby eliminating themselves from serious competition. Lemonade offers won't work, unless they are perceived by the seller, as a fair exchange for whatever he's selling! Remember, value is "in the eye of the seller," too. What seems like junk to you, might be viewed as a potential gold mine to even the most desperate seller. You must be careful to avoid insulting or intimidating owners with ridiculous proposals, otherwise you run the risk of not being taken seriously!

Lemonade offers can be the perfect solution to cleaning out your overstuffed garage. Old weightlifting sets, ski boats, unused pianos and the extra car that no one drives anymore, are suddenly worth *TOP DOLLAR* as trade items. Let me tell you about an actual deal that worked very well for me several years ago. I think you'll see the potential benefits of acquiring profitable investment properties using something other than cash.

A LEMONADE OFFER THAT WORKED

The property I acquired consisted of a junky four-unit apartment and two detached single-family houses, all on one large city lot. A local property management company was renting the units for an out-of-town owner. At the time I purchased the property, vacancies were high in my town, and only three of the units were rented. One reason for this was the management company had allowed junky non-running cars and trash to accumulate, creating a very unsightly mess, and making it harder to attract decent tenants. When the owner finally showed up to see what was wrong, he immediately fired the manager and listed the property for sale. That's when I first found out about the deal. The asking price was $120,000, with $20,000 cash down payment.

The classified ad drew lots of "looky-loos"—those are the folks who just look at properties, without making offers. I never heard if any of them made offers, but I suspect they didn't. The property was a terrible mess, with dismantled cars everywhere. Still, there was potential value for someone with enough vision to picture six attractive rental units with a nice positive monthly income, once the clean-up work was done. The situation was ideal for my *lemonade technique!* Here's the offer I presented, which was slightly modified and then accepted.

TOTAL PURCHASE PRICE $105,000

DOWN PAYMENT to be..20,000

Consisting of the following combination of cash and
personal property:

$4,500 Cash
$4,000 1968 Barracuda—318 engine—in top condition
$1,500 1981 Suzuki SS1100 motorcycle—low mileage, in
        showroom condition
$7,500 Self-contained 1971 Winnebago—60,000 miles,
        new paint
$1,550 E2500 Honda generator—used 15 hours
$ 950   Cherrywood hutch—beautiful condition

$20,000 Total down payment

Buyer to assume existing mortgage with a balance of approximate $61,700—with payments of $718 per month, including principal and interest. Seller to carry back second deed of trust and promissory note in the amount of $23,300—with payments of $221.87 per month, including 8% interest, for a 15-year term.

With No Other Buyers—Sellers Have Few Options

At first the owner agreed to take just the automobile, motor bike and travel trailer. To which I said, "That's fine, but I don't have anymore cash to give you, and I can't afford higher monthly payments, either." Originally, I proposed interest-only payments on the seller's equity, with $155.33 per month payments for 15 years. My compromise was to amortize the note and pay the higher monthly payments. I told the seller I would need to switch back to interest-only payments if he didn't take the generator and hutch. Finally, he agreed and we signed the deal.

Obviously, the real key to making this deal work was the **high level of seller motivation!** To my knowledge, there were no other offers on the property. I also learned the seller was really strapped for cash. With only three units rented at

$275 each, he just barely collected enough rents to pay the mortgage payment, and the management fees and maintenance costs were extra. He was paying those out of his pocket each month. Living out of town, with a regular job to attend, made it impossible for the owner to help himself. Selling the property, to cut his losses, was the only logical choice he could make.

Fixing Problems Builds Wealth Faster

The day I took over the apartments, my mortgage debt was higher than the rental income! However, it didn't stay that way for long. The apartments were easy to rent for $300 each, once the property was cleaned up, which took me and my helper about three months to complete. Today these apartments rent for $450 each. The property never fails to produce at least $1,000 net income every single month. That's not a bad deal for ONLY $4,500 cash—plus all that stuff I was not using anyway—wouldn't you agree?

EASY PATH TO CASH FLOW

The benefits for a buyer using the lemonade technique can be quite impressive once the motivation level of the seller can be determined! Obviously, if the owner simply wants to sell, but is in no serious bind, lemonading looses its clout.

However, when there is a serious need to sell, I've seen a $2,000 truck traded for $5,000 value. Used furniture, weight-lifting sets, boats and almost any personal property can be "marked up" three and four times the value of a yard sale clearance. Sellers who need to sell will make very heavy concessions to get whatever cash they can! Stated another way—they will take a lot of junk they don't want to get what they do want. When you consider the discounted selling price of $105,000—plus my personal property items traded at $15,500—with a yard sale value of $6,400 tops!—you can begin to see how cash flow status is much easier to achieve with lemonade deals!

The Deal Begins With Personal Property Inventory

The best way to become a "LEMONADER" is to make a list of everything you own that you don't mind parting with. Consider everything in the garage and, especially, those "faddish" recreational items that you only used one or two seasons, and are now packed away. I've found almost everything has value to someone. If one item seems sort of odd and appears to have no redeeming value, mix it with something else! For example, if you trade a boat, it's easy to toss in fishing gear, skis, tents, outdoor cooking gear and scuba diving equipment. Add another $1,000, or so, to the value of the boat, and call it "the deluxe package." Tools and campers can always go with trucks at premium trading prices.

HOW TO MAKE LEMONADE OFFERS WORK

Quite obviously, not everyone who owns real estate will go for this kind of deal! *The good news is you don't need "everyone."* You just need one or two who will; and, believe me, they're out there. Most folks foul up in two areas—they misjudge the seller's motivation level, and they try to acquire properties that don't have very much wrong with them! On a scale of one to ten, with ten being top-notch properties and one being little more than stacked kindling—try for two through five deals. You'll find the right seller somewhere in that range.

There's another point worth mentioning here—don't use up all your good stuff on one single offer unless it's necessary to close the deal. In other words, don't give away anymore than you absolutely have to. You may well have enough junk stored away to complete several of these transactions. A good rule of thumb for splitting up the down payment is about 33% cash, 33% for some major item of value—like a car or boat, and the balance can be filled in with miscellaneous junk items or accessories—like water skis or travel trailer. This one technique can save you a lot of down payment money. Try it, you'll be very pleased with your increased buying power—I promise!

CHAPTER 24

The Big Picture and Long-Term Wealth

REMEMBER—PLANNING IS NECESSARY FOR SUCCESS! ACCORDING TO ROBERT SCHULLER: "YARD-BY-YARD, LIFE IS TOO HARD, BUT INCH-BY-INCH, IT'S REALLY A CINCH." AND, BASEBALL GREAT, YOGI BERRA, SECONDS THE IDEA—"IF YOU DON'T KNOW WHERE YOU'RE GOING, YOU COULD WIND UP SOMEPLACE ELSE."

Seldom do I begin a chapter with poetry. However, I think it's particularly appropriate here, since we'll be discussing "a tree-top view" of investing and landlording. It's good to step back and add up the score every now and then, for the purpose of self-examination. Are we investing according to our plan? Do we even have a plan? If so, are we working the plan on schedule?

DON'T GET BOGGED DOWN WITH ROUTINE STUFF

I find that even experienced operators need some kind of measurement plan, otherwise, it's difficult to stay on course. Without direction, or a means of making necessary investing adjustments, it's nearly impossible to keep yourself "fine-tuned." It's quite easy to head down a non-productive pathway to nowhere! For this reason, it's helpful to pretend for a moment, that you're high up on a mountain top, above the daily struggles! Up there, it's much easier to look down and see everything that's happening. That's what we're after here, a "top-down" view—or *the big picture*, if you will!

It's very easy to inadvertently allow yourself to get "bogged down" with daily routine and completely lose sight of your investment goals! I've said this many times—you'll never get to where you'd like to be without a GOOD PLAN, or without the means and skills necessary to make your plan work. Making lots of money in real estate is deliberate! It's not an accident and it certainly has very little to do with luck, believe me!

IT TAKES A WORKABLE AND REALISTIC PLAN

I hate complicated plans and formulas! The way I figure, if I can't easily understand what I'm supposed to do—or how I'm supposed to do it—chances are, it won't get done. Most folks in the real estate business, including myself, have no trouble whatsoever with visions and dreams! It's easy to picture myself with an overflowing bank account, driving a high-priced sports car and living in a million-dollar mansion. The problem is, there's an awfully long stretch between my dreams and reality.

Don't misunderstand me here, I'm not knocking dreams. They're important, and we all need a "dangling carrot" to keep us in the race; but, don't forget, dreams are always at the far end of the rainbow. Therefore, we need a reliable vehicle to take us there—*to that workable and realistic plan!*

What is a Workable-Realistic Plan

By nature, I'm a very positive person! Therefore, I don't want to sound like I'm raining on your Sunday picnic, but let me start out on a somewhat negative note and tell you what WORKABLE and REALISTIC are not!

It is not realistic for *start-out* or *part-time* investors to waste their time, thoughts and energies figuring out how to buy properties in some distant location, far from where they live. I could offer you at least ten convincing arguments why you shouldn't do this, but we're talking about the "BIG PIC-TURE" here, so I'll sum them up by saying—you lose the *home field advantage*. That's far too much to give away, in my judgment. People who determine the gambling odds in football know, from experience, the hometown players have a three point win advantage because they are playing in their own ballpark. The advantage comes from the local fan support and familiarization with the playing field. There's really not much difference in the real estate investment game! You must ask yourself—Why in the world would you ever want to bet against the odds?

THE *WALL STREET JOURNAL* DREAM

Reading newspaper ads—we all do it and it's likely we'll all end up with about the same results! Would you care to guess how much money I've made from reading the *Wall Street Journal* and *USA Today* real estate ads! On a scale of one to ten (ten being high), how much help (investment-wise) would you suppose I've gotten over the years by reading articles about national real estate trends and so-forth? You'll know the answer by the end of this page.

I must confess, I enjoy reading Friday's *Wall Street Journal* special real estate section. I love ads that say—5 TIMES GROSS IN THE OZARKS, $7,000 PER UNIT IN HOUSTON, FAR BELOW THE REPLACEMENT COST IN SAN FRANCISCO. These are exciting ads to us real estate junkies. Reading national papers makes it easy to stay "high." Why, because it's like reading about Fantasy Island—far away places always makes for "SWEETER DREAMS."

I can easily picture myself as a rich tycoon in some far-away place, sipping "mint juleps," while collecting big real estate profits! Unfortunately, as with most dreams, I always wake up to find I'm still in Redding—and I'm still telling my tenants how to flip the handle on their running toilets, still listening to mumbled excuses about "the check's in the mail"—my dream is suddenly gone as quickly as it started.

Over the years, I've made some "hefty profits" with my real estate projects. However, looking back, I can't think of a single time when reading national real estate articles has helped me financially. Oh yes, on the scale of one to ten, my answer is perhaps two or three. The only reason it's even that high is because real estate stories fascinate me. They write the kind of things that dreams are made of!

Join the Real World of Investing

Most folks who follow my teachings and investment strategies, see me as a very practical person, I think I've learned to separate my dreams from the real world stuff that earns money. I also learned a valuable lesson years ago, and that is—*it pays big dividends to find a good mentor early on in your investment career*. You can save yourself a lot of time and effort—especially efforts that are taking you in the wrong direction! Obviously, you should thoroughly check out a potential mentor before settling on them. Find someone who is successful at doing what you want to do, then learn everything you can from that person.

I listen to many people who say, "I'm the kind of person who needs to do things my way." Just remember there's lots to learn about investing, and you'll end up light years ahead if you copy successful people! Don't worry, there's still plenty of wide open territory for individuality and creativity. In fact, there's no end to learning, when it comes to real estate investing. That's the reason it's so exciting, in the first place.

THE SERIOUS SIDE TO INVESTING

Over the years, I've had lots of fun as an investor! However, many things I do are definitely not fun! The fun part of investing is not the same as fishing, taking an exotic vacation or a night on the town. It's a more serious type of fun as in the kind of fun you have putting deals together, watching transactions close and turning a scum-bag property into a cash-flow investment. It's also a great deal of fun receiving a $50,000 check as compensation for using my specialized investment skills. This is what I mean by *"serious fun."*

In my opinion, you must have some fun investing, otherwise the hard work part will discourage you quickly—and, it's likely you'll get burned out and give up! Still—part of the "BIG PICTURE" is understanding that real estate investing is not all fun. Landlording can be a disheartening task, especially when you first start out. In fact, there are many who feel you must possess masochistic tendencies in order to even consider doing it!

There are basically two ways to learn landlording and neither one is fun! You can learn it from your tenants, or learn it from people like me! I always recommend a combination of "on-the-job" training (with real live tenants) and formal classroom-type education (schooling) from an experienced landlord-teacher (like me). People often ask, "Why should I pay $500 for a seminar to hear about renting my own property out?" For every person who asks, I can introduce them to a "beat-up" landlord who will tell them exactly why, very convincingly and very quickly!

MY BEST PROFITS COME FROM DUMB LANDLORDS

Landlording is not a question of having good luck, bad luck or picking all the right tenants. Certainly, that's a part of it. However, during any reasonable investment career (over several years) you'll need all the expertise you can get! Forking out $500 for a seminar from an expert will seem like peanuts compared to what tenants will cost you for lessons. The point is—landlording is a business! You must accept it as such and treat it very seriously. Remember, some of my best real estate purchases (discount prices and liberal owner financing terms) are from sellers who have been driven completely "bonkers" (over the edge) by their tenants—they simply can't handle them. They are forced to give up their property for a pittance. Don't sell landlording short! Knowing the job well pays big bucks! I'll assure you—education is cheap, compared to the lack of it when it comes to the business of landlording.

What's a Nice Guy Doing in a Place Like This?

Remembering back (seems like 100 years now), when I traded my "old" traditional corporate job for my *current career* as investor, horse trader and landlord, I had "lofty" dreams and visions of "sugar plums!" The bottom line was—I wanted to make a lot of money, instead of being stuck in a job where my future earnings would always be limited to those meager cost of living increases. Of course, back then, I didn't even think about the possibility of my corporate job disappearing; however, several years later it did!

I felt REAL ESTATE INVESTING offered the best opportunity for me to earn more money without restrictions, BASED ON HOW HARD I WAS WILLING TO WORK! I must tell you—I don't mind working long hours, or weekends, nights, or whatever else it takes if the rewards are in line with my extra efforts! What I've always been opposed to is working at some "fixed-pay-level job," regardless of my initiative or the extra efforts I contributed. Naturally, that's one of the BIG differences with my real estate career—being my own boss has been a very rewarding experience. Besides, there's absolutely no limit to how big my paydays can be anymore!

THE DREAM ALONE IS NOT ENOUGH

WORKABLE and REALISTIC, that's what we're talking about here! If you already have a plan for yourself, make certain it includes both essential ingredients. *THEY ARE NOT OPTIONAL.* Earlier, I said investing away from home is not a good idea. I mean this seriously! It's also not a good idea to invest your money in a property when you have very limited knowledge about it.

I friend of mine bought an old hotel 60 miles away from where he lives. I advised him not to do it, unless he moved there and operated it himself. He showed me some impressive income figures and, I must admit, they looked great if he could keep all the rooms rented! The problem was he couldn't or didn't! He frequently changed managers and completely wore out his new Toyota truck driving back and forth on evenings and weekends.

His original plan was to convert the small sleeping rooms to regular monthly rentals. The plan was to remodel the hotel, combining two sleeping rooms to create larger more desirable efficiency apartments. Each new unit would have cooking and plumbing facilities, which would be much easier to rent to permanent-type tenants. Efficiency apartments are always in short supply.

YOU MUST LEARN TO WALK BEFORE YOU RUN

My friend's plan was very workable, but not realistic for him! His regular 40-hour weekday job was naturally his main priority. Although he had purchased the hotel for an excellent price and terms, he was "tapped out" for additional money, after the down payment. He had no other funds available for remodeling the way he wanted to. It's quite obvious, in hindsight, that even if he had the money to remodel, he still didn't have the time to do the job. Almost every manager he hired was a drunk. Obviously, he needed a bit more practice interviewing perspective employees! Eventually, the inevitable happened, and he lost the hotel. What had seemed like a genuine cash flow bargain at the beginning, turned out to be a $46,000 loss at the end. The saddest part to the story is the building had all the potential for making $100,000 net profit in the hands of an experienced operator. My friend had owned only one single rental house before leaping into a management-intense, 33-unit hotel, which was 60 miles from home. Simply put, the property was much more than he was capable of handling at the time.

DON'T WANDER AROUND WITH BLINDERS ON

Spirited horses are often fitted with blinders. They're sort of like curtains or shades mounted on each side of their head to prevent them from seeing objects off to the side. The blinders completely block their side view, allowing the horse to only see what's straight in front of him. This is done to prevent distractions, which often "spook" or panic the horse. By limiting vision to straight ahead, the horse is unable to see any potential danger that might spook him. The blinders create a very limited view. Real estate investors don't wear blinders on their heads like horses, but they can develop them mentally! When they do, they work exactly the same way as the real ones.

Not long ago, a young man came to me with a problem. He had an opportunity to buy six dumpy houses for an extremely cheap price. Just when he was set to close the deal, a local real estate agent advised him that—"He would lose his shirt." "The reason," he said "Is the location! The houses were located directly behind a soap factory. They'll never appreciate in value, plus there's a better than average chance you'll be stuck with them forever!"

That's typical salesman advice and, who knows, he might be absolutely right! The problem, however, is he's got blinders on! He's already thinking about *future sales* and *appreciation*. Those things might not be too important. In fact, in this situation, as you will see, they're really not. Many folks, including real estate professionals, are quite good when it comes to finding shiny gold nuggets that sparkle brightly in plain view. However, their detection abilities quickly diminish when the golden nuggets get slightly tarnished or buried beneath the mud!

Remember this—It's not the location or how they look that makes gold mines valuable. Most of them are smelly holes in the ground. Most look rather ugly! The point I'm making is that value isn't always apparent when you first look at something!

HIGH RENT-TO-VALUE RATIO IS TIP-OFF TO PROFITS

The six houses behind the soap factory are really "hidden gold." They had a 2.0 percent *rent-to-value factor* at the close of escrow. Rent-to-value is a number that expresses the monthly rent in relation to the value of the property. Old-timers used to talk about the *one-percent rule* for renting. It means a $70,000 house should rent for $700 per month. With a 2.0% rent factor, the same house would rent for $1,400 per month. The RENT-TO-VALUE FACTOR is calculated by dividing the monthly rent by the total value of the house. In this particular deal, the young man bought the entire property (all six houses) for a total price of $135,000.

Let's look at the big money picture here. The six houses are earning $450 each, or a $2,700 per month total. Four out of the six are rented to HUD tenants with guaranteed rents; the other two could easily be the same if the owner wants to. Here's a good way to analyze profitability. Each house is earning 24% of it's total value *annually*. The value or purchase price of each house is $22,500. Rent equals $450 per month times 12 months, equals $5,400 annual ($5,400 divided by $22,500 equals 24%). Can you see that a 24% rent return means that each house will earn it's entire cost ($22,500) back in just slightly over four years time—it's 4.16 years to be exact.

If you have a nose for making money, you should at least start sniffing about now! Many folks get cold feet when they think about the soap factory—they suddenly develop blinders. They fail to scratch the surface to find the shiny gold! Anytime an asset generates enough income to completely pay itself off in just four short years, you should be very interested in that asset! They're not all that easy to find. When you do, don't pass it up without a thorough investigation.

SELLING FOR WHAT YOU PAID, AND STILL MAKING A PROFIT

When I was telling this story to an investor group, a lady asked how you can tell if a deal is good or not. She said, "You haven't told us what the down payment was or what the monthly expenses and the mortgage payment cost!" The answer is—IT DOESN'T MATTER MUCH, unless, of course, something is terribly out of whack! In this particular case, I happen to know the cash down payment was $18,500 (13%), which means the mortgage balance was $116,500.

Even with high leverage deals like this, it's quite easy to structure the seller financing (quite often interest-only) in a way that will allow the owner to enjoy a very respectable cash flow, starting with the first day of ownership! It's also very common, with these type of rental units, to earn handsome profits, even if the property doesn't appreciate at all! The reason is because the property is a ***cash flow machine***. You've actually acquired your own private gold mine, so to speak. Every month you'll be able "mine out" fresh green cash!

Let's suppose you're able to net-out $500 each month from the "soap factory houses." That adds up to $6,000 the first year, and your return on cash invested is 32%. You have also acquired approximately $100,000 worth of depreciable property (income shelter).

Forget about the future of the houses for a moment and only consider the income stream. For the next ten years, even with very modest "cost of living" rent increases, it's a very good bet you'll be "netting out" $10,000 annually, by the end of the term. It's not the least bit difficult to visualize this one small property generating $100,000 worth of cash and tax benefits in just ten years time. Even if you sell the houses ten years from now—at the same price you paid—that's not all bad! I will assure you, much worse things can happen to real estate investors.

An important thing to remember when you are lucky enough to locate income properties that have a rent-to-value factor of 1.5, or above, it's like the gold miners say—*"You're beginning to see some very good colors!"* Stick with the deal and figure it out. There's a very good chance you're standing real close to a cash flow spigot!

I DIDN'T GROW UP TO BE A LANDLORD

Investing in rental houses and being a landlord are not my goals, and they never have been! Rental houses and landlording are the vehicles that are taking me to my goals! In order to keep my focus on the BIG PICTURE, I must keep the vehicles separated from the goals.

Stated another way, I'm in the housing business to make money, not to simply own a bunch of rental properties! In fact, as much as I like owning houses, my underlying motive sounds almost selfish. Houses are the best vehicles I've found to take me where I wish to go. But my houses are certainly not the end—*they are the means to the end*.

Understanding this makes it much easier to make sound investment decisions, I believe! Let me explain it this way— my goals and personal dreams are probably quite similar to those of every other investor I know. *Without mincing any words*—MY MAIN GOAL IS MAKING MONEY!

I have discovered that all properties are not the same, when it comes to making money! For example, I own several properties that just sit there and cost me money. They are not actively participating as vehicles taking me toward my investment goals! In fact, one property is pulling me backwards—away from my money-making goals. The problem is, I paid too much for looks! Every month it costs me $335 for the privilege of being the owner!

YOU NEED A CLEAR UNDERSTANDING OF WHERE YOU'RE GOING

When you understand goals are the objective—*not the vehicle*, it helps you "zero-in" on an investment plan that makes the best use of your time and resources (money). For example, in my case I needed to quickly develop monthly cash flow without paying out a ton of cash (which I didn't have) for my properties. Only certain types of properties will provide cash flow, so that's where I directed my energies. Also, another one of my personal goals was to quit my regular salaried job. That put some tight restrictions on my time limits because I had to start earning a living on the income from my houses within three years from the day I started buying properties.

Avoid Getting Your Investment Cart Before The Horse

A major failing on the part of many investors I talk with is they develop tunnel vision! They have unknowingly created blinders for themselves. They are concerned about the vehicles more than the benefits. If you do that, it's very difficult to achieve your goals in a reasonable period of time. That's the main reason investors can struggle along for many years buying properties and never have any cash flow. They can only brag about imagined equities—which won't buy very many groceries at the supermarket!

It's my feeling that all investment portfolios should be diversified! You need some properties that provide good cash flow and some that just occupy the lot waiting for appreciation or higher and better usage. By setting goals, schedules, time limits and minimum cash return requirements, you'll quickly determine what kind of properties you need and how many of each it will take to get you where you're going.

In case you don't already know—landlording is not fun, and it's not supposed to be. However, if you learn it well—it could easily end up being the best paid job you've ever had! It's strictly up to you.

Change Brings On New Opportunity

Opportunities for serious-minded real estate investors—who want to earn a bigger share of the "American Dream," create a more rewarding life for their families and ultimately achieve financial independence—are just as good today as they've ever been!

Passive tax laws, variable-rate mortgages and a record numbers of foreclosed properties does not mean the real estate bubble has finally burst. What it means is the rules are constantly changing! Changing rules almost always deal a death blow to the inflexible. However, the same changes offer new hope and opportunity to those who are willing to meet the challenge.

INVESTING IN REAL ESTATE IS LIKE KISSING FROGS

Taking advantage of new opportunities means knowing what to do and how to do it. You must be willing to roll up your sleeves and get started! You alone are the person who must make it happen. Making things happen is a lot like "kissing frogs in order to find your prince." Chances are, you'll have to kiss a lot of frogs before you finally meet a prince. You cannot stay inside the comforts of your home and wait for the frogs to hop in and kiss you. Frogs don't do that, so nothing will ever happen.

You've got to spend many hours of your time with the frogs. Remember, most frogs are generally hiding deep in the muddy marshes. Kissing frogs is a contact sport. You cannot ask your friends, your real estate broker or secretary to do it for you—they haven't got the stomach for it. You can't keep changing kissing styles because one seems easier than the other, and you can't spend all your time in kissing school. That's not where you'll find a prince. Don't kiss the same frog forty times either. It shouldn't take you that much smooching to tell a frog from a prince.

Kissing Frogs is a whole lot like investing in real estate—the more time and effort you put in, the more experience you gain. After a while you'll be able to tell if you have a prince on the very first kiss. The point I'm making here is simple—forget shortcuts, because there are none! Don't worry about how fast the folks on TV are getting rich—just remember, that's "SHOW BIZ."

Avoid Doom and Gloom Like the Plague

> *"THE DAYS OF OPPORTUNITY ARE OVER. THERE'S NO LONGER ANY USE TRYING TO SAVE FOR INVESTMENT. THE BEST YOU CAN HOPE FOR IS TO KEEP A STEADY JOB AND STAY OFF WELFARE. NOBODY WILL EVER AGAIN BE ABLE TO BUILD AN ESTATE LARGE ENOUGH TO PRODUCE AN INDEPENDENT INCOME."*

Those are the words as they were spoken in a speech given by the economics professor at Fresno State College. Over three hundred graduating students were in the audience.

Folks familiar with California's economy would have little quarrel with the professor's bleak assessment! With the huge layoffs at IBM and the telephone company, thousands of lost jobs in the defense industry—and nearly everywhere you look, there's downsizing by the state's largest corporations. Indeed, it would be hard to find fault with the professor's reasoning! However, there's just one thing wrong! The professor wasn't talking about the economy today! He was making his speech to the graduating class of 1931, in the middle of the Great Depression!

William Nickerson was in that class! Fortunately for him, he didn't pay much attention to the professor's advice. Bill recalls the speech in his best selling real estate book, "HOW I TURNED $1,000 INTO FIVE MILLION IN REAL ESTATE" (Simon and Schuster, 1980 Rev.). According to Nickerson, the professor really didn't mean any harm. He simply didn't know any better. And obviously, it's a good thing for self-made millionaire investors like Bill that most *DOOM AND GLOOM* predictions are merely opinions of the misinformed. It's unfortunate that many who are charged with teaching others have great difficulty seeing beyond the ends of their noses.

History has clearly proven that economic opportunity for enterprising students didn't end with the class of 1931—and there's certainly no end in sight today. Unfortunately, most teachers don't encourage students to develop their entrepreneurial skills and rely on themselves to earn a living! Hopefully, these 24 chapters have convinced you there are numerous possibilities.

ROADBLOCKS—YOUR MOMENTUM WILL CARRY YOU AROUND THEM

The single, most dangerous roadblock facing every new investor, or career changer, is PROCRASTINATION. There is no doubt that many folks, with the best intentions and even a good workable plan, will procrastinate forever. Look around you—how many people are financially independent compared to those who just talk about it? "Very few" is the answer. However, don't let the numbers discourage you! Remember, financial success is not some wild stroke of luck. *It's a solid workable plan and it's you working the plan.* Your success will come almost automatically when you do the things I've suggested.

POSITIVE CASH FLOW MAKES IT ALL WORTHWHILE

Over the years I've discovered that investing in *cash-flow*-producing real estate is even better than I dreamed it would be when I first started out—financially speaking, that is! People often ask me, "Isn't it hard work renting your houses to Section 8 (HUD) tenants?"; "Don't you have a lot of trouble with people who don't pay on time?"; and, "Don't you ever get 'fed-up' cleaning filthy properties when trashy tenants move out?" The answers are "YES," "YES," "YES"—but just remember, those are very "narrow vision" items. They're simply part of the vehicle that takes me where I'm going!

THE BIG PICTURE IS that my rewards are very generous when you compare them to the tasks I perform. That's exactly what I asked for when I first started investing—as you may recall!

I once had 218 houses that produced average rents of $388 per month. Even though rents are relatively low in my area— still, when you do the math, you can see how quickly they add up to substantial annual income. As my mortgages are paid off, obviously, I get to keep a bigger share of the rent money each month! And "Who pays off the mortgages?," you ask. Why it's the same folks who pay all my other expenses too! *Is this a great program, or am I still just dreamin'?*

APPENDIX

Exhibits:
Income Property Analysis Form
Typical Property Sketch
Co-Ownership Agreement

Resources:
Jay's Seminars and Fixer Camps
Trade Secrets Newsletter—"Why You Need it"
Fixer Jay Training Products

INCOME PROPERTY ANALYSIS FORM

Property Name _____   Date _____

LINE
NO.   INCOME DATA (MONTHLY)   PER MONTH

1	Total Gross Income (Present)	$ _____
2	Vacancy Allowance Min. 5% LN-1 Attach copy of 1040 Schedule E or provide past 12 months income statement for verification	$ _____
3	Uncollectable or Credit Losses (rents due but not collected)	$ _____
4	Net Rental Income	$ _____

EXPENSE DATA (MONTHLY)

5	Taxes, Real Property	$ _____
6	Insurance	$ _____
7	Management, Allow Min. 5%	$ _____
8	Maintenance	$ _____
9	Repairs	$ _____
10	Utilities Paid by Owner (Monthly)	$ _____

	Elec	$ _____
	Water	$ _____
	Sewer	$ _____
	Gas	$ _____
	Garbage	$ _____
	Cable TV	$ _____
	Totals =	$(_____)

| 11 | Total Expenses | $ _____ |
| 12 | OPERATING INCOME (LN 4 - LN 11) | $ _____ |

Existing Mortgage Debt

1st Bal Due	_____	Payments	(Montly)	Due Mo/Yr
2nd Bal Due	_____	Payments $ _____		_____
3rd Bal Due	_____	Payments $ _____		_____
4th Bal Due	_____	Payments $ _____		_____
5th Bal Due	_____	Payments $ _____		_____

13	Totals	(13A)	$ _____	_____
14	MONTHLY CASH FLOW AVAILABLE	$ _____	
	(LN-12—13A) (Pos or Neg)		_____

NOTE: Line 14 shows available funds to service new mortgage debt from operation of property.

REMARKS: All lines must be completed for proper analysis. Enter the actual amount on each line or Ø.

TYPICAL PROPERTY SKETCH

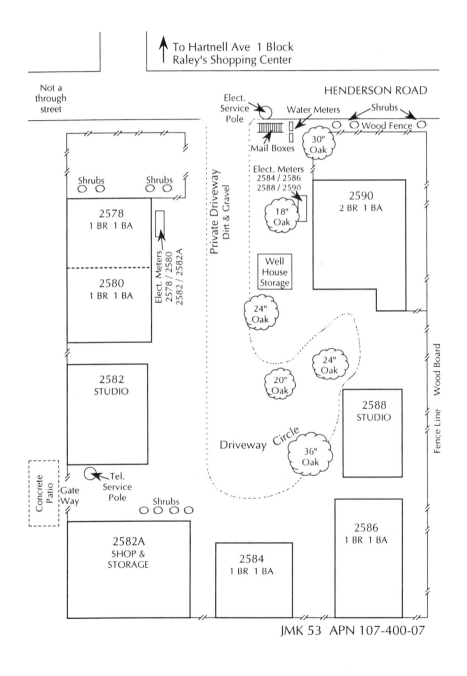

To Hartnell Ave 1 Block
Raley's Shopping Center

Not a through street

HENDERSON ROAD

Elect. Service Pole

Water Meters

Shrubs

Mail Boxes

Wood Fence

30" Oak

Shrubs   Shrubs

Elect. Meters
2584 / 2586
2588 / 2590

Private Driveway
Dirt & Gravel

2578
1 BR 1 BA

Elect. Meters
2578 / 2580
2582 / 2582A

2590
2 BR 1 BA

18" Oak

2580
1 BR 1 BA

Well House Storage

24" Oak

2582
STUDIO

24" Oak

20" Oak

2588
STUDIO

Wood Board

Fence Line

Driveway Circle

36" Oak

Concrete Patio

Tel. Service Pole

Gate Way

Shrubs

2582A
SHOP & STORAGE

2584
1 BR 1 BA

2586
1 BR 1 BA

JMK 53 APN 107-400-07

CO-OWNERSHIP AGREEMENT
1234 Easy Street
Uglyville, CA 96001

THIS AGREEMENT is made effective as of the 23rd of March, 1997 between Jay P. Decima dba Fixer Jay and Ivan M. Smith dba, Investor Ivan.

1. Transaction: Investor Ivan (Ivan) and Fixer Jay (Jay) will join together as co-owners for the purpose of owning and operating that certain real estate located at 1234 Easy Street, Uglyville, CA. herein called (the "Property") for the mutual benefit and profit of each. Each party agrees to perform fully under this Agreement for the success of both parties herein.

2. Acquisition of Profit: Ivan and Jay have purchased the Property for a purchase price of Three Hundred Fifty Thousand Dollars ($350,000) pursuant to Escrow Instructions dated February 25, 1997 to North State Title, a copy of which is attached hereto as Exhibit "A." The cash down payment of Fifty Thousand Dollars ($50,000) was paid equally.

3. Cash Distributions from Rental: All excess cash derived from rental of the Property, after payment of all expenses and debt service, shall be divided fifty percent (50%) to Ivan and fifty percent (50%) to Jay.

4. Cash Proceeds From Sale or Refinancing of the Property: Net cash proceeds derived from sale or refinancing of the Property shall be shared as follows: First, each party shall receive back all of his capital invested in the Property by way of initial down payment, fix-up expenditures and operating expenses made pursuant to paragraphs 2, 11 and 12 hereof. Thereafter, all remaining proceeds derived from sale or refinancing shall be shared fifty percent (50%) Ivan and fifty (50%) Jay.

5. Management: All decisions regarding the management of the Property shall be made upon the joint approval of both Ivan and Jay; provided, however, it is agreed that Jay will have primary responsibility for the day-to-day management operations, such as rent-ups, property maintenance, repairs, cleaning and the like in order to conduct an efficient rental business. Jay shall receive a management fee of ten (10%) of the gross rents collected from the Property for management of the Property and shall be reimbursed for his actual out-of-pocket costs and expenses incurred in connection with such management.

6. Books & Records: All books and records will be kept at the office of Jay. A statement of operations will be provided to Ivan on a monthly basis. This statement will be prepared by Jay as part of his management duties.

CO-OWNERSHIP AGREEMENT

7. Bank Accounts: Jay shall maintain a commercial checking account at North Valley Bank, 2930 Bechelli Lane, Redding, California or at such other banking institution that shall be approved by Ivan, for the purpose for operating the Property.

8. Indemnification: Each party shall indemnify and hold harmless the other party and the Property from and against all separate debts, claims, and demands of said party.

9. Termination: This Agreement shall terminate upon sale of the Property or by mutual consent of Ivan and Jay. Ivan shall have the sole right to determine when the Property is to be sold, provided, however, that Ivan shall first offer Jay the right to purchase the Property for the same amount and upon the same terms and conditions as Ivan is willing to sell the Property pursuant to a bona fide offer received from any third party. Jay shall consummate the transaction within ninety (90) days after exercise of his right of first refusal.

10. Death of Parties: Upon the death of Ivan, Jay shall have the right to either purchase Ivan's interest in the property in the manner described in Paragraph 9 hereof based upon a bona fide offer received by Ivan's estate or, in the absence of such an offer, Jay shall have the right to cause the Property to be sold and the proceeds divided in accordance with Paragraph 4 of this Agreement. In the event liquidation is elected, Jay shall proceed with reasonable diligence to liquidate the Property within six (6) months after Ivan's death.

11. Initial Fix-up Expenditures: Initial fix-up funds for rehabilitation of the Property will be contributed equally. All work will be performed by employees of Jay, ONE STOP HOME RENTAL COMPANY.

12. Operating Funds: All expenses, improvements, taxes, insurance, maintenance and other operating expenses deemed necessary for the operation of the Property shall be paid first from rental income derived from the Property and thereafter from additional funds to be contributed equally.

13. Business Address: The official management office for the Property will be ONE STOP HOME RENTAL COMPANY, located at 2551 Park Marina Drive, Redding, California 96001. Mailing address is c/o JMK Traders, P.O. Box 493039, Redding, California 96049-3039.

14. No Partnership or Joint Venture: The relationship between Ivan and Jay under this Agreement shall be solely that of co-owners of real estate and under no circumstance shall said relationship constitute a partnership or joint venture.

CO-OWNERSHIP AGREEMENT

IN WITNESS WHEREOF, the Parties have executed this Agreement as of the day and year first above written.

By_____   By_____
        Investor Ivan   Investor Jay

Location: _____   Date:_____

Resources

JAY'S SEMINARS AND FIXER CAMPS

CASH FLOW is the life blood for real estate investors. Therefore, that's exactly what you'll learn the most about at my seminars. I have developed many successful techniques from building my own personal real estate wealth during the past 30 years. What you will learn from me is not some theory about investing—but it's tried, proven and tested methods that will work very well for **today's real estate investors.**

The central theme of my "How To" training seminars is structured around four basic ingredients that every investor must use in order to succeed.

1. **WORKABLE PLAN** – One that will teach you how to evaluate the profit potential of any investment property, from start to finish! One test is to explain your deal to an uninterested party and convince them, beyond a reasonable doubt, it's profitable.

2. **PROPER COURSE OF ACTION** – You'll learn the facts about buying and selling versus long-term ownership. No need to guess which is right—*it's easily proven*. Investors must learn to properly evaluate "hear-say" and concentrate their energies toward a strategy that pays off. At Jay's seminar, you'll learn how to judge for yourself.

3. **INVESTING RIGHT** – The first rule of investing is knowing the value of returns! You'll learn the difference between investing and speculating, and how to measure the benefits.

4. **COMPOUNDING** – The magic that multiples your wealth! Learning how to make a few dollars do the work of many is the secret to real estate riches. Jay teaches you how to consistently earn 25 to 50%, annually.

Discover for yourself how to find the kinds of properties that earn the highest cash flow, starting on the day you acquire them; how you can easily move up from one house deals to larger properties with *higher profits* and *less work*; where to get the best financing and how to sell for the biggest profits; how to balance-out passive income from rents—and carry back mortgage payments from installment sales so you don't lose your profits to the tax man; and how pyramiding small start-out investments can make you extremely wealthy in a relatively short period of time! Obviously, you must apply all four basic *wealth-building ingredients* we've been discussing.

My long-time friend and personal real estate broker, Fred Quigley, can never pass up an invitation to enlighten my students about finding the right agent—and how investors and agents can truly benefit each other. Fred shows licensees how to create big paydays—and how commissions can be structured to minimize cash down payments. Fred is a "Buyer's Broker" who specializes in buying and selling investment properties.

PART-TIME INVESTORS and **CAREER CHANGERS** alike will learn a great deal from this personalized *hands-on training*. Jay's 400-page plus workbook—including many forms and special wealth-building examples is provided to each student who attends a three-day seminar or fixer camp. For more information call **800-722-2550**. Ask for Kathy. She will send you seminar flyers and advise you of current schedules.

TRADE SECRETS NEWSLETTER—WHY YOU NEED IT

JUST ONE EXCUSE CAN STOP YOU FROM MAKING $1,000,000

Nearly everyone I talk to about real estate investing claim they could have made a fortune 10 years ago when they had extra money! Some say they could have done it easily if only they hadn't been laid-off at the rocket factory.   Others admit they should have used the Las Vegas money for a down payment instead.   Each of these stories can be boiled down to a single word... *EXCUSES*. I'm sure we've all heard the same excuses and others like them a thousand times. The reason they sound so familiar is because they're repeated month after month—*year after year*. The only thing that ever changes is the date.

SACRIFICES ARE TINY WHEN COMPARED TO BIG REWARDS

Understand the facts—real estate investing requires a bit more from its participants than merely wishing to make lots of money. Certainly, I grant you—**You can make big money all right, but you'll have to earn it**. Also, you'll need to make some tough financial decisions! For most of us these decisions often mean sacrifice. I can tell you from personal experience, it's not much fun when you're forced to give up a trip to Disneyland—and spend the money on a fixer house.

INVESTING TODAY MAY BE EASIER THAN YOU THINK

Many folks have the mistaken idea that investing today is tougher than it used to be! That, my friend, is simply another excuse. If investing has changed at all in the past 40 years, I'd have to say—*IT'S EASIER NOW*. My good friend Bill Nickerson agrees! As you may already know, if you've read Nickerson's book—It was no small task turning a $1,000 investment into 5 million dollars worth of real estate. In case you're wondering if Nickerson and his wife, Lucille, ever made sacrifices—I can tell you this much: They began investing with a small savings from Bill's telephone company paycheck when he was earning less than $20 a week! *I'll let you be the judge!*

WE CAN ALL HELP EACH OTHER MAKE MONEY

A student of mine in Bakersfield lost his job several years ago.   Alan had just started investing when the ax fell. The bad news was—*His three rental houses were all negative cash flow*. He needed a job to support them! When Alan attended my Los Angeles seminar, he was already in serious trouble with his rental houses. He was certain that without his job he'd lose them. I showed him three different ways to save the properties. He only needed one to make things work. ***Today there's good news!***

He now owns twelve more income properties. Alan doesn't need a regular job anymore because he works full-time on his houses. He still shows up at my seminars and he still keeps buying more properties. I'm sure by now Alan has discovered the same thing I learned years ago... *The quickest path to financial success is to learn the ropes from people who are already successful.*

IT'S A BRIGHT FUTURE—MAKING BIG MONEY WITH HOUSES

Investing is not for everyone! Some people just don't like the pressure of extra mortgage payments or the idea of fixing toilets on the weekend no matter how much money they can make. Obviously, those folks should not be investors. However, for the rest of us, the future is indeed a very bright one! Buying *income producing real estate* offers you an opportunity to acquire the personal wealth you've always dreamed about.

If my message sounds like I'm trying to motivate you, *you're right, I am*. But the reason is because I'm absolutely sure I can teach you how to make big profits with real estate. I've already done it for myself, but more importantly, *I've taught many other people just like you.*

HOW YOU WILL BENEFIT YOURSELF THE MOST

Real estate investing can often turn out quite well in spite of doing many things wrong! But the problem is—*it's much too risky* and is likely to take you twice as long. The biggest profits are always made by investors who act quickly and make the right decisions... *the kind that make things happen.* If you follow my advice, you will learn to act quickly when opportunities are available! Your decision to read *TRADE SECRETS* will help you because it will provide you with the "knowledge" to make fast, educated decisions. I'll teach you how to "size-up" the best deals quickly, long before your competition has a chance to act!

If your goal is to earn more family income for college or retirement, let me show you how! If you plan to start a new career... that's my specialty too! I'm living proof because I've done it! Here's where *TRADE SECRETS* will really help you: Whether you invest part-time or full-time, I'll teach you how to make the *right kind of investments*—the kind that earn you cash flow money to buy groceries and survive. I'm sure you'll agree with me—*that's valuable training.*

ONE SINGLE DECISION CAN MULTIPLY YOUR WEALTH

Failure to make small decisions stops far more people from becoming successful than lack of money! Some decisions mean making changes! Change runs against the grain of human nature! *We all resist them*—still each of us is blessed with the ability to make decisions if we choose! Waiting for things to happen rather than *making them happen is often* the big difference between people who are successful—and those who miss out. When you observe wealthy people, you can't help but notice how quickly they make decisions. They know what they want and they do it!

TO MAKE A MILLION DOLLARS YOU NEED SPECIFIC DIRECTIONS

TRADE SECRETS is a monthly "How To" newsletter which I personally write. Each new issue gives you *specific information* about the *Cash Flow Techniques* I've used to acquire over two-hundred investment houses. Obviously, *TRADE SECRETS* alone can't make you rich. However, I can promise you this much—*TRADE SECRETS* will play a big part in helping you get there, if that's your goal.

When you order my *TRADE SECRETS,* today—you'll be making a very smart business decision to improve your financial health. *Here's what you get for only $4 an issue*—you'll learn new *high profit strategies* every month. I'll show you the details and numbers for actual *cash flow transactions.* As always, *I guarantee TRADE SECRETS* will help you or you get your money back. There's no way you can lose! *I'll take the risk!*

WEALTHY INVESTORS ARE DECISIVE AND THEY ACT QUICKLY

Rich people have learned to recognize good deals quickly when they find them. *Watch a wealthy friend, you'll see!* Also, observe how quickly they act. They don't wait for the crowd—Instead, they're always way out front. Once you learn to make important decisions quickly for yourself—you'll start to see very positive results. *When you can make things happen instead of watch them happen*—you'll be well on your way to greater wealth. *TRADE SECRETS* will keep you *upbeat and positive* because that's what I write about! Negative thoughts will stop you from achieving your goals. Don't waste your time thinking about "doom and gloom." *If the sky is really falling... it will fall on the heads of the losers who predict it!*

MAKING THE FINAL DECISION . . . IT'S ALWAYS UP TO YOU

I'm asking you to order my *TRADE SECRETS—Do It Right Now Without Delay! You have my money back guarantee with the deal.* Every month I'll send you new "High Profit" strategies to help you make more money. I'll also keep you advised about my latest training materials and seminars.

I'm sure you will agree with me—making a quick $48 decision should be EASY AS pie if you intend to make the big ones—Pick up the phone and call **800-722-2550**. That's how easy it is.

Fixer Jay's Training Products
How To Build Real Estate Wealth and Manage for Profits

FIXING HOUSES & RUNDOWN APARTMENTS DELUXE HOME STUDY COURSE

Big Book (361 pages) plus 8 full 1-hour audio tapes.

Jay's most popular home study course is designed for every serious minded investor who wants to learn more about making money with HIGH PROFIT FIXER HOUSES. Jay shares a wealth of personal "How-to" knowledge about **buying the Ugly Ducklings...and turning them into beautiful White Swans.** He shows you how to generate cash flow and build profits quickly. Jay's *miracle-like* ADDING VALUE TECHNIQUES show you how to double your investment dollars in the shortest possible time. **Jay walks you through the buying and selling numbers so you completely understand why economical fix-up houses are the most profitable game in town.** This self-help home study course is complete with 8 full hours of audio training...plus Jay's Best Selling Fix-Up Book. You get 361 pages, crammed full of **Money making ideas and strategies**...perfect for the do-it-yourself investor. This course is ideal for beginners and "Old Salts" alike if fixing rundown houses for profit is your goal. National syndicated real estate columnist, Robert J. Bruss says, *"Jay's tape package is filled with practical 'Nuts and Bolts,' 'How -To-Examples of profiting from rundown fixer-upper properties.'—* Bruss Real Estate Tape Review."

Bonus—Free Subscription Trade Secrets (12 issues) when you order product #2100 (Free $48 value)

Order Product #2100 ....Price $195

Book Only #2105 ....$49

TRADE SECRETS TRAINING BINDER -

Jay's complete 3-day seminar course plus 6 "Quick Start" training tapes.

Includes 400 pages of cash flow instruction plus six 1-hour audio tapes

Making Money Strategies - covers all material from Jay's intensive 3-day "Trade Secrets" Seminar. **Big binder, Over 2 inches thick, filled with money-making instructions** and all the forms you need, *Section 8 strategies, free grant funds,* record keeping, lease option contracts, how to hire handyman contractors are only a few of the techniques you'll learn. Tapes are studio type...not recorded at the seminar. They are crammed full of Jay's proven success secrets. **This binder and audio tape course is your "millionaire training kit."** You get the same insider information and profit making techniques Jay has personally used to build his own multi-million dollar investment property empire. **This Step-by-Step training binder will be your personal road map to wealth.**

Order Product #2103 ....Price $175

CASH FLOW FROM CITY HALL-GUARANTEED RENTS

Special HUD Grant Funds for Property Owners

How to obtain Fix-Up Grants up to $8,500 per unit...plus how to qualify your houses for Section 8 rent subsidies for all your tenants. **The city pays your rents.** You get 120 pages of "how -to" information plus four 1-hour audio tapes. Jay explains how to upgrade your rundown properties with FREE HUD REHAB GRANT MONEY. Also, how you can qualify for **FIX-UP LOANS** from your local housing authority. This course is designed to show you how to profit using LOANS and GRANTS from your local Public Housing Authority (PHA). Jay tells you which properties work best and how to obtain guaranteed rents for all eligible tenants. **Special HUD forms and instructions to qualify your property are included.**

Order Product #2113 ....Price $95

JAY'S MONTHLY NEWSLETTER "TRADE SECRETS"

You'll get insider **Wealth Formulas... An exciting new Money-Maker Topic every month...** You'll learn Financing, Fix-Up Techniques, How to buy right, Fast evictions, Where to borrow money, How to estimate your profits, How to protect yourself with extra collateral on carryback notes, Building wealth with money partners, Where to buy houses...plus many more high profit strategies. **Jay's newsletter is like money in the bank for every do-it-yourself Real Estate investor who wants to build personal wealth faster. Remember...64% of all Jay's subscribers make their living with Real Estate.**

Order Product #2130 ....Price $48 Annually

Jay's Success Secrets Revealed
TO ORDER NOW CALL 1-800-722-2550

LANDLORDING– MANAGING TENANTS & TOILETS

Includes: • Big Book (250 pages)
Plus • 8 Full 1-hour audio tapes • Dozens of Forms you need

Jay's best selling 'Landlord' Study course will guide you through the complexities of property management. **You'll learn how to manage your tenants with very little person contact.** How to prevent rent collection hassles before they start. This home training course is filled with the same 'street smart' techniques Jay uses to manage more than 200 tenants and all his houses. Smart Management is often the difference between success and failure. *Knowing what to do... and how to avoid costly problems is worth thousands of dollars to every real estate investor.* The landlording knowledge you'll receive with this course is worth many times what it costs. Jay teaches you how to qualify low-income tenants so you don't lose rents. He also explains the forms you'll need from every tenant who occupies your houses...*and how to write up the contracts to protect yourself.* Jay's book is 250 pages thick filled with many **specialized landlording forms** you'll need to manage you properties and your tenants. You get 8 full hours of **insider information** designed to teach you the landlording ropes.

Bonus—Free Subscription Trade Secrets (12 issues) when you order product #2101

Order Product #2101 ....Price $195
Book only #2106 ....$49

INVESTOR'S PARTNERSHIP GUIDE...Includes Jay's special contract you'll need to satisfy money partners and protect yourself. Full 60 minute audio tape and "How to" booklet - You need this information if you intend to purchase properties using a down payment investor. Jay clearly explains the benefits you can expect...and how to avoid the common pitfalls. Samples of THE ARTICLES OF A PARTNERSHIP are included...*also copies, examples and explanations of Jay's special contracts are included. You'll learn why tenants-in-common could be the smartest way for you to hold title with your investment partner- many important tips to save you Big Bucks and keep money partners coming back.*

Order Product #2110 ....Price $49

(HANDYMAN) HOW TO HIRE INDEPENDENT CONTRACTORS AND AVOID PERSONAL LIABILITY
If you hire workers to help fix your properties - **you need this booklet fast!**... you get 60 minutes of audio instructions...plus **Jay's Special Rubber Stamp you need for your handyman paychecks.** *(spells out your workers tax liability)*—you must understand the Independent Contractor Rules to protect yourself. Jay tells you how to determine the difference between *employees* and *independent contractors*...**also why you must always use an Independent Contractor Agreement.** Jay's **LEGALLY APPROVED** contractor agreement is included with your instructions... *how to fill it out line-by-line.*

Order Product #2111..Price $69

JAY'S LEASE OPTION PLAN - IMPROVE YOUR CASH FLOW - SELL FOR 110% OF VALUE
You get Jay's instruction booklet...plus two 60 minute audio "how-to" tapes. You'll also receive Jay's special ready-to-use **OPTION AGREEMENT** - with filled in example. You get instructions about **TERMS AND CONDITIONS** that make the option work. *Jay explains how to initiate Lease Options with credit worthy tenants for 3 year terms. Jay's lease-option is different.* It provides for annual price adjustments to protect you from inflation. You'll learn why lease-options are one of the slickest techniques for *increasing cash flow and reducing your monthly operating expenses.*

Order Product #2112..Price $49

JAY'S 90/10 INVESTOR PLAN DEAL STRATEGY WHEN YOU HAVE THE RIGHT PLAN...BUT LITTLE MONEY
For **Aggressive Investors** who need financial help. **How to find partners with cash.** Shows you a *non-traditional method* to invest with very little money. Jay tells you how to attract money investors. What to offer them in return for a profit opportunity. **This Strategy will keep your investment plans on the Fast Track using other people's money...** you furnish the skills and creativity. This powerful information about making money...**includes how-to booklet, forms, agreements and a 60-minute audio instruction tape to make sure you understand how it works.**

Order Product #2115 ....Price $49

"EVICTIONS" HOW TO REMOVE UNWANTED TENANTS QUICKLY - WITHOUT A HASSEL
Step by step instructions (California rules). However, Jay offers personal strategies and techniques used by Street Smart Investors to speed up the process. *You get a training booklet with* **forms and how-to** *information*—plus a 60 minute audio tape with a **step-by-step eviction** procedure. Also, Jay tells you what to do when your tenants leave a house full of furniture and personal junk behind. Knowing the rules will save you a ton of money. **Remember this...deadbeat tenants are no match for knowledgeable property owners who know the rules...and use them.**

Order Product #2116 ....Price $49

Ordering is easy! ☎ 1-800-722-2550
As always your order is 100% guaranteed-If you can't make it work, send it back for a refund.

Index

A

Accounts Receivable   371
Action Mode   98
Adding Value   15, 16, 105, 107, 161, 182, 202
Adding Value Strategy   20, 52, 202, 204
Ads For Sale   226, 227, 229
Agent Fred   76, 77, 132, 195, 401
Agreement Partnership   269, 286
   Co-owner   272
All-Inclusive Mortgage   252
All-Inclusive (Wrap)   35, 252, 258
Amateur Landlord   359
American Dream   432
Appraisers   41, 315
Appreciation   164, 211, 413
Archimedes-*Greek Mathematician* 297
As-Is Condition   150
Asset (Liability)   322

B

Balloon Notes   203
Bank (Savings) Loan   53
Beautiful Swan   31
Beirut Airport   43
Beneficial Finance   308, 319, 320
Beneficiary   344, 351, 354
Bid-Up   145, 194
Bidding War   148
Big Picture   422, 424, 430, 435
Big Wealth Builders   295
Blind Faith   118
Blinders   427
Bloated Estimates   118
Block Funding (Grant)   328, 342
Blue Ribbon Deals   58
Borrowing (Equity)   11, 222
Bruss, Robert *Newsletter*   192
Business Decision   382

Business Sense   72
Buying Debt   260
Buying Power   420
Buying Skills   145

C

California Civil Code   395
Camp Fixer Jay's   72, 443
Capitalization   201
Carryback Notes (Paper)   14, 29, 164
Cash Equivalent Skills   176, 301
Cash Flow   12, 22, 61, 62, 128, 163, 203, 215, 285
Cash Flow Machine   429
Cash Flow Mode   17, 37, 403
Cash-Less Offer   247
Charm (*Homeyness*)   44, 106, 169
Cheap Utilities   189
Christopher Columbus *Technique* 56, 65, 142
Christmas Letter   356, 357
City Hall Motivation   340
Classified Ads   23, 129, 142, 227
Clean-Up   41
Clint Eastwood   51
Clobbered Financing   30
Co-Investors   176, 272, 289
Co-Signer   387
Cold Call Letters   129, 134, 138, 140
Columbo Lieutenant   131, 231, 356
Combat Zone   184
Compounding (*Brain*)   6, 9, 212, 219, 220, 415
Computer (Printout)   116
Contractor Building   155, 159
Contributions   248
Corrected Problems   192
Cost Fix-Up   153
Cost Per Unit   81
Cream-Puff Fix-Up   171

Creative Thinking 415
Credit Cards-Losses 82, 97
    Rating 319
Customers (Tenants) 377
Cyclical Business 211, 369

D

Deadbeat Tenants 17, 22, 206, 376, 380
Deferred Maintenance 229
Depreciation 174, 215
Detective *House* 72, 141, 163, 193
Diamonds (Rough) 44, 145
Dirt Bag (Property) 30
Discount Mortgages 223, 354
Discrimination 398
Disneyland 140
Distressed 248
Do-It-Yourself 100, 409
Dog Properties 133
Don't Wanters 67, 142
Doom and Gloom 433, 434
Down Cycles 325
Dry-Rot 181
Due-On-Sale (Clause) 18, 345, 355
Dumb *Landlords* 425
Dumbest (Question) 372, 385

E

Easy Money 220
El Dumpo Villa 285
Entrepreneurial Skills 434
Epley Richard 360
Equal Benefits 161, 266
Equity Buildup 216
Eviction (Laws) 371, 392, 393
Exterior Paint 158, 177, 180
Extract Profits 304

F

Fairness 401
Fast Forward Mode 74
Fast Track Investing 60, 62
Favorable Terms 97
Fed-Up Factor 27

Feeling (Homey) 44
Fences 179
Filthy Rich 252
Financial Independence 99
    Documents 321, 324, 337
    Information 29, 109
    Status 30, 57
    Tools 319
Fire Sale 66
First Impressions 158, 160
First Main Street 298, 300
First Property 59
Fix-Up 41, 103
Fix-Up Bids 96
Fix-Up Plan 41
    Revolution 170
Fixed Interest 308
Fixed Cost 115
Fixer Camp 152, 167, 443
Fixer Jay 90
Fixer Jay Seminars 72, 160
Fixer Uppers 225
Fixing Houses 50, 241, 247
    Wrong Things 106
Flat Roofs 174
Foo-Foo (Fix-Up) 167, 168, 185
For Sale 64, 141
Forced Appreciation 113
    Equity 305
Foreclosure 60, 244, 246
Form *1040 Schedule* 46, 64, 82, 87, 88
Formica Countertops 108, 171
Fred Quigley 79, 130, 444
Free Money 326
Frills Fixing 178
Front Yards 176

G

Garages 180, 190
Garden Variety *Transaction* 97
Get Rich Gurus 143
Godfather Loans 308
Gold Mine 109, 145, 194, 220
Golden Partner 297
Good Looks 158
    Colors 430
Grade A Apartment 415
Great Depression 434

Gross Rent Multiplier   45, 110, 205, 227, 282
Grunt Work   25, 32
Gypsies   400, 401

H

Habitability Laws   65, 395
Hamburger Helper   297
Hamilton Avenue   7
Hands-On (Investing)   21
Handyman College   36, 174, 270
   Stores   174
Handyman Skills   270
Hardball Terms   60
Hauling Junk   159
Haywood Houses   23, 26, 31, 32
Help Wanted   265
Hidden Profits   36, 193
High-Tech Programs   109
Highest Paid   259
Hillcrest Cottages   3, 7, 144, 244, 247, 248
Home Field Advantage   423
Homey Atmosphere   44
Horror Stories   380
House Calls   372
House Detective   72, 193, 195
Housewise *BRANGHAM*   156
Housing Authority   65
   Property   330
Housing Rehabilitation   326
HUD-*Dept. of Housing & Urban Development*   2, 47, 71, 194, 326, 327
HUD Rehab.   329
   Loans   335, 338
   Tenants   386

I

Improvement   59
Income Property Analysis Form   61, 63, 81, 87, 89, 235, 242, 313
Inherited Owners   95, 162
Installment Sales   259
Institutional Lenders   53
Internal Rate Return (IRR)   117
Installment Sale   412

Investment Plan   72, 125

J

Jay's 60/40 Rule   264, 268, 303, 402
   30/30 Plan   307, 309, 310, 313
   90/10 Plan   293, 299
Jay's Rundown Houses   82
John Housman   405
John Schaub   120, 255
Jump-Start   97
Junker-Type Property   99

K

Keeper Rental Units   12
Killer Contract   389
Kissing Frogs   432, 433
Knowledge Builds Fortunes   192

L

Land Contract   284
Landlord Memos   372
Landlord Skills   369, 375
Landscaping   176
Lease W/Option   95, 162
Legwork   81
Lemonade Offers   16, 415, 417, 420
Leper Property   270, 279
Letter Cold Call   138, 140
Leverage   6, 9, 212, 217
Loan To Value   224
Loan Over Basis   253
   To Value   30, 224
Loans (HUD)   335
Locations   47, 48, 124
Long-Term (Profits)   59
Looky-Loos   417
Low-Ball (Shotgun) Offers   80
Low Profile Techniques   401
Lump Sum Cash   14

M

Magic Compounding   212
Making Money   5, 101, 431
Management Property   84

Mark-Up Profits   43
Marketplace   119
Maximum Control   203, 228
MBA Degree   81
McDonald's Corp.   317
Mentor   424
Merv, *Real Estate Agent*   80, 132
Middleman Buffer   79
Million Dollar *Mansions*   50
Millionaire   97
MLO (Money Left Over)   280
Mom & Pop Owners   79, 229
Money Partner   265
Monthly Rental Income   11
Moratorium (Payments)   150, 152
Mortgage   29
Motivated Sellers   19, 98, 198,
    141, 143, 418
Motor Lodge   145
Mr. & Mrs. Right   75, 77
Multiple Listings (Service)   79,
    80, 129, 132
Multiple Skills   103
Multiplier Gross Rent   45
Murphy's Law   153

N

Napier, Jimmy   218, 358
Negative Numbers   105
Neglect   25
Negotiating Terms   235
Net Spendable Income   137
Nickerson, William   1, 71, 434,
    445
Nit-Wit (Boss)   218
No Cash Down (Money)   56, 97,
    312, 409
Nob Hill   112
Non-Performing *Mortgages*   147
Non-Cash Values   296
Non-Marketable   355
Non-Traditional (*Properties*)   44,
    145
Nothing Down (Also, *No Cash
    Down*)   100, 250, 304

O

Obnoxious Agents   76

One-On-One Counseling   166
One Percent Rule   428
One Stop Home Rental Co.   323
Operating Income   86
OPM (Other People's Money)   298
Opportunities High Profit   349
Option Purchase   271
Over (Fixing)   106
    (Paying)   209, 210
Owner Financing   61
Owner's Shadow   362

P

Pain Relief   198, 199
Painting Fumes   173, 191
Paperwork (Tenants)   396, 399
Partnerships   261, 264, 265
Payback Fix-Up   185
People-Problems   26, 82
Perfect Timing   245
Perry Mason   393
Personal Efforts   52
    Skills   91, 244, 303, 304
    Property   419
Picket Fences   39, 377
Pier-Type Foundations   25
Playing (*20 Questions*)   367
Pocket Listings   132
Pool Hall Analogy   160
Positive Cash Flow   123, 128
Potential Market Value   193
Pre-Hung (Doors)   172
Pride Ownership   62, 203
Problem Solving   99
    Tenants   94, 199
Profit Makers   98, 405
    Mode   16
    Bulbs   220, 259
    Fix-Up   49, 186
Promissory Notes   29
    (*Also See Mortgage*)
Property Management   377
    Sketches   135, 339
Property Detective   162
Property Analysis Form   79, 80,
    82, 89, 239, 313, 315
Proposition   13, 83
Proprietorship   364, 365
Public Housing   327

Nuisance   146
Puffery (*Classified Ads*)   227
Purchase Notes   343, 355
Purple Toilets   39

Q

Qualified Buyers   251
Quick Returns   49, 108
Quickie Cost Estimate   87

R

Ray Kroc   12, 317
Reader's Digest   219
Ready-Made (Fixtures)   172
Real Estate Agents   75, 78, 88, 131
Real Estate *Guru*   93
REO (Real Estate Owned)   146
Real World Experience   104
Red Adair   99
Red-Tagged Houses   146, 147
Red Tape (Rules)   337, 338
Relief Pain   198, 199
Remodelers   45, 105, 155, 156, 176
Renovators   45, 155
Rent Multiplier   45, 81, 282
Rent-To-Value Factor   428
    Ratio   121
Rental Income   59, 82
Rental *Marketplace*   101
Right Properties   96
    Stuff   132
Ring-Wise Investors   127
Risk (No Down Sales)   71, 250
Robert Schuller   70
Rules (Of Thumb)   80
    Leadership   274

S

Sam *Moneybags*   288
Schedule Gross Rents   46
    1040 Tax Form   46, 64
Schaub, John *Seminars*   120, 255
Schuller, Robert-Minister   70
Scumbag Property (Villa)   112

Second Opinion   192
Section 8 *(also see HUD)*   327, 386, 388, 435
Section, *Code IRS 1031*   287
Selection (Partners)   262
Self Employment   164
Seller Financing   18, 161, 221, 325
Seller Motivation   4i8
    Problems   55, 259
Seller Carryback *Mortgages*   19, 150, 165, 257, 353
Seller's Urgency   245
Separate Meters (Utility)   83
Serious Money   145, 147
    Fun   424
Shirley McLaine   120
Short-Term Cash Flow   59
    Tenants   401
Shotgun Offers   88, 93
Single-Party Control   362
Sizzle *Fix-Up*   43, 185
Skin-Deep Beauty   158
Skills *Fix-up*   41, 103
Skinny Deals (Transactions)   100
Skip-Outs   82
Smith-Barney Investments   406
Snob Hill (Location)   112
Soggy Paper (Mortgages)   355
Sonneborn, Harry   317
South Side Houses   315
Specialization (Fix-Up)   55, 101
    Knowledge   41
Speculating   73, 403, 404, 405
Squeaky Clean (Properties)   34
Strategy *Adding Value*   16, 52
Street Savvy Investor   376
Subordination   307
Subsidized Tenants   326
Substantial Profits   194
Successful *Fix-Up*   175
Suzanne Brangham   156
    Also, *See Housewise*
Swan Beautiful (Houses)   31

T

Target Areas   335

Tax Benefits-*Deductions* 135, 269
 Bills 135
 Form 237
 Free 73
 Returns 64
Techniques *Lt. Columbo* 231, 233
Tenants-In-Common 287, 288, 300
Tenant Cycling 32, 199, 301, 400
 Sob Stories 372
 Urgency 391
Terms Conditions 17, 130
 Negotiating 22, 150
Three Day Notice 378
Thrift Leader - Banks 308
Tightening Up 98
Time Life (Books) 173
TLC (Tender Loving Care) 298
Top Dollar 417
Track Record (Investor) 276
Trade Accounts 324
Trade Secrets Newsletter 104, 105
Trash-Outs (Trashed) 96, 146, 393
Trees and Shrubs 176, 185
Trust Deeds 35, 197
Tunnel Vision 47
Turn-Around Properties 207
Two-Fer One (Sales) 251

U

Ugly Ducklings (Houses) 13, 31, 148, 314
Ugly Bath-Kitchens 177
 Tenants 159
Uncle Sam (Government) 82
Uncollectable Rents 82
Under-Performing Property 20, 161
Under Market Rents 201
Underlying Mortgages 254, 257, 258
University (Fixer) 174
Unlawful Detainer 397

Unsightly (Mess) 417
Up-Cycles (*House Market*) 13
Upgrading Houses 61
USA Today (Newspaper) 423

V

Vacancies 68, 417
Variable Rate *Mortgages* 101
Viola Cottages 334, 335
Vision Investor 143

W

Wall Street Journal (Newspaper) 423
Wanta-Be Investor 50
Weak Offers 416
Wealthy Tycoons 115
White Picket *Fences* 7, 43, 169, 179, 199, 377
Wholesale Prices 209, 227
Wild Goose Chase 80
William Nickerson 1, 71, 434, 445
Willingness (*Investor*) 92
Window Coverings 174, 177
Women's Advantage 42
Workable Plan 410, 422, 434
Working Investors 276
 Partners 267
Worry-Free Investments 203
Wrap-Around Financing 35, 252, 254-255
Written Contracts 227

Y

Yellow Court Project 146, 151, 153
Yellowed Squares (*City Housing*) 331, 341
Yogi Berra 117, 421
Yucks (Dirty Houses) 12

Z

Zenderdorff, William 143, 144
Zero Cash Down 295
 Also See, *No Cash Down*

FREE FIXER CAMP TAPE

If you wish to know more about what goes on at Jay's Fixer Camps in Redding, California— we'll send you a 60 minute audio tape with details. Ask for Tape #6100.

Write to **KJAY PUBLISHING** at the address shown on the reverse side or call toll free 1-800-722-2550.

Order Form – *Fixer Jay's Training Products*

KJAY Publishing Co • P.O. Box 491779 • Redding, CA 96049

How To Build Real Estate Wealth and Manage for Profits

Product #	Description of Fixer Jay's Training Product		Price	Order
2100	Fixing Rundown Houses and Small Apartments	(Full course)	$195	
2105	Fixing Rundown Houses and Small Apartments	(Book only)	49	
2101	Managing Tenants and Toilets	(Full course)	195	
2106	Managing Tenants and Toilets	(Book only)	49	
2103	Trade Secrets - Seminar Training Course	(Book only)	175	
2102	Full Time Real Estate investing & Career Changer	(Full course)	195	
2110	Investors Partnership Guide		49	
2111	Handyman - How To Hire Independent Contractors		69	
2112	Jay's Lease Option Plan		49	
2113	Cash Flow from City Hall - Guaranteed Rents	(Full course)	95	
2114	Jay's Loan Kit - Forms and Audio Tape Instructions		49	
2115	Jay's 90/10 Investor Plan		49	
2116	Evictions – How to Remove Unwanted Tenants		49	
2117	Fixin' Ugly Houses For Money - 12 Audio Tapes - Full Course		195	
2118	Successfully Manage Tenants with Postage Stamps - 4 Tapes		95	
2119	Bank Repossessed Houses – Yale Court REOs		49	
2130	Jay's Monthly Newsletter – Trade Secrets	(12 issues)	49	
2131	Strategies I – Real Estate Investor Techniques	(Full course)	149	
2132	Strategies II – Real Estate Investor Techniques	(Full course)	149	
2133	Strategies III – Real Estate Investor Techniques	(Full course)	149	
2134	Strategies IV – Real Estate Investor Techniques	(Full course)	149	
2135	Strategies V – Estate Investor Techniques	(Full course)	149	

PLEASE PRINT   100% Money-Back Guarantee

NAME_____	TOTAL AMOUNT OF ORDER
ADDRESS _____	DISCOUNT AMOUNT
CITY _____	SALES TAX (CALIF ONLY)
STATE_____ ZIP _____	Under $50 - Add $3.00 / Over $50 - Add $5.00 SHIPPING
TEL NO (_____)_____	TOTAL AMOUNT DUE $
CARD NO _____	EXP DATE _____
SIGNATURE _____	❑ VISA ❑ MC ❑ Check

VISA MasterCard

Ordering is easy! ☎ 1-800-722-2550
Fax Orders (anytime) 1-530-223-2834

FREE FIXER CAMP TAPE

If you wish to know more about what goes on at Jay's Fixer Camps in Redding, California—we'll send you a 60 minute audio tape with details. Ask for Tape #6100.

Write to **KJAY PUBLISHING** at the address shown on the reverse side or call toll free 1-800-722-2550.

Order Form – *Fixer Jay's Training Products*

KJAY Publishing Co • P.O. Box 491779 • Redding, CA 96049

How To Build Real Estate Wealth and Manage for Profits

Product #	Description of Fixer Jay's Training Product		Price	Order
2100	Fixing Rundown Houses and Small Apartments	(Full course)	$195	
2105	Fixing Rundown Houses and Small Apartments	(Book only)	49	
2101	Managing Tenants and Toilets	(Full course)	195	
2106	Managing Tenants and Toilets	(Book only)	49	
2103	Trade Secrets - Seminar Training Course	(Book only)	175	
2102	Full Time Real Estate investing & Career Changer	(Full course)	195	
2110	Investors Partnership Guide		49	
2111	Handyman - How To Hire Independent Contractors		69	
2112	Jay's Lease Option Plan		49	
2113	Cash Flow from City Hall - Guaranteed Rents	(Full course)	95	
2114	Jay's Loan Kit - Forms and Audio Tape Instructions		49	
2115	Jay's 90/10 Investor Plan		49	
2116	Evictions – How to Remove Unwanted Tenants		49	
2117	Fixin' Ugly Houses For Money - 12 Audio Tapes - Full Course		195	
2118	Successfully Manage Tenants with Postage Stamps - 4 Tapes		95	
2119	Bank Repossessed Houses – Yale Court REOs		49	
2130	Jay's Monthly Newsletter – Trade Secrets	(12 issues)	49	
2131	Strategies I – Real Estate Investor Techniques	(Full course)	149	
2132	Strategies II – Real Estate Investor Techniques	(Full course)	149	
2133	Strategies III – Real Estate Investor Techniques	(Full course)	149	
2134	Strategies IV – Real Estate Investor Techniques	(Full course)	149	
2135	Strategies V – Estate Investor Techniques	(Full course)	149	

PLEASE PRINT   100% Money-Back Guarantee   TOTAL AMOUNT OF ORDER

NAME_____   DISCOUNT AMOUNT

ADDRESS _____   SALES TAX (CALIF ONLY)

CITY _____   Under $50 - Add $3.00   SHIPPING
                                              Over $50 - Add $5.00

STATE_____ ZIP _____   **TOTAL AMOUNT DUE** $

TEL NO (_____)_____   **VISA** MasterCard.

CARD NO _____   EXP DATE _____

SIGNATURE _____   ❑ VISA ❑ MC ❑ Check

Ordering is easy! ☎ 1-800-722-2550
Fax Orders (anytime) 1-530-223-2834